The Dark Side of Silicon

Amir M. Rahmani • Pasi Liljeberg
Ahmed Hemani • Axel Jantsch • Hannu Tenhunen
Editors

The Dark Side of Silicon

Energy Efficient Computing in the Dark Silicon Era

Springer

Editors
Amir M. Rahmani
University of Turku
Turku, Finland

Pasi Liljeberg
University of Turku
Turku, Finland

Ahmed Hemani
Department of Electronic systems
School of ICT, KTH
Royal Institute of Technology
Kista, Sweden

Axel Jantsch
Vienna University of Technology
Vienna, Austria

Hannu Tenhunen
KTH Royal Institute of Technology
Stockholm, Sweden

ISBN 978-3-319-31594-2 ISBN 978-3-319-31596-6 (eBook)
DOI 10.1007/978-3-319-31596-6

Library of Congress Control Number: 2016936374

Printed on acid-free paper

This Springer imprint is published by Springer Nature
The registered company is Springer International Publishing AG Switzerland

Contents

Part I
Architecture and Implementation Perspective

Chapter 1
A Perspective on Dark Silicon

Anil Kanduri, Amir M. Rahmani, Pasi Liljeberg, Ahmed Hemani, Axel Jantsch, and Hannu Tenhunen

The possibilities to increase single-core performance have ended due to limited instruction-level parallelism and a high penalty when increasing frequency. This prompted designers to move toward multi-core paradigms [1], largely supported by transistor scaling [2]. Scaling down transistor gate length makes it possible to switch them faster at a lower power, as they have a low capacitance. In this context, an important consideration is *power density*—the power dissipated per unit area. Dennard's scaling establishes that reducing physical parameters of transistors allows operating them at lower voltage and thus at lower power, because power consumption is proportional to the square of the applied voltage, keeping power density constant [3]. Dennard's estimation of scaling effects and constant power density is shown in Table 1.1. Theoretically, scaling down further should result in more computational capacity per unit area. However, scaling is reaching its physical limits to an extent that voltage cannot be scaled down as much as transistor gate length leading to failure of Dennardian trend. This, along with a rise in leakage current, results in increased power density, rather than a constant power density. Higher power density implies more heat generated in a unit area and hence higher chip temperatures which have to be dissipated through cooling solutions, as increase in temperature beyond a certain level results in unreliable functionality, faster aging, and even permanent failure of the chip. To ensure a safe operation, it is

A. Kanduri • A.M. Rahmani • P. Lijeberg
University of Turku, Turku, Finland
e-mail: spakan@utu.fi; amirah@utu.fi; pasi.liljeberg@utu.fi

A. Hemani • H. Tenhunen
KTH Royal Institute of Technology, Stockholm, Sweden
e-mail: hemani@kth.se; hannu@kth.se

A. Jantsch
TU Wien, Vienna, Austria
e-mail: axel.jantsch@tuwien.ac.at

A.M. Rahmani et al. (eds.), *The Dark Side of Silicon*,
DOI 10.1007/978-3-319-31596-6_1

Table 1.1 Practical scaling in post-Dennardian era

Parameter	Symbol	Ideal scaling	Practical scaling
Transistor dimensions	L, W, t_{ox}	1/k	1/k
Area	$A = 1/L \times W$	$1/k^2$	$1/k^2$
Capacitance	C	1/k	1/k
Gate delay	RC	1/k	1/k
Frequency	$F = 1/RC$	k	k
Supply voltage	V_{dd}	1/k	1/s
Power	$P = CV^2F$	$1/k^2$	$1/s^2$
Power density	$D = P/A$	1	k^2/s^2

essential for the chip to perform within a fixed power budget [4]. In order to avoid too high of a power dissipation, a certain part of the chip needs to remain inactive; the inactive part is termed dark silicon [5]. Hence, we have to operate working cores in a multi-core system at less than their full capacity, limiting the performance, resource utilization, and efficiency of the system.

The Dark Silicon Phenomenon

Too high of a power density is the chief contributor to dark silicon. It is to be noted that trends in computer design that are technology-node-centric have a significant impact on area, voltage, frequency, power, performance, energy, and reliability issues. Most of these parameters are cyclically dependent on one another in a way that improving one aspect will deteriorate the other. Many-core systems face a new set of challenges at the verge of dark silicon to continue providing the expected performance and efficiency. Computational intensity of future applications such as deep machine learning, virtual reality, big data, etc., demands further technology scaling, leading to further rise in power densities and dark silicon issue. Increase in power density leads to thermal issues [6], hampering the chip's performance and functionality. Subsequently, issues of reliability and aging have come up, in addition to limitation in performance. ITRS projections have predicted that by 2020, designers would face up to 90 % of dark silicon, meaning that only 10 % of the chip's hardware resources are useful at any given time when high operating frequencies are applied [7]. Figure 1.1 shows the amount of usable logic on a chip as per projections of [5, 7]. Increasing dark silicon directly reflects on performance to a point that many-core scaling provides zero gain [8]. Thermal awareness and dark silicon-sensitive resource allocation are key techniques that can address these challenges.

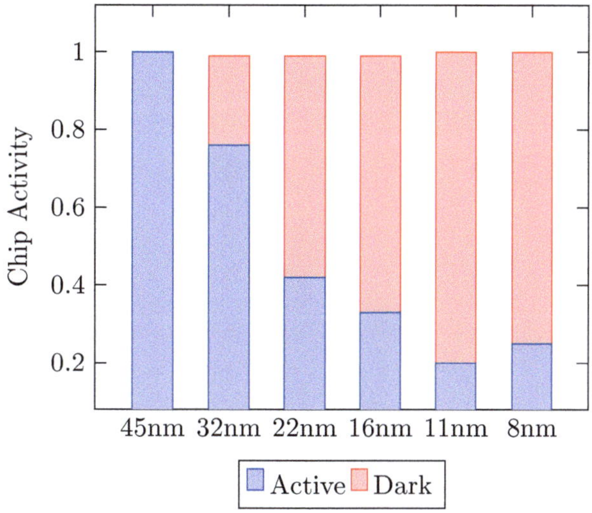

Fig. 1.1 Projections of dark silicon [5, 7]

Power Density

Working components of a chip consume power and generate heat on the chip's surface area that increases chip's temperature. Higher chip temperatures adversely affect the chip's functionality, induce unreliability, and quicken the aging process. Heat accumulated on the die has to be dissipated through cooling solutions in order to maintain a safe chip temperature. The amount of heat generated on the chip's surface area depends on its power consumption. Most of the computer systems must function within a given power envelope, since heat dissipated in a given area is restricted by cooling solutions. The power budget of a chip is one of the principal design parameters of a modern computer system. With every technology node generation, power densities have been steadily increasing, while the areas and cooling solutions are stale. Increase in power density that comes with integration capacity is shown in Fig. 1.2, which is estimated based on 2013 ITRS projections [9] and conservative projections by Borkar [10]. We see an exponential rise in power density, while the ideal scaling principle of Dennard states that it would remain constant. Increased power density is currently handled by unwanted dark silicon.

The scaling conclusions from Dennard's principle [3] are summarized in Table 1.1. Scaling transistor's physical parameters, viz., length (L), width (W), and oxide thickness (t_{ox}), translates into reduced voltage and increased frequency. This means that scaling physical parameters by a constant factor k would also scale down voltage and capacitance and scale up frequency by k. This reduces power consumption and chip area by the same factor of k^2, keeping the power density constant. This argument held only until a reaching certain point, called the utilization wall [11]. The major causes for increase in power density are discussed in following sections (Fig. 1.3).

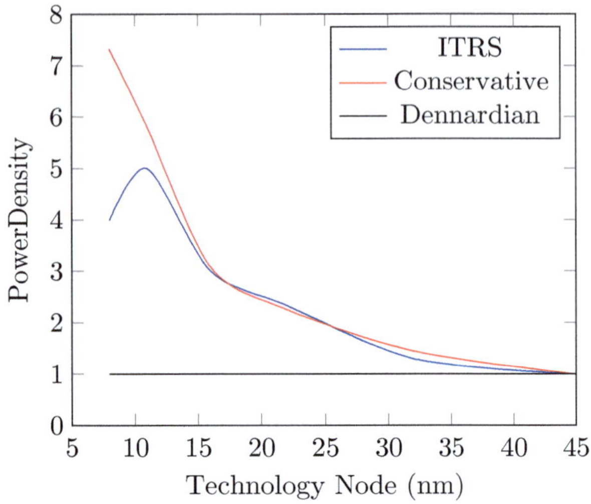

Fig. 1.2 Increase in power density with scaling

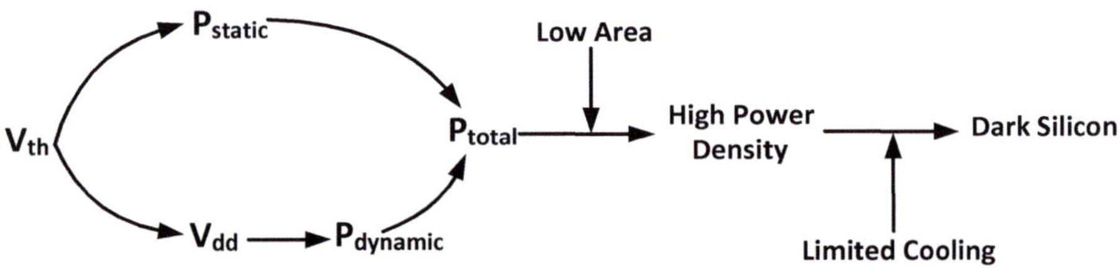

Fig. 1.3 Chief contributors to dark silicon

Power Consumption in CMOS Chips

Power consumption in CMOS-based chips mainly consists of *dynamic power*—the power consumed when transistors are switching between states and *static power*—the power drawn by leakage currents, even at an idle state, as expressed in Eq. (1.1).

$$P_{total} = P_{dynamic} + P_{static} \tag{1.1}$$

The dynamic power component, as in Eq. (1.2), is the power consumed to charge or discharge capacitive loads across the gate. Dynamic power is proportional to the capacitance C, the frequency F at which gates are switching, squared supply voltage V, and activity A of the chip (the number of gates that are switching). In addition, a short circuit current I_{short} flows between the supply and ground terminals for a brief period of time τ, whenever the transistor switches states [12]. Dynamic power consumption can be minimized by reducing any of these parameters.

$$P_{dynamic} = ACV^2F + \tau AVI_{short} \tag{1.2}$$

With transistor scaling, the voltage and frequency of the chip can also be scaled, as proposed in [3] and shown in Table 1.1. Frequency varies with supply and threshold voltages as

$$F \propto \frac{(V - V_{th})^2}{V} \tag{1.3}$$

Reducing physical parameters reduces voltage, capacitance, and hence the *RC* delay of a transistor [13]. Noticeable aspect here is that frequency varies almost as voltage, assuming a smaller threshold voltage. With higher level of integration, reduced frequency can be compensated with parallel multi-processor architectures which yield better performance [14]. Voltage scaling can reduce power by factor of a square, encouraging transistor scaling. Unlike a perfect switch, a CMOS transistor draws some power even when it is idle (i.e., not switching between the states) [15]. This can be attributed to the leakage current arising at lower threshold voltages and gate oxide thickness. More details on leakage power are presented in section "Leakage Power".

Slack Voltage Scaling

As predicted by Gordon Moore, the semiconductor industry has been successfully improving transistors of increasing speed with decrease in gate length for every technology node generation [2]. Moore's prediction has obtained validation through Dennard's scaling law [3] which establishes an empirical relation between technology node scaling and power density such that power density remains constant even with aggressive integration of transistors in a given area. The reason behind validity of Dennard's scaling is that reduction in gate length of transistors also lowers operating voltage and load capacitance—which in turn reduces dynamic power consumption significantly [14]. Dennardian scaling held as long as voltage and transistor gate length were scaled down at a similar rate. However, voltage scaling is reaching its limits, and it is increasingly challenging to scale down further. Figure 1.4 shows projections of voltage and gate length scaling from ITRS reports of 2011 [7], 2013 [9], and conservative approach [10]. (Data of [7] and [10] retrieved from [5].) All the projections clearly indicate that voltage scaling is almost stuck in comparison with integration capacity, which continues to scale gracefully. Increasingly wide gap between voltage scaling and transistor scaling results in higher power density and dark silicon. This can also be considered a corollary on failure of Dennard's scaling, where it is expected that voltage and technology nodes could be scaled at a reasonably similar rate. Scaling the supply voltage is pegged with scaling the threshold voltage. The requirements on performance and reliability determine minimum and maximum limits of the supply voltage. Along with them, the risk of scaling down threshold voltage is the reason for limitation in

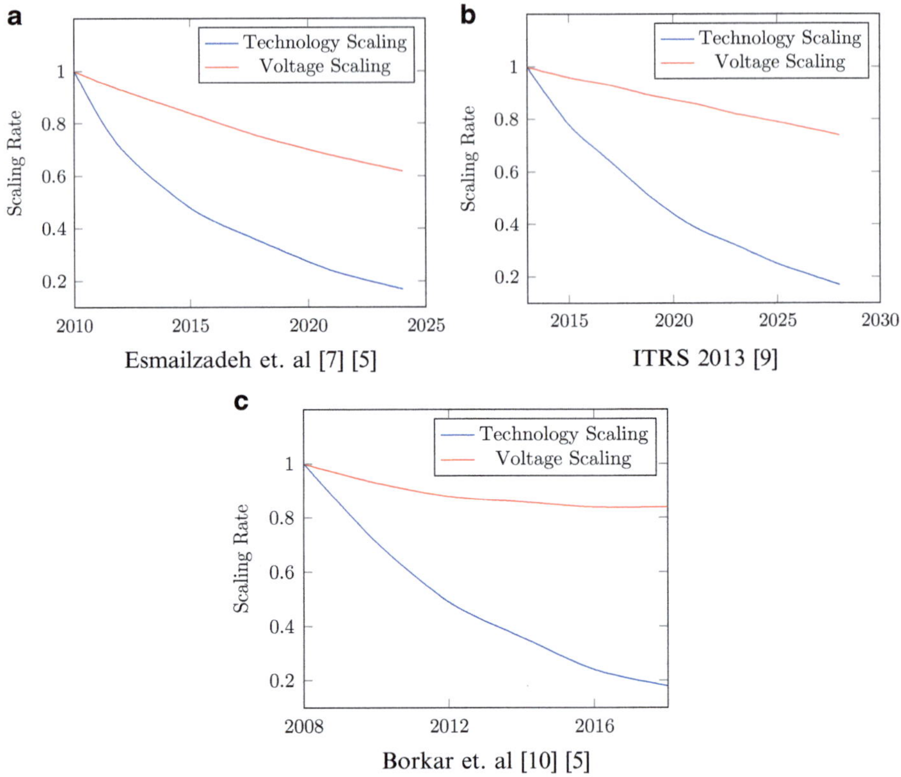

Fig. 1.4 Technology node versus voltage scaling. (**a**) Esmailzadeh et al. [5, 7]. (**b**) ITRS 2013 [9]. (**c**) Borkar et al. [5, 10]

voltage scaling [16]. Since the voltage no longer scales with the transistor size, ideal Dennardian scaling will no longer hold. Instead, the power density starts increasing with every new technology node generation, as indicated in [11].

Table 1.1 shows realistic scaling in the post-Dennardian era. Assuming that voltage scales as a factor s instead of transistor scaling factor k, the power density becomes k^2/s^2, a nonconstant and greater than unity quantity. This indicates that aggressive technology scaling (high k) and/or slack voltage scaling (low s) put together results in a high k^2/s^2 value, which is nothing but higher power density. Safer limit on power consumption can be achieved by operating only a section of the entire chip's resources, resulting in dark silicon [14]. From this viewpoint, accumulated heat has to be dissipated by controlling power consumption, which practically is not possible, unless there is dark silicon.

Leakage Power

Power has to be viewed as summation of static and dynamic components, as in Eq. (1.1). Reducing voltage and frequency would obviously reduce dynamic power consumption and also total power, as long as the static component is negligible. However, an increasingly critical component is the static power that is drawn by the leakage current I_{leak}, which is increasing with scaling [17]. Leakage current is split into two components, subthreshold leakage and gate oxide leakage, as in Eq. (1.4).

$$P_{static} = VI_{leak} = V \times (I_{sub} + I_{ox}) \tag{1.4}$$

Leakage current is the static power drawn even at an off state (not switching), meaning that a chip can still consume appreciable power without being active [14]. Off-state leakage current is experimentally driven in [18] and is represented as in Eq. (1.5).

$$I_{off} = K_1 W e^{\frac{-V_{th}}{nV_\theta}} (1 - e^{\frac{-V}{V_\theta}}) \tag{1.5}$$

where K_1 and n are experimentally derived constants , W is transistor width and V_θ is 25 mV. With V_θ being much smaller than V, this is further simplified in [12] as in Eq. (1.6).

$$I_{off} \propto e^{\frac{-qV_{th}}{kT}} \tag{1.6}$$

where k is a constant and T is temperature. This indicates that leakage current is low when the exponential component is small, meaning that threshold voltage has to be high. In other words, leakage current increases exponentially with a decrease in threshold voltage (negative exponential component). Reduction in transistor physical parameters reduces voltage and thus also reduces threshold voltage. This results in higher leakage current and thus higher static power being drawn. It can be said that with every technology node generation, leakage current increases exponentially with a decrease in threshold voltage [13]. Also, leakage power is a cumulative of all transistors' leakage, while the number of transistors more or less doubles for every technology node generation, contributing to increasing leakage power [14]. Reduction in transistor physical dimensions also reduces oxide thickness, to an extent that electrons start to leak also through the oxide resulting in leakage current [15]. Oxide thickness is reaching around 1.5 nm, result in significant leakage current (I_{ox}) [19]. Gate oxide leakage current is shown in Eq. (1.7).

$$I_{ox} = K_2 W (\frac{V}{T_{ox}})^2 e^{\frac{-\alpha T_{ox}}{V}} \tag{1.7}$$

where K_2 and α are experimental constants. Analogous to off-state leakage, gate oxide leakage depends exponentially on oxide thickness. The squared denominator and negative exponential component of Eq. (1.7) shows that gate oxide leakage increases exponentially with decrease in oxide thickness. Since oxide thickness depends on other transistor parameters, it has to be scaled down with technology node scaling, thus increasing I_{ox}. While dynamic power was a lone significant contributor to total power, static power is alarmingly increasing and would exceed dynamic component with further technology scaling [12]. In addition, leakage increases exponentially at higher temperatures, implying that a chip would have far more leakage if it operates to its full potential [20]. Power density would not have been constant with scaling; however, it would have been in a relatively controllable reach, had there been no static power.

Thermal Issues

CMOS-based chips are limited by battery technology and cooling solutions that can dissipate the chip's heat. Therefore, a safe peak operating temperature and the corresponding power that can be dissipatable within a given volume are used as design time parameters. The upper limit on power consumption, and thus temperature, is popularly known as thermal design power (TDP) [21]. Safe and reliable operation of a chip is guaranteed as long as power consumption and heat dissipation stay within TDP guidelines [21]. However, TDP is not the maximum power that can be consumed by a chip, but it is only a safe upper bound. TDP is calculated assuming a worst-case application running at worst-case voltage and frequencies of components, although a typical application might consume power that is less than TDP and yet runs under TDP constraint. Performance has to be sacrificed through voltage and frequency reduction and clock and power gating in order to stay within TDP. This is acceptable for applications that are power hungry; however, other applications that may never exceed TDP would still end up running at lower than full compute capacity of the chip. In this sense, current systems are rather conservative. Such conservative estimates of the upper bound can be met through inactivating part of the chip, contributing to dark silicon. Thus far the most common practice of ensuring safe chip functionality has been by estimating TDP [21] in a conservative manner. Recent upgrades on AMD and Intel Corporations' CPUs have the option of a configurable TDP; however, they are limited to a maximum of three modes, without any fine-grained control [22, 23].

Challenges and Consequences of Dark Silicon

Increase in dark silicon reduces the number of simultaneously active components on a chip and hampers performance, energy efficiency, reliability, aging, and effective

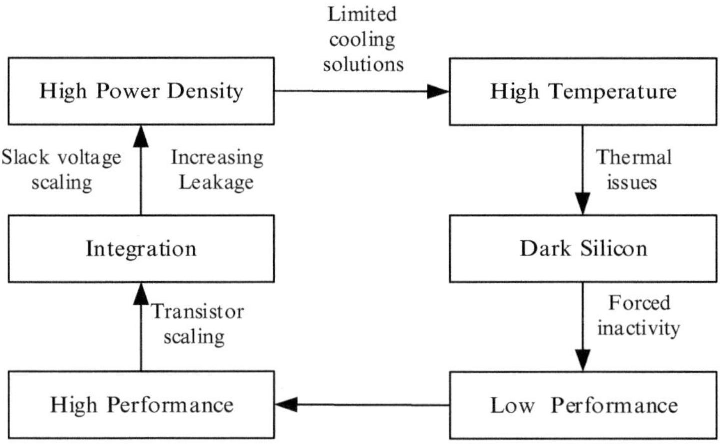

Fig. 1.5 Performance degradation due to dark silicon

resource utilization of computer systems [5]. With the dark silicon phenomenon being a hardware and device-level issue, other layers of the computing stack are unaware of the sources of inefficiency. This makes computer design challenging, considering the lack of support from workload characteristics, programming languages through compilers. The key challenges and consequences of dark silicon are summarized in the following sections.

Performance

With a section of the chip being inactive, expected performance from many-core systems can never be realized in practice. In the quest for higher performance, designers opt for denser chips, with the expectation of increased performance. Transistor scaling has paved the way to integrate more cores at a lower power consumption, effectively increasing compute capacity per area [2]. Inherent parallelism, if present, in workloads could take advantage of multi, and many-core computers to show improved performance to a certain extent [24]. Dark silicon changes this consensus, since we cannot power up all the available resources at any given time [25]. Figure 1.5 shows cyclic dependency of performance requirements and dark silicon, where integration—the solution to high performance—becomes the subsequent cause for low performance. Consequently, there is no significant performance gain from technology node scaling, and moving into the future, there is no clear direction on attacking this problem [26].

Energy Efficiency

Extensive work has been done toward energy efficiency in computer systems [27–30], although they are restricted to dynamic power component. Deterioration in performance due to dark silicon also affects energy efficiency of the system [31]. With static power dominating the total power consumption, achievable performance in a fixed energy budget, i.e., *performance per watt*, is declining. Around half of the energy budget is wasted toward static power, literally implying spending energy for doing nothing. Assuming that transistor scales down by 30 % in physical dimensions, voltage and area being scaled the same way, energy savings should be as much as 65 % [14]. However, voltages and subsequently power densities are no longer scaling better, thus also limiting the possible energy gains.

Resource Allocation and Utilization

Applications that are designed at a higher level (e.g., algorithms, programming languages, compilers) and run-time and operating systems that allocate resources to these applications are blindsided since dark silicon is only visible at a lower level. Run-time resource allocation can influence performance and energy efficiency of many-core systems significantly [32]. A naive allocation can result in unbalanced load across the chip, which potentially might lead to hotspots. State-of-the-art run-time resource management techniques for many-core systems target benefits in terms of network performance, minimizing congestion, system throughput, power optimization, etc. [33–35], without considering the dark silicon scenario at all. Subsequently, schedulers allocate tasks to hardware expecting them to finish efficiently, while inefficiency is inevitable. As a result, most of the system resources go underutilized and poorly allocated, limiting both performance and energy gains.

Thermal Management

High power density and accumulation of heat results in an increase of average and peak temperatures of the chip. This opens the window for manifesting soft errors and unreliability in computation [36], in addition to decrease in lifetime of the chip [37]. This ages the chip faster and puts an additional penalty on manufacturing cost per reliable chip. Another important aspect is the exponential dependence of gate oxide leakage on temperature. A chip working at full throttle can dissipate more heat and thus high temperature, which in turn increases leakage too [20]. Again, the omnipresent cyclic dependence of temperature, leakage, performance, and energy efficiency surface here. Further, dissipating heat requires a heat sink and a fan that blows air through it, adding to manufacturing costs. For a data-centric level compute platform, more sophisticated cooling solutions have to be used, further increasing maintenance costs.

Solutions for Dark Silicon

Architectural and run-time management techniques can mitigate dark silicon to a large extent and improve energy efficiency. In this book, some of the potential directions and solutions to attack, mitigate, and exploit dark silicon are discussed from an architectural and run-time management perspectives. Figure 1.6 summarizes a typical heterogeneous multi-core/many-core computing platform implementing design time and run-time optimization techniques tailored for dark silicon era.

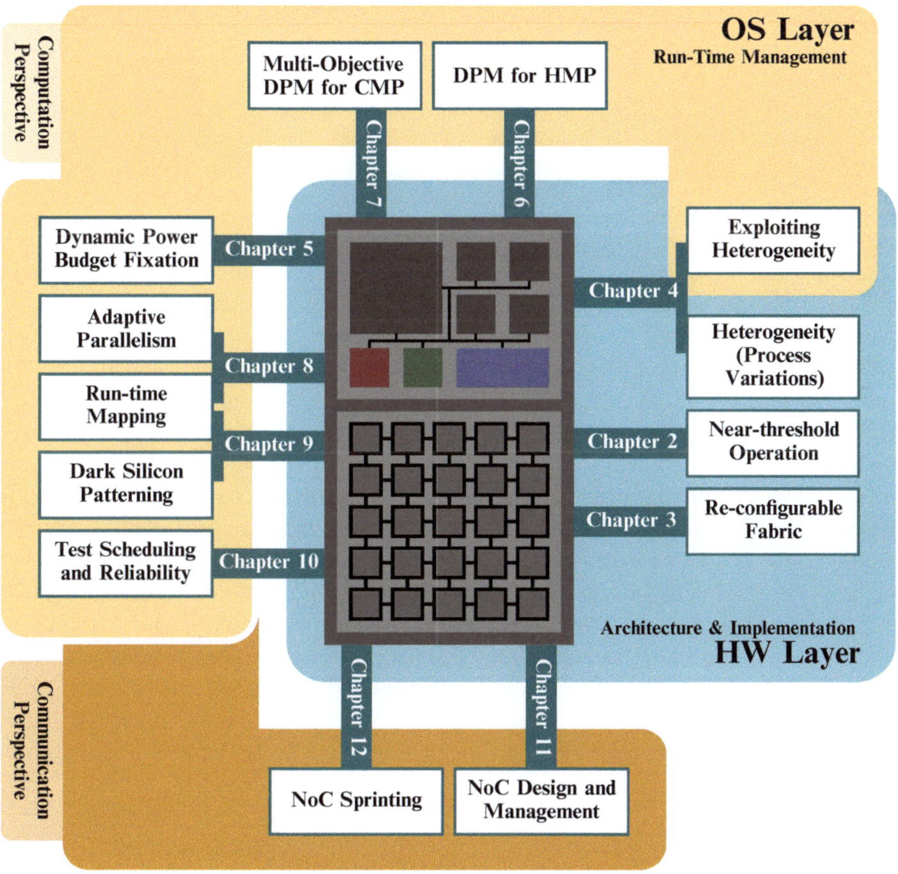

Fig. 1.6 Design and run-time solutions to dark silicon

A NoC-based many-core system is chosen to cater wider implementation scenarios with future application characteristics in view, while traditional bus-based multi-core system is used to better represent heterogeneity. The *architecture and design* phase is a design time approach, based on architectural asymmetry and heterogeneity of cores and task-specific accelerators that are expected to run energy efficiently. All the three phases augment each other in avoiding, mitigating, or minimizing dark silicon.

Architecture and Implementation Perspective

Operating cores with same instruction set architecture (ISA) but different micro-architectures, possibly at different voltage and frequency levels, results in asymmetric cores with different power-performance characteristics. Such an asymmetric multi-core system offers the choice of executing specific application on a core with suitable power-performance characteristics that can improve energy efficiency.

As of now, multi-core systems have been using more common conventional methods such as DVFS and clock and power gating for power management. As an advancement, near-threshold computing (NTC), where the chip's supply voltage is scaled down to as low as threshold voltage, offers ultralow power consumption, at the loss of performance. The cores that operate at low power and performance have penalty on per-core performance, yet the overall throughput of the chip is better than in case of dark cores that are simply idle. Near-threshold computing leads to better per-chip-throughput cores, also termed dim silicon. Energy efficiency can be improved by selectively downscaling the voltage, inducing dim silicon over dark silicon. Chapter 2 focuses on aspects of near-threshold computing and dim silicon for mitigating dark silicon. It provides *Lumos*, a framework for design space exploration of future many-core systems quantifying dark silicon effect. The framework analytically models technology node scaling effects at near-threshold operation.

Current compute platforms are still largely homogeneous, calling for innovations from an architectural standpoint. Designing cores that are different at micro-architectural as well as instruction set architectural (ISA) levels provides core-level heterogeneity. Heterogeneous systems offer highly customized processing units that can run specific tasks and applications at high performance and low energy consumption. Emerging accelerators like neural network, fuzzy, quantum, and approximation paradigms are expected to dominate future many-core systems. Both asymmetric and its super set, heterogeneous cores, provide customized hardware that can run certain tasks or applications more efficiently at the loss of generality. Yet, appropriate design and choice of asymmetric and/or heterogeneous cores can mitigate dark silicon significantly, over than conventional homogeneous systems that use same cores for even different tasks. Further, variable quality of service (QoS) and throughput requirements of certain applications presents a chance for coarse- and fine-grained (i.e., per-cluster and per-core) dynamic voltage and frequency scaling to enhance overall energy efficiency. Chapter 3 presents a special case of heterogeneous design leveraging process variations induced by technology scaling. Further, they provide architecturally synthesized heterogeneous platform for minimizing dark silicon and analyzing the parameters of area and power budgets, workloads to be executed, and a library of cores.

Design of reconfigurable logic at different levels of granularity lends adaptability to the architecture. The compute platform can be reconfigured on the fly subject to power consumption and workload characteristics, improving performance and energy efficiency along with area. Further, domain-specific custom accelerators implemented on reconfigurable fabric can be tightly coupled with traditional cores, making it a special case of heterogeneity. Dark silicon can be minimized by adjusting the granularity and the extent of reconfigurability. Further, dynamic compilation techniques that control degree of parallelism and choice of specialized hardware make the decision of reconfiguring hardware more comprehensive. Chapter 4 proposes use of a coarse-grained reconfigurable fabric as an alternative to traditional core design for mitigating dark silicon. Hardware architecture, storage, and interconnects are reconfigured dynamically in view of workload and resource availability through dynamic compilation and run-time decision making.

Run-Time Resource Management: Computational Perspective

Run-time management operates in an observe-decide-act loop to enhance the energy efficiency of underlying hardware. Observe phase involves monitoring critical parameters of chip's performance including instantaneous power consumption, active and dark cores, network congestion, temperature accumulation, etc. The decide phase uses a pro-active strategy, relying on application mapping, resource allocation, and scheduling to reduce potential dark silicon. Act phase employs a reactive strategy that applies different actuators, such as clock and power gating, voltage and frequency scaling, hardware reconfiguration, and task migration, to mitigate dark silicon.

Current computer systems use a conservative estimate of upper bound on power budget, the thermal design power (TDP). However, a fixed TDP value for variable application scenarios results in poor utilization of available power budget. Run-time estimation of power budget subject to the set of applications currently executing on the chip and their spatial alignment can provide a thermally safe upper bound on power budget, variable at run-time. This variable upper bound, thermal safe power (TSP), improves utilization of available power budget and minimizes dark silicon that otherwise manifests from conservative design time estimate of TDP. Chapter 5 introduces the concept of TSP, the variable yet safe upper bound on power budget, evaluated dynamically based on current set of applications being executed. It also provides a lightweight C instrumentation library that can compute the safe power budget, thermal safe power (TSP), on the fly as a function of spatial alignment of working cores.

Individual actuators like voltage and frequency scaling including near-threshold operation, clock and power gating, task allocation and migration, etc., can manage power consumption, occasionally by sacrificing performance. However, a suitable combination of more than one of these knobs will guarantee an acceptable level of performance within a given energy budget. Dynamic power management (DPM) techniques that use the right combination of these knobs can monitor and manage power consumption with a fine-grained control. Observation, decision making, and control of power by considering all possible knobs and actuators would yield a comprehensive power management platform tailored for the dark silicon era. Chapter 6 present dynamic power management (DPM) framework for asymmetric and heterogeneous multi-processor systems (HMPs), employing per-cluster dynamic voltage and frequency scaling. Subject to TDP and QoS demands of applications, it further enhances energy efficiency using run-time task migration and task allocation techniques.

Traditional power management techniques were dominantly triggered by instantaneous power consumption and actuating voltage and frequency. With dark silicon scenario, a combination of several critical parameters should be used as knobs to monitor, and similarly, a combination of several actuators are needed for fine-grained power management. Knobs that are needed to be monitored in addition to power consumption are upper bound on power budget (TDP/TSP), critical temperature, thermal profile, network congestion, and applications' characteristics such as arrival, QoS demands, criticality, etc. Adaptive actuators, like adjustable degree of parallelism (DoP), computational and interconnect sprinting, application mapping (mapping/queuing the application), contiguity and dispersion, and coarse- and fine-grained re-configurability, can be used in addition to traditional actuators like voltage and frequency scaling and clock and power gating. Chapter 7 proposes a comprehensive power management framework for chip multi-processors (CMPs) for the dark silicon era that considers workload characteristics, network congestion, QoS demands, and per-core and per-chip power requirements. They use actuators like voltage and frequency scaling, clock and power gating, and mapping decisions, within available power budget, maximizing utilization. It is to be noted that resource allocation though run-time application mapping lends a fine-grained control on several critical chip parameters and impacts performance and energy efficiency.

Dynamic application mapping involves allocating task per core at run-time, given an unpredictable sequence of incoming applications. An efficient mapping affects network parameters like congestion, dispersion, spatial availability, power consumption, etc. With varied limitations of parallelism and sequential bottleneck, different applications run efficiently at a different degrees of parallelism. Also considering heterogeneous systems, it is advantageous to consider task per core from the degree of parallelism standpoint. Chapter 8 presents a run-time mapping approach controlling the degree of parallelism of applications to minimize average energy consumption. In addition, they provide reliability and soft error awareness leveraging process variations to find optimal set of cores for mapping an application.

Dark silicon agnostic techniques greedily strive for performance via contiguous mappings. Given variable workload characteristics and application scenarios, aforementioned TSP can be maximized with appropriate dark silicon-aware application mappings. Mapping applications sparsely can avoid potential hotspots and distribute heat accumulation evenly across the chip area, in turn increasing the safer limit of power budget (TSP). Gain in power budget may be used to activate more cores improving performance and resource utilization, over compensating the loss from dispersed mapping. Chapter 9 proposes dark silicon patterning, an efficient run-time mapping scheme that maximizes power budget and utilization. It employs sparsity among mapped applications to balance heat distribution across the chip and thus maximize utilizable power budget. Inevitable dark cores are patterned alongside active cores providing the necessary cooling effect.

Despite the loss in performance and energy efficiency caused by dark silicon, an alternative approach to attack dark silicon is to exploit it for chip maintenance tasks like testing, reliability, and balancing the lifetime. Testing a core consumes power resources and interrupts applications' execution. Test procedures can be scheduled on specific set of cores based on chip activity such that available power budget is efficiently utilized for both computation and testing, which otherwise is wasted as dark cores. Chapter 10 exploits dark silicon to schedule test procedures on idle cores. With new application to be mapped, they leverage the possibility of few cores being left idle due to limited power budget and schedule test procedures on such cores. They propose a mapping technique that selects working cores to evenly balance the idle periods for testing.

Design and Management: Communication Perspective

Interconnection network forms the other major components of a computing platform, accounting for performance latency and power consumption. Power management actuators, like power gating and voltage and frequency scaling, can also be applied to interconnect components at both design time and run-time, giving fine-grained control. An optimized network with suitable energy-efficient and reliable interconnect components can increase overall energy efficiency of the system. Chapter 11 proposes *darkNoC*, an adaptive design time energy-efficient interconnect for exploiting dark silicon, over conventional voltage and frequency scaling. They also present *SuperNet*, a combinatorial run-time optimizable interconnection approach for executing in either energy efficiency, performance, or reliability modes, through voltage and frequency scaling.

Operating a section of the chip at highest possible voltage and frequency for a brief period of execution time is known as computational sprinting. Sprinting can accelerate performance with an increase in power consumption, possibly violating the safe upper bound. However, the sprinting time is kept low enough so that increase in power does not reflect directly in temperature accumulation. Since thermal safety is maintained, relatively high performance and utilization can be achieved despite dark silicon. Sprinting can be employed to specific tasks or applications particularly on requirements and demands. Chapter 12 proposes fine-grained interconnect sprinting that activates only a selected number of routers for short burst of heavy communication, instead of activating all the communication resources. Specific communication channels are chosen for this purpose based on application and energy demands.

Summary

Dark silicon has become a growing concern for computer systems ranging from mobile devices up to data centers. Architecture, design, and management strategies need rethinking when moving into the future exascale computing. Although dark silicon scenario is a challenge, it still presents opportunities to innovate and pursue new research directions. An overview of dark silicon regime has been presented in this chapter. State-of-the-art solutions to minimize, mitigate, and attack dark silicon are organized into the subsequent chapters of this book. Several interesting discussions and key insights leading to open research directions addressing energy efficiency are covered.

References

1. H. Sutter, The free lunch is over: a fundamental turn toward concurrency in software. Dr. Dobb's J. **30**(3), 202–210 (2005)
2. G.E. Moore, Cramming more components onto integrated circuits. Proc. IEEE **86**(1), 82–85 (2002)
3. R.H. Dennard, V. Rideout, E. Bassous, A. Leblanc, Design of ion-implanted mosfet's with very small physical dimensions. IEEE J. Solid-State Circuits **9**(5), 256–268 (1974)
4. M.-H. Haghbayan, A.-M. Rahmani, A.Y. Weldezion, P. Liljeberg, J. Plosila, A. Jantsch, H. Tenhunen, Dark silicon aware power management for manycore systems under dynamic workloads, in *Proceedings of IEEE International Conference on Computer Design* (2014), pp. 509–512
5. H. Esmaeilzadeh, E. Blem, R.S. Amant, K. Sankaralingam, D. Burger, Dark silicon and the end of multicore scaling, in *Proceedings of IEEE International Symposium on Computer Architecture* (2011), pp. 365–376
6. J. Kong, S.W. Chung, K. Skadron, Recent thermal management techniques for microprocessors, in *ACM Computing Surveys* (2012), pp. 13:1–13:42
7. Semiconductor Industry Association, International technology roadmap for semiconductors (ITRS), 2011 edition (2011)
8. H. Esmaeilzadeh, E. Blem, R.S. Amant, K. Sankaralingam, D. Burger, Dark silicon and the end of multicore scaling. IEEE Micro **32**(3), 122–134 (2012)
9. Semiconductor Industry Association, International technology roadmap for semiconductors (ITRS), 2013 edition (2013)
10. S. Borkar, The exascale challenge, in *Proceedings of the VLSI Design Automation and Test* (2010)
11. G. Venkatesh, J. Sampson, N. Goulding, S. Garcia, V. Bryksin, J. Lugo-Martinez, S. Swanson, M.B. Taylor, Conservation cores: reducing the energy of mature computations, in *ACM SIGARCH Computer Architecture News*, vol. 38(1) (2010), pp. 205–218

12. T. Mudge, Power: a first-class architectural design constraint. Computer **34**(4), 52–58 (2001)
13. S. Borkar, Getting gigascale chips: challenges and opportunities in continuing Moore's law. Queue **1**(7), 26 (2003)
14. S. Borkar, A.A. Chien, The future of microprocessors. Commun. ACM **54**(5), 67–77 (2011)
15. H. Sasaki, M. Ono, T. Yoshitomi, T. Ohguro, S.-I. Nakamura, M. Saito, H. Iwai, 1.5 nm direct-tunneling gate oxide Si MOSFET's. IEEE Trans. Electron Devices **43**(8), 1233–1242 (1996)
16. S. Borkar, T. Karnik, S. Narendra, J. Tschanz, A. Keshavarzi, V. De, Parameter variations and impact on circuits and microarchitecture, in *Proceedings of ACM Design Automation Conference* (2003), pp. 338–342
17. N.S. Kim, T. Austin, D. Baauw, T. Mudge, K. Flautner, J.S. Hu, M.J. Irwin, M. Kandemir, V. Narayanan, Leakage current: Moore's law meets static power. IEEE Comput. **36**(12), 68–75 (2003)
18. A.P. Chandrakasan, W.J. Bowhill, F. Fox, *Design of High-Performance Microprocessor Circuits* (Wiley-IEEE Press, New York, 2000)
19. C.-H. Choi, K.-Y. Nam, Z. Yu, R.W. Dutton, Impact of gate direct tunneling current on circuit performance: a simulation study. IEEE Trans. Electron Devices **48**(12), 2823–2829 (2001)
20. G. Sery, S. Borkar, V. De, Life is CMOS: why chase the life after? in *Proceedings of ACM Design Automation Conference* (2002), pp. 78–83
21. Intel Corporation. Intel Xeon Processor - Measuring Processor Power, revision 1.1. White paper, Intel Corporation, April 2011
22. Intel Corporation. Fourth Generation Mobile Processor Family Data Sheet. White paper, Intel Corporation, July 2014
23. AMD. AMD Kaveri APU A10-7800. Available: http://www.phoronix.com/scan.php?page=article&item=amd_a_45watt&num=1. Accessed 28 Feb 2015 [Online]
24. G.M. Amdahl, Validity of the single processor approach to achieving large scale computing capabilities, in *Proceedings of ACM Spring Joint Computer Conference* (1967), pp. 483–485
25. G. Venkatesh, J. Sampson, N. Goulding-Hotta, S.K. Venkata, M.B. Taylor, S. Swanson, Qscores: trading dark silicon for scalable energy efficiency with quasi-specific cores, in *Proceedings of IEEE/ACM International Symposium on Microarchitecture* (2011), pp. 163–174
26. H. Esmaeilzadeh, E. Blem, R.S. Amant, K. Sankaralingam, D. Burger, Power limitations and dark silicon challenge the future of multicore. ACM Trans. Comput. Syst. **30**(3), 11:1–11:27 (2012)
27. Y. Almog, R. Rosner, N. Schwartz, A. Schmorak, Specialized dynamic optimizations for high-performance energy-efficient microarchitecture, in *Proceedings of International Symposium on Code Generation and Optimization: Feedback-directed and Runtime Optimization* (2004), pp. 137–148
28. K. Khubaib, M.A. Suleman, M. Hashemi, C. Wilkerson, Y.N. Patt, MorphCore: an energy-efficient microarchitecture for high performance ILP and high throughput TLP, in *Proceedings of IEEE/ACM International Symposium on Microarchitecture* (2012), pp. 305–316
29. G. Semeraro, G. Magklis, R. Balasubramonian, D.H. Albonesi, S. Dwarkadas, M.L. Scott, Energy-efficient processor design using multiple clock domains with dynamic voltage and frequency scaling, in *Proceedings of IEEE International Symposium on High-Performance Computer Architecture* (2002), pp. 29–40
30. D.M. Brooks, P. Bose, S.E. Schuster, H. Jacobson, P.N. Kudva, A. Buyuktosunoglu, J.-D. Wellman, V. Zyuban, M. Gupta, P.W. Cook, Power-aware microarchitecture: design and modeling challenges for next-generation microprocessors. IEEE Micro **20**(6), 26–44 (2000)
31. M.B. Taylor, Is dark silicon useful?: harnessing the four horsemen of the coming dark silicon apocalypse, in *Proceedings of ACM Design Automation Conference* (2012), pp. 1131–1136
32. E.L. de Souza Carvalho, N.L.V. Calazans, F.G. Moraes, Dynamic task mapping for MPSoCs. IEEE Des. Test Comput. **27**(5), 26–35 (2010)
33. C.-L. Chou, R. Marculescu, Contention-aware application mapping for network-on-chip communication architectures, in *Proceedings of IEEE International Conference in Computer Design* (2008), pp. 164–169

34. M. Fattah, M. Daneshtalab, P. Liljeberg, J. Plosila, Smart hill climbing for agile dynamic mapping in many-core systems, in *Proceedings of IEEE/ACM Design Automation Conference* (2013)
35. M. Fattah, P. Liljeberg, J. Plosila, H. Tenhunen, Adjustable contiguity of run-time task allocation in networked many-core systems, in *Proceedings of IEEE Asia and South Pacific Design Automation Conference* (2014), pp. 349–354
36. J. Srinivasan, S.V. Adve, P. Bose, J.A. Rivers, The case for lifetime reliability-aware microprocessors, in *Proceedings of IEEE International Symposium on Computer Architecture* (2004), pp. 276–287
37. J. Srinivasan, S. Adve, P. Bose, J. Rivers, The impact of scaling on processor lifetime reliability, in *Proceedings of International Conference on Dependable Systems and Networks* (2004)

Chapter 2
Dark vs. Dim Silicon and Near-Threshold Computing

Liang Wang and Kevin Skadron

Introduction

Threshold voltage scales down more slowly in current and future technology nodes to keep leakage power under control. In order to achieve fast switching speed, it is necessary to keep transistors operating at a considerably higher voltage than their threshold voltage. Therefore, the dwindling scaling on threshold voltage leads to a slower pace of supply voltage scaling. Since the switching power scales as CV^2f, Dennard Scaling can no longer be maintained, and power density keeps increasing. Furthermore, cooling cost and on-chip power delivery implications limit the increase of a chip's thermal design power (TDP). As Moore's Law continues to double transistor density across technology nodes, total power consumption will soon exceed TDP. If high supply voltage must be maintained, future chips will only support a small fraction of active transistors, leaving others inactive, a phenomenon referred to as "dark silicon" [6, 26].

Since dark silicon poses a serious challenge for conventional multi-core scaling [6], researchers have tried different approaches to cope with dark silicon, such as "dim silicon" [11] and customized accelerators [23]. Dim silicon, unlike conventional designs working at nominal supply, aggressively lowers supply voltage close to the threshold to reduce dynamic power. The saved power can be used to activate more cores to exploit more parallelism, trading off per-core performance loss with better overall throughput [7]. However, with near-threshold supply, dim silicon designs suffer from diminishing throughput returns as the core number increases. On the other hand, customized accelerators are attracting more attention due to their orders of magnitude higher power efficiency than general-purpose processors [5, 9]. Although accelerators are promising in improving performance

L. Wang (✉) • K. Skadron
University of Virginia, Charlottesville, VA, USA
e-mail: lw2aw@virginia.edu; skadron@cs.virginia.edu

A.M. Rahmani et al. (eds.), *The Dark Side of Silicon*,
DOI 10.1007/978-3-319-31596-6_2

with less power consumption, they are built for specific applications, have limited utilization in general-purpose applications, and sacrifice die area that could be used for more general-purpose cores. Poorly utilized die area would be very costly if there are more efficient solutions, in terms of opportunity cost. Therefore, the utilization of each incremental die area must be justified with a concomitant increase in the average selling price (ASP).

To investigate the performance potential of dim cores and accelerators, we have developed a framework called *Lumos*, which extends the well-known Amdahl's Law, and is accompanied by a statistical workload model. We investigate two types of accelerators: application-specific logic (ASIC) and RL. When we refer to RL, we generically model both fine-grained reconfigurability such as FPGA and coarse-grained such as Dyser [8]. We also assume the reconfiguration overhead is overwhelmed by sufficient utilization of each single configuration. In this chapter, we show that

- Systems with dim cores achieve up to 2× throughput over conventional CMPs, but with diminishing returns.
- Hardware accelerators, especially RL, are effective to further boost the performance of dim cores.
- Reconfigurable logic is preferable over ASICs on general-purpose applications, where kernel commonality is limited.
- A dedicated fixed logic ASIC accelerator is beneficial either when: (1) its targeted kernel has significant coverage (e.g., twice as large as the total of all other kernels), or (2) its speedup over RL is significant (e.g., 10×–50×).

Related Work

The power issue in future technology scaling has been recognized as one of the most important design constraints by architecture designers [15, 26]. Esmaeilzadeh et al. performed a comprehensive design space exploration on future technology scaling with an analytical performance model in [6]. While primarily focusing on maximizing single-core performance, they did not consider lowering supply voltage, and concluded that future chips would inevitably suffer from a large portion of dark silicon. In [2], Borkar and Chien indicated potential benefits of near-threshold computing with aggressive voltage scaling to improve the aggregate throughput. We evaluate near-threshold in more detail with the help of Lumos calibrated with circuit simulations. In [11], Huang et al. performed a design space exploration for future technology nodes. They recommended dim silicon and briefly mentioned the possibility of near-threshold computing. Our work exploits circuit simulation to model technology scaling and evaluates in detail the potential benefit of improving aggregate throughput by dim silicon and near-threshold computing. In short, Lumos allows design space exploration for cores running at near-threshold supply factoring in the impact of process variations.

Another set of studies has evaluated the benefit of hardware accelerators as a response to dark silicon. Chung et al. studied the performance potentials of GPU, FPGA, and ASIC in [5], regarding physical constraints of area, power, and memory bandwidth. Although a limited number of applications were studied in their work, our work corroborates that reconfigurable accelerators, such as FPGA, are more competitive than dedicated ASICs. Wu and Kim did a study across several popular benchmark suites regarding the targets to be accelerated in [27]. Their work suggested a large number of dedicated accelerators are required to achieve a moderate speedup due to the minimal functional level commonality in benchmark suites like SPEC2006. This is consistent with our observation that dedicated fixed logic accelerators are less beneficial due to limited utilization across applications in a general-purpose workload, and the importance of efficient, reconfigurable accelerators. In [24], Tseng and Brooks build an analytical model to study tradeoffs between latency and throughput performance under different power budgets and workloads. However, the model lacks the support of voltage and frequency scaling, especially near threshold and the capability of hardware accelerator performance modeling, limiting the design space explorations of future heterogeneous architectures. In [28], Zidenberg et al. propose MultiAmdahl model to study the optimal allocations to each type of computational units, including conventional cores and u-cores. The MultiAmdahl model cannot support voltage scaling and near-threshold effects. Therefore, its design space exploration capability is limited to heterogeneous systems under dark/dim silicon projections. Lumos takes a broader view of these issues, by providing a highly configurable tool for exploring the design space composed of conventional cores, various forms of accelerators, and low-voltage operation, with various power constraints and various workload characteristics.

Lumos Framework

The design space of heterogeneous architectures with accelerators is too huge to use extensive architectural simulation in a feasible way. To enable rapid design space explorations, we build Lumos, a framework putting together a set of analytical models for power-constrained heterogeneous many-core architectures with accelerators. Lumos extends the simple Amdahl's Law model by Hill and Marty [10] with physical scaling parameters calibrated by circuit simulations with a modified Predictive Technology Model (PTM) [4]. Lumos is composed of:

1. a technology model for inter-technology scaling;
2. a model for voltage–frequency scaling;
3. a model evaluating power and performance of conventional cores;
4. a model evaluating power and performance of accelerators;
5. a model for workloads and applications.

Technology Modeling

Technology scaling is built based on the modified PTM, which tunes transistor models with more up-to-date parameters extracted from recent publications [4]. Two types of technology variants are investigated: a high-performance process and a low-power process. The primary difference is that low-power processes have a higher threshold, as well as a higher nominal supply voltage, than high-performance processes for every single technology node, as shown in Table 2.1.

Inter-technology scaling mainly deals with scaling ratios for the frequency, the dynamic power, and the static power. From the circuit simulation, we compare the frequency, the dynamic power, and static power at the nominal supply for each of technology nodes. We assume CPU cores will have the same scaling ratio as the circuit we simulated. As shown in Table 2.2, the scaling factors of high-performance variants are normalized to 45 nm, while the scaling factors of low-power variants are normalized to their high-performance counterparts.

Table 2.1 Nominal supply voltage (V_{nom}) and threshold voltage (V_t) for each PTM technology variants

		45 nm	32 nm	22 nm	16 nm
High	V_{nom}	1.0	0.9	0.8	0.7
Perf.	V_t	0.424	0.466	0.508	0.505
Low	V_{nom}	1.1	1.0	0.95	0.9
Power	V_t	0.622	0.647	0.707	0.710

Voltage units are volt (V)

Table 2.2 Technology scaling for high-performance (HP) processes and low-power (LP) processes

Tech. (nm)	Freq.	Switch power	Static power
(a) HP, normalized to 45 nm			
45	1.0	1.0	1.0
32	0.95	0.49	0.31
22	0.79	0.21	0.12
16	0.66	0.09	0.13
(b) LP, normalized to HP counterparts			
45	0.27	0.30	0.0084
32	0.28	0.32	0.042
22	0.26	0.34	0.23
22	0.26	0.40	0.49

The area scaling is assumed to be ideal so that the area of a single-core shrinks by 2× per technology node

Frequency Modeling

Voltage-to-frequency scaling is modeled by interpolating empirical results from circuit simulations. We use a 32-bits ripple carry adder (RCA) to emulate the core frequency changes subject to various supply voltages. The adder is synthesized with a standard-cell library [16] at 45 nm. Sizes of transistors in the post-synthesized netlist are scaled to other technology nodes. These circuit netlists are then simulated using a modified version of PTM for the bulk devices. Note that Lumos is not limited to the bulk devices, but can be extended to other technologies by applying the same RCA characterization with the target technology library. More specifically, Lumos can be extended with minimal efforts to support FinFET devices by using the latest PTM from [22].

The switching speed of transistors scales exponentially to the threshold voltage when it is operating at near-threshold voltage; therefore, the working frequency of a circuit in the near-threshold region is extremely sensitive to the threshold voltage. To investigate the fluctuation of frequency subject to process variations, the test circuit has been simulated with Monte-Carlo analysis, assuming a standard Gaussian distribution with 3-σ on the threshold voltage of transistors.

The results for 16 nm are shown in Fig. 2.1. For the high-performance process plotted in Fig. 2.1a, the maximum performance penalty due to process variations is less than 10 % when the circuit is operated at higher voltages around 1 V. However, when supply voltage approaches the threshold, frequency penalty increases significantly to more than 90 %, which means an order of magnitude slowdown on frequency. A similar trend is observed in Fig. 2.1b for the low-power process, though the low-power process suffers from a larger percentage of frequency loss than the high-performance process even at a higher supply voltage, e.g., 1 V.

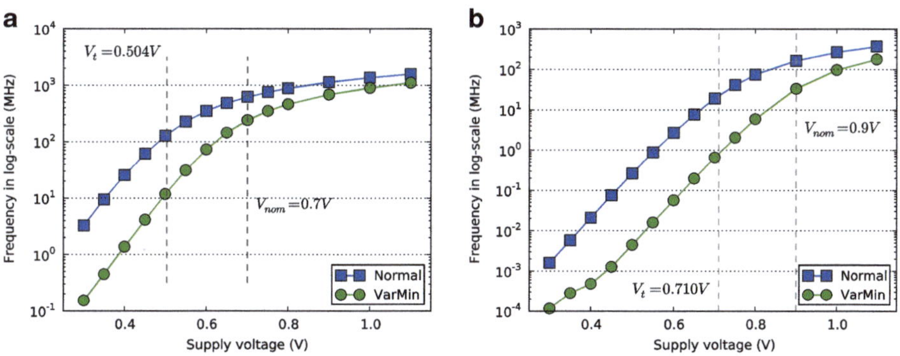

Fig. 2.1 Voltage–frequency scaling characterization on a 32-bits ripple carry adder (RCA) simulated at 16 nm, subjecting to process variation (VarMin). V_t is the threshold voltage, and V_{nom} is the nominal supply voltage. (**a**) High-performance with $V_t = 0.504$ V. (**b**) Low-power with $V_t = 0.710$ V

Core Modeling

Power

Core power consumption is modeled as two major sources: dynamic power due to transistor switching and static power due to leakage. Therefore, per-core power is given by

$$P_{\text{total}} = P_{\text{dynamic}} + P_{\text{leakage}} \tag{2.1}$$

Generally speaking, dynamic power is given by

$$P_{\text{dynamic}} = \alpha \cdot C_{eff} \cdot V_{dd}^2 \cdot f \tag{2.2}$$

where α is the activity factor, C_{eff} is the effective capacitance, V_{dd} is supply voltage, and f is the core running frequency. We assume a constant activity factor and effective capacitance when the core frequency is scaled with various supply voltages. Therefore, dynamic power changes quadratically to supply voltage and linearly to frequency. According to [3], the static power of a system is given by

$$P_{\text{leakage}} = V_{dd} \cdot N \cdot k_{design} \cdot \hat{I}_{leak} \tag{2.3}$$

where V_{dd} is supply voltage, N is the number of transistors, k_{design} is a device-specific constant, and $\hat{I}_{leak}$ is the normalized per-transistor leakage current. We assume the normalized per-transistor leakage current is proportional to the leakage current of a single transistor, which, according to [21], is given by

$$I_{leak} = I_0 \cdot 10^{\frac{V_{gs}-V_t+\eta V_{ds}}{S}} \cdot \left(1 - e^{-\frac{V_{ds}}{V_{th}}}\right) \tag{2.4}$$

where V_t is the threshold voltage, η is the drain-induced barrier lowering factor (DIBL), V_{th} is the thermal voltage, and n is a process-dependent constant. The thermal voltage at room temperature is around 28 mV, which is far less than the supply voltage of interest in this project, therefore, $e^{-V_{ds}/V_t} \approx 0$. Because the transistor is at its static state when considering the static leakage power, V_{gs} and V_{ds} are roughly proportional to the supply voltage. As a result, the above equation can be reduced to

$$\hat{I}_{leak} \propto 10^{\frac{V_{dd}}{\hat{S}_{leak}}} \tag{2.5}$$

where $\hat{S}_{leak}$ is the aggregate scaling coefficient derived from fittings to the simulated results.

Performance

Lumos uses aggregate throughput under physical constraints as the primary performance metric. For the aggregate throughput performance (pf_s), we model it with Amdahl's Law as shown in Eq. (2.6)

$$\text{Speedup} = \frac{1}{\frac{1-\rho}{S_{\text{serial}}} + \frac{\rho}{n \cdot S_{\text{parallel}}}} \tag{2.6}$$

where ρ is the parallel ratio of the studied workload, S_{serial} is the serial part speedup over a baseline core, S_{parallel} is the per-core speedup when the system works in parallel mode, and n is the number of active cores running in parallel. For S_{serial}, only one core is active. In this case, the core is boosted to 1.3× nominal supply to achieve the best single-core performance, denoted as $\widehat{pf}_c$. When the system is in parallel mode, both n and S_{parallel} are determined by supply voltage. First, Lumos picks the optimal frequency with a given supply voltage; Second, it calculates per-core power consumption including both the switching power and the leakage power according to the frequency and the supply. Finally, the number of active cores is the minimum restricted by the overall power and area budgets, which is given by

$$n(v) = \min(\frac{P}{p(v)}, \frac{A}{a}) \tag{2.7}$$

where $p(v)$ is the per-core power as a function of supply voltage, P and A are system budgets of power and area, respectively. As a result, Eq. (2.6) can be rewritten as

$$\text{Speedup} = 1 \Bigg/ \frac{1-\rho}{\frac{\widehat{pf}_c}{pf_0}} + \frac{\rho}{n(v) \cdot \frac{pf(v)}{pf_0}} \tag{2.8}$$

where pf_0 is the performance of a single baseline core, $pf(v)$ is the per-core performance as a function of supply voltage,

When considering process variation, performance loss of a single core leads to the poor throughput of a symmetric many-core system, in which the slowest core dictates the overall performance of the whole system. Adaptive Body Bias (ABB) is not considered here, but is important future work. However, the worst-case variation we used here sets an upper bound on the benefits of ABB. As a result, the parallel performance of a system is confined to the core with the worst performance, denoted as $pf(v)_{min}$. As for the power, we assume no fine-grained per-core level power management mechanisms; therefore, all cores contribute to the total power consumption through the whole period that the system is in parallel mode. We use the total mean power ($\overline{p}$) to estimate the per-core power, so Eq. (2.7) is rewritten to

Table 2.3 Characteristics of a single Niagara2-like in-order core at 45 nm, obtained from McPAT [14]

Frequency (GHz)	Dynamic power (W)	Leakage power (W)	Area (mm^2)
4.20	6.14	1.06	7.65

$$n(v) = \min(\frac{P}{\overline{p}}, \frac{A}{a}) \tag{2.9}$$

When considering process variations, the overall speedup in Eq. (2.8) is rewritten to

$$\text{Speedup} = 1 \Bigg/ \frac{1-\rho}{\frac{\widehat{pf}_c}{pf_0}} + \frac{\rho}{n(v) \cdot \frac{pf(v)_{min}}{pf_0}} \tag{2.10}$$

Baseline

A Niagara2-like in-order core is chosen as a single-core baseline design. The characteristics of this baseline core at 45 nm are obtained from McPAT [14] and summarized in Table 2.3.

Accelerator Modeling

We model the speedup and the power consumption of RL and customized ASIC for a given kernel by u-core parameters (η,ϕ). "u-core" is a term proposed by Chung et al. [5] to represent any single computational unit that is not a conventional CPU core. Examples of "u-core" include dedicated hardware accelerators and reconfigurable accelerators. η[1] is the power relative to a single, basic in-order core, and ϕ is the performance relative to the same baseline in-order core. We assume that u-cores are only used to accelerate kernels that are ideally parallelized. Therefore, we model the relative performance of a u-core proportional to its area:

$$\text{Power}_{ucore} = \eta \cdot \frac{A_{ucore}}{A_0}, \quad \text{Perf}_{ucore} = \phi \cdot \frac{A_{ucore}}{A_0}$$

where A_{ucore} is the area of a u-core, and A_0 is the area of a single baseline in-order core.

[1]In the original paper of [5], μ is used to denote relative performance of u-cores. However, μ is commonly used as the mean in statistics. Therefore, we use η as an alternative to avoid confusions.

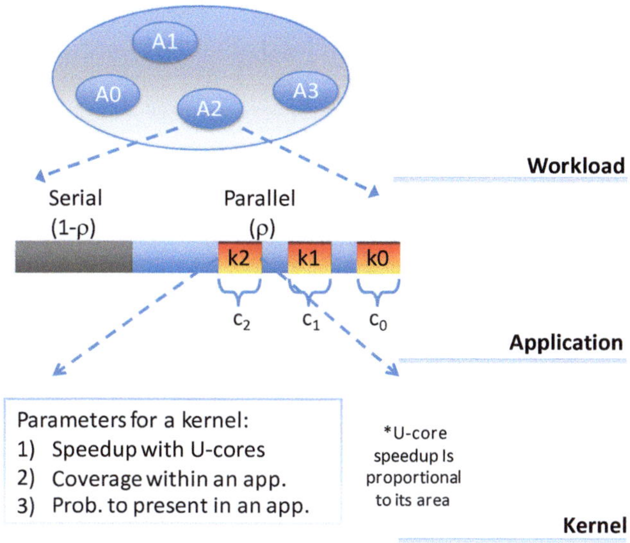

Fig. 2.2 A workload is composed of applications. An application may have several computing kernels, which can be accelerated by many-core parallelization, RL, and ASIC

Workload Modeling

Figure 2.2 illustrates the workload model, where a workload is composed of a pool of applications. Each single application is divided into two parts: serial and parallel. Part of an application can also be partitioned into several computing kernels. These kernels can be accelerated by various computing units, such as multi-core, possibly dim CPU cores, an RL accelerator, or a dedicated ASIC accelerator. We add two more parameters to model the relationship between applications and kernels:

Presence (λ) A binary value to indicate whether or not a kernel present in an application.

Coverage (ε) The time consumption in percentage for a kernel when the whole application is running with a single base line core.

To model these two parameters, we have profiled the PARSEC benchmark suite [1] using Valgrind [18]. We use native input set for all applications except for x264, due to that Valgrind failed to capture source annotations for x264. We extract top kernels for every single application. Their presence and coverage, which are plotted in Fig. 2.3, show trends:

- Kernels, such as library calls, have a larger probability of presence, but lower coverage in a single application;
- Kernels, such as application-specific routines, have very limited presence, but higher coverage once they present;
- The majority of kernels are application-specific.

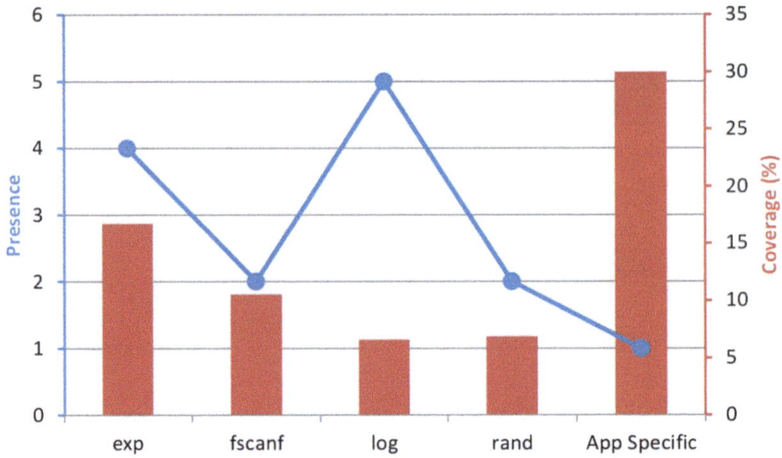

Fig. 2.3 Kernels coverage and presence. Kernels are extracted from PARSEC applications, excluding x264. The coverage percentages of *exp*, *fscanf*, *log*, and *rand* are averaged across their all occurrences. The coverage percentage of the "app specific" is the average of all application-specific kernels

In addition to general-purpose benchmark suites like PARSEC, we have observed a similar trend on domain-specific benchmarks suite such as SD-VBS [25] and Minebench [17]. Based on these observations, we propose a statistical approach to model kernels' speedup (η), presence (λ), and coverage (ε), which is shown in Eq. (2.11).

$$\begin{cases} \eta \sim \mathcal{N}(\mu_s, \sigma_s^2) \\ \lambda = B(1,p) \qquad \text{where } p = PDF(\eta) \\ \varepsilon \sim \mathcal{N}(\mu_c, \sigma_c^2) \end{cases} \tag{2.11}$$

Lumos samples the speedup (η) of kernels from a normal distribution. Speedup values of common kernels (e.g., library calls) are close to the mean of the distribution. On the other hand, speedup values of application-specific kernels are sampled from both tails of the distribution. The left tail represents kernels that are hard to accelerate, such as control-intensive ones. The right tail denotes kernels that are highly efficient on accelerators, such as compute-intensive streaming ones. Lumos samples the presence (λ) of these kernels from a Bernoulli distribution where the distribution parameter equals to $PDF(\eta)$. Lumos then samples the coverage (ε) from an independent normal distribution. We justify coverage parameters using values extracted from PARSEC. We extract speedup values of RL accelerators by normalizing reported data from recent publications in the RL research community. Finally, we assume a fixed performance scaling ratio of 5× between an ASIC accelerator and its counterpart RL implication. Table 2.4 summarizes values of all parameters used in Lumos's workload model.

Table 2.4 Summary of statistical parameters for kernel modeling

μ_s RL	μ_s ASIC	σ_s RL	σ_s ASIC	μ_c	σ_c
40×	200×	10	50	40 %	10 %

Table 2.5 System configurations

System type	Area (mm^2)	TDP (W)	P. density (W/mm^2)[a]	# of cores[b]
Small	107	33	0.308	14
Medium	130	65	0.5	17
Large	200	120	0.6	26

All numbers are for core components, assumed to be 50 % of the total TDP and the die area
[a]P. density is defined as maximum power allowed for a unit area
[b]Computed based on the single-core area at 45 nm from Table 2.3

System Configuration

Lumos models the basic system as a symmetric multi-core system composed of in-order cores. We assume the power and the area of the last-level cache (LLC) remains a constant ratio to the whole system. According to [14], we take the assumption that un-core components attribute to 50 % of both the total thermal design power (TDP) and the die area. The bandwidth of the memory subsystem is assumed to be sufficient for future technology nodes by applying advanced techniques such as 3D-stacking, optical connections, and multi-chip modules. In the rest of this section, the power and the area of a system only refer to core components.

We have modeled three systems with different configurations of power and area by extracting data from commodity processors fabricated in 45 nm. They include a small system representing desktop level processors [12], a medium system for normal server processors [13], and a large system that represents the most aggressive server throughput processors [20]. One more parameter is introduced as the power density of a system, reflecting the maximum power allowed for a unit area. It is calculated by dividing the area from the TDP. Table 2.5 summarizes detailed parameters of system configurations for Lumos.

Within an application, we assume that each kernel takes a considerable amount of computation time for a single run. Therefore, we can ignore various overhead in reconfiguring accelerators, as well as setting up the execution context. We leave the detailed overhead modeling and the modeling of transient kernel behaviors as future work.

Discussions

Instead of detailed cycle-accurate simulation, Lumos takes an analytical approach with abstractions to model heterogeneous computer architectures composed of cores and accelerators. Due to its fast evaluation time, Lumos is able to explore a large design space that is prohibitively expensive for simulation-based approaches. However, the drawback is that it is not as accurate as simulation-based approaches for a specific design. Therefore, the best use case of Lumos is to explore design implications at early stages or provision future designs subject to the technology scaling. Be complementary to each other, Lumos can help to prune down the design space for detailed simulation, while characterizations from simulation-based approaches benefit Lumos with better modeling accuracy.

Lumos Release

Lumos is composed of a set of Python scripts that read in technology-specific values described in previous sections, as well as results from circuit simulations. Typically, Lumos needs a workload description as its input for analysis. This description is encoded in XML format, enumerating all applications that compose the workload. Each application is described by a parallel fraction, as well as possible kernels with their coverage in percentage. Each kernel is described by u-core parameters described in section "Workload Modeling". We have pre-compiled a couple of descriptions for kernels and workloads following distributions as described in section "Workload Modeling". Lumos provides APIs to generate these XML descriptions by user-specified distribution parameters. Lumos requires the user to specify the physical constraints of a system, such as TDP and area, then define systems by explicitly allocating area resources to any accelerators. Lumos allocates the rest of area to conventional cores to form a heterogeneous many-core system. When all these are ready, Lumos evaluates workload performance of the defined system or brute-force search system configurations to find the one with the best TDP-constrained performance. More details on its usage can be found at http://liangwang.github.com/lumos/tree/v0.1. Lumos is released under a BSD open-source license.

Design Space Exploration

Effectiveness of Dim Silicon with Near-Threshold Operation

Unlike dark silicon, which we define as maximizing the frequency of cores to improve the overall throughput at the expense of potentially turning some off, dim silicon aims to maximize chip utilization to improve overall throughput by exploiting more parallelism. Dark silicon systems apply the highest supply voltage, which is assumed to be 1.3× the nominal supply voltage in Lumos, to cores in parallel mode; while dim silicon systems scale down supply and pick the optimal voltage according to the overall throughput. Figure 2.4 shows the comparisons between dark silicon (dark Si.) and dim silicon (dim Si.).

For performance, dim silicon beats dark silicon with up to 2× throughput improvement at 16 nm. When variation comes into play, however, both systems experience throughput loss compared to non-variation cases, respectively. Dim

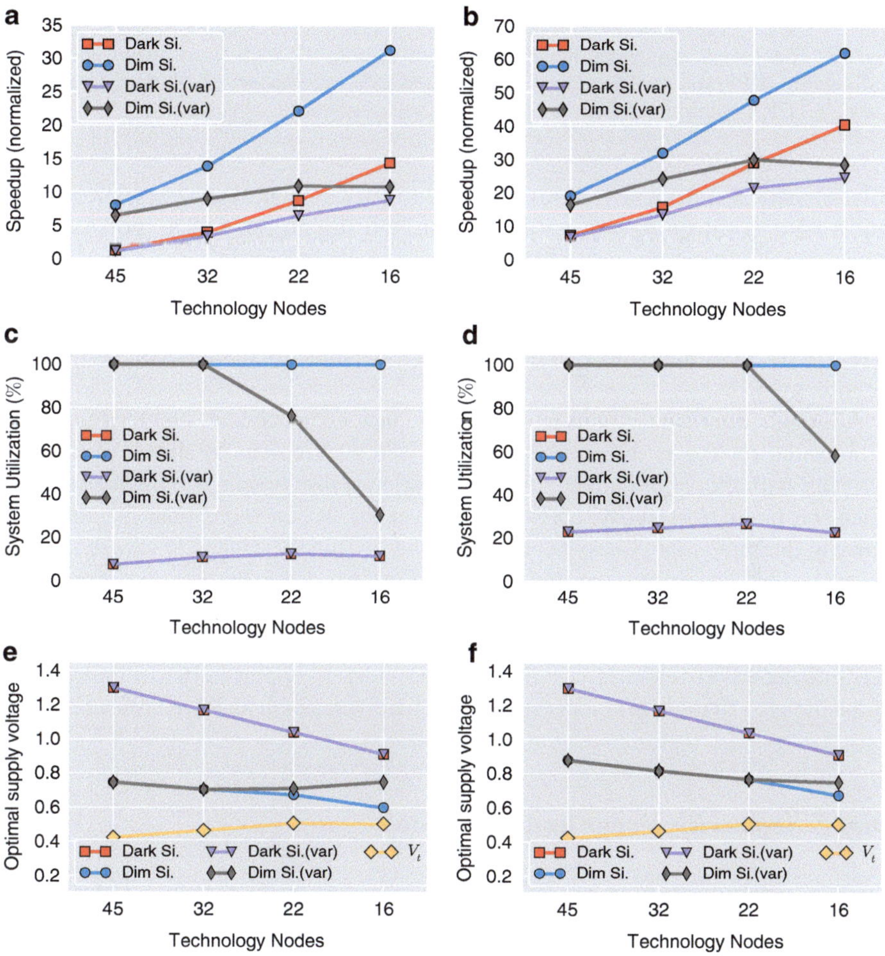

Fig. 2.4 Dark silicon vs. dim silicon, regarding the throughput-based speedup, die utilization, and the supply voltage with optimal throughput. Two system budgets (see Table 2.5) are compared: a small-budget system plotted on the *left* and a large-budget system plotted on the *right*. (**a**) Speedup (small). (**b**) Speedup (large). (**c**) Utilization (small). (**d**) Utilization (large). (**e**) Supply voltage (small). (**f**) Supply voltage (large)

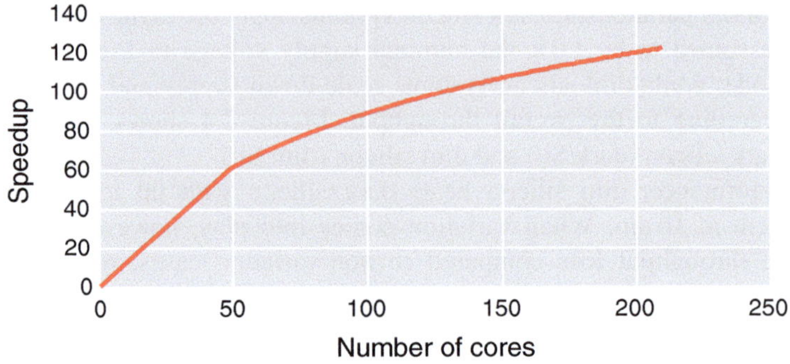

Fig. 2.5 The overall throughput-based speedup of in-order (IO) cores at 16 nm with high-performance (HP) process. The system is configured with large budget of 200 mm^2 in area and 120 W in TDP. Diminishing returns are observed starting around 50 active cores

silicon systems suffer even more compared to dark silicon, revealing more vulnerability of dim silicon to process variations. Dim silicon can utilize more cores to achieve high utilization, up to 100 % for the small system and the large system. Similarly, utilization of dim silicon systems is sensitive to the process variation. Being aware of variation, the small system with dim silicon sees the utilization drop to as low as 20 %, quite close to the utilization obtained by dark silicon at the same technology node.

Although dim cores manage to deliver higher throughput, they suffer from diminishing returns as the number of active cores increases. As shown in Fig. 2.5, a system with the large budget starts to experience diminishing performance gains when the number of active cores passes over 50. When more than 50 cores are active at the same time, the system has to lower its supply voltage to stay within the power budget. Therefore, the lower per-core frequency of dim cores compromises the performance improvement from more cores. As a result, it is not cost-effective to lower supply voltage aggressively to reach the optimal throughput. The limited performance improvement from dim cores provides an opportunity to allocate some of the die area to more efficient customized hardware, such as RL or ASICs.

Dim Silicon with Reconfigurable Logic

In this section, we explore the design space that includes conventional CPU cores and accelerators built on reconfigurable logic. Our goal is to find the optimal area allocation for RL that achieves the highest overall throughput on a given application. We start by looking at the scenario that every application in the workload has only a single kernel with a considerable amount of coverage, e.g., an application-specific kernel. To do this, we generate a pool of kernels from a population of distribution described in Table 2.4, as well as applications associated with each of

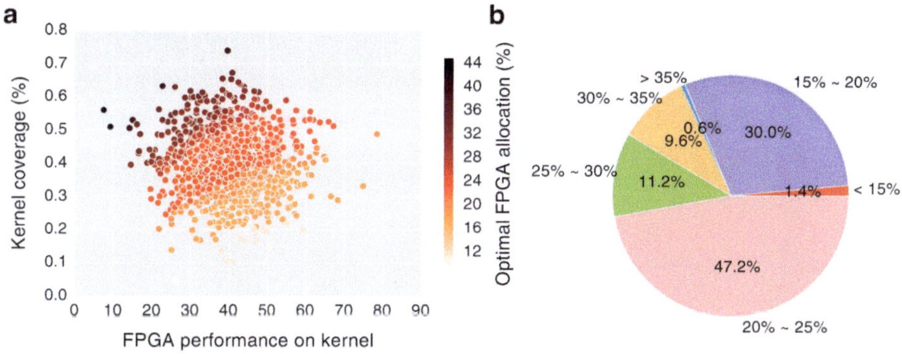

Fig. 2.6 Optimal area allocations on reconfigurable logic (RL) accelerators. Derived from exhaustive search, optimal RL allocations for each application are plotted in (**a**). In this plot, each of dots represents an application that has been characterized by the coverage of the kernel (*Y*-axis) and the kernel's u-core performance parameter (*X*-axis). The *colors of dots* indicate the optimal RL allocation in percentage for the given application, the darker the color the higher the optimal allocation. On the other hand, the distribution of optimal allocations is illustrated by a pie-chart in (**b**). Percentage numbers within the pie-chart suggest the distribution of each category, while the percentage numbers around the pie-chart indicate the range of optimal RL allocations within each category. (**a**) Optimal area allocations on RL. (**b**) Distribution of optimal RL allocations

generated kernels. To search for the optimal RL allocation, we adopt a brute-force approach by sweeping RL allocations with a step size of 1 % of the total area budget. As shown in Fig. 2.6, despite various kernel characteristics of each application, a majority of applications flavor RL allocations that are less than 30 % (Fig. 2.6a and b). This is because RL manages to accelerate the kernel significantly even with a small area allocation. As a result, the application performance is limited by the performance that dim silicon delivers on the rest part of the application, according to Amdahl's Law.

Considering a system with certain die area dedicated to RL, some applications may suffer from performance penalties when the allocation is not large enough. We compare the performance of a system to the best speedup ever achieved by varying the area allocation of RL. Figure 2.7a–d plots the normalized speedup of systems with RL allocation of 15 %, 20 %, 25 %, and 30 %, respectively. Systems with smaller RL allocations, such as 15 %, 20 %, and 25 %, have close-to-optimal performance for a majority of applications, with a worst-case performance loss of 35 % with 15 % allocation. On the other hand, the system with 30 % RL allocation only experiences less than 5 % penalties for most applications, and less than 10 % for the worst-case. This indicates that the speedup is somewhat insensitive to RL allocation within a reasonable range.

We vary statistical parameters of speedup and coverage, with five other alternatives listed in Table 2.6. The sensitivity study on these alternatives is summarized in Fig. 2.8. As the coverage drops down from high to low profile, the optimal area allocation drops down as well. This is because the lower the coverage of a single

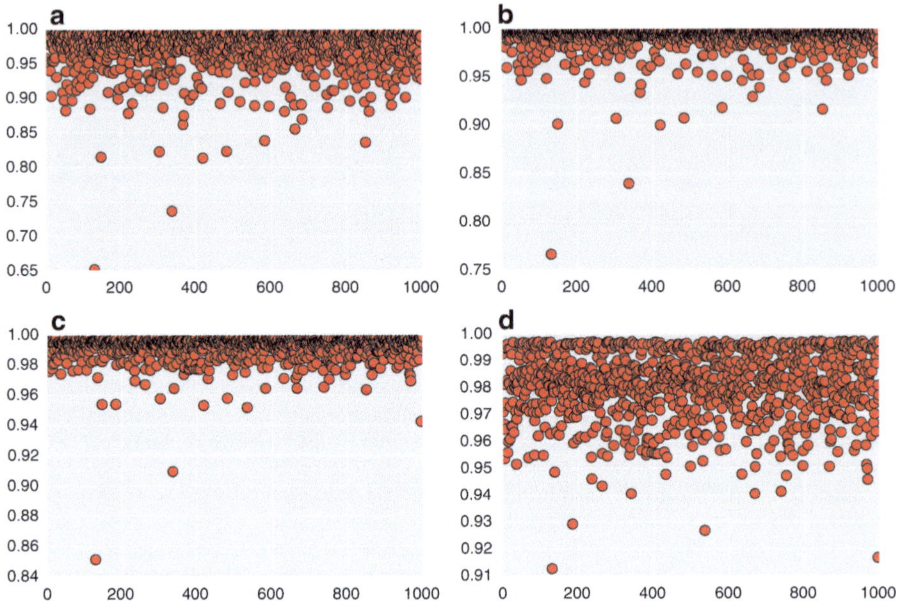

Fig. 2.7 The performance of systems with reconfigurable logic accelerators. For each application, we plot the performance of the given system relative to the system of the optimal RL allocation for the application. Four systems with alternative RL allocations are shown at 15 % in (**a**), 20 % in (**b**), 25 % in (**c**), and 30 % in (**d**). *X*-axis is the index number of an application within the workload. *Y*-axis is the relative performance achieved by the given system configuration relative to the performance achieved by the system with the optimal RL allocation. Note that each of dots is normalized to different baseline since the optimal RL allocation is application-specific (see Fig. 2.6). (**a**) 15 % RL allocation. (**b**) 20 % RL allocation. (**c**) 25 % RL allocation. (**d**) 30 % RL allocation

Table 2.6 Parameters for alternative distributions of speedup and coverage

	Speedup			Coverage	
	Fast	Medium	Slow	High	Low
μ	80	40	20	0.4	0.2
σ	20	10	5	0.1	0.05

kernel, the less significant the kernel speedup at any allocation. As the RL speedup goes from fast to slow, the optimal area allocation goes up. This is because the less the speedup of the RL on kernels, the more area it requires to achieve a comparable performance.

Then we take a look at a more realistic scenario, where applications follow characteristics defined by Eq. (2.11). We choose ten synthetic kernels with speedup performance sampled evenly from $\mu - 3\sigma$ through $\mu + 3\sigma$, representing various kernel performance and kernel's probabilities of presence in a specific application. We refer to these kernels with number 0 through 9. The accelerator speedup increases from kernel 0 through kernel 9, while the probability of presence peaks at

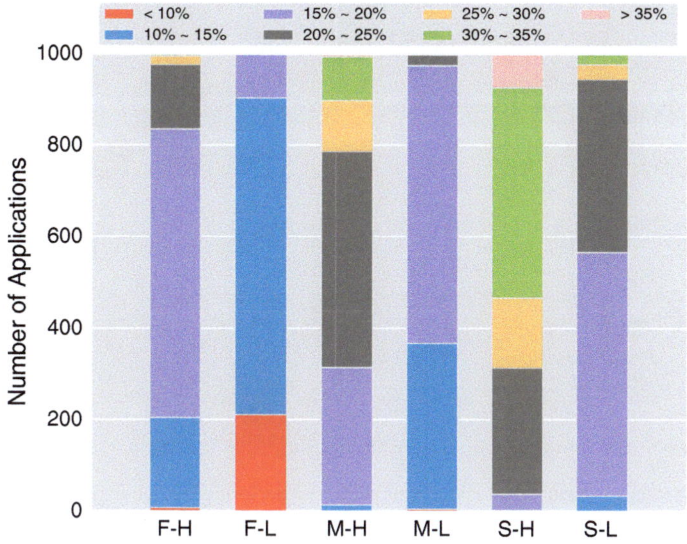

Fig. 2.8 Optimal RL allocations with alternative distributions for speedup and coverage. Distributions for speedup cover three scenarios from slow to fast, while distributions for coverage cover two scenarios from low to high. Parameters for these distributions are detailed in Table 2.6

kernel 4 and 5 and bottoms at kernel 0 and 9 (i.e., normal distribution). We draw a sample population of 500 applications. For each application, the sum of all kernels' coverage follows the given coverage distribution. We use the mean of speedups on all applications as the performance metric. We do a brute-force search to find the optimal RL allocation.

The result suggests an optimal RL allocation around 20 %, in Fig. 2.9. A larger allocation on RL limits the number of available dim cores to accelerate the non-kernel parts of the application. Therefore, it delivers worse overall performance according to Amdahl's Law. For net speedup comparison of RL vs. dim silicon, see Figs. 2.11 and 2.12 in following sections.

In summary, across a range of kernel characteristics, a relatively small area allocation on RL, e.g., 20–30 %, is good enough to achieve the optimal throughput that is almost 3× larger than a non-RL system of only dim cores.

Dim Silicon with ASICs

We use the same synthesized workload as described in the last subsection to study the performance implications of systems equipped with various fixed logic ASIC accelerators. Although specialized accelerators can achieve tremendous speedup for application-specific kernels, limited overall coverage of its targeted kernel within

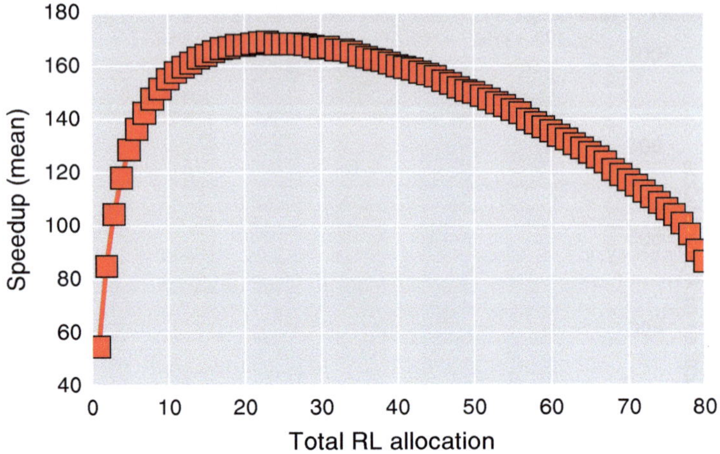

Fig. 2.9 Speedup of various RL allocations. Optimal performance achieved at around 20 % RL allocation

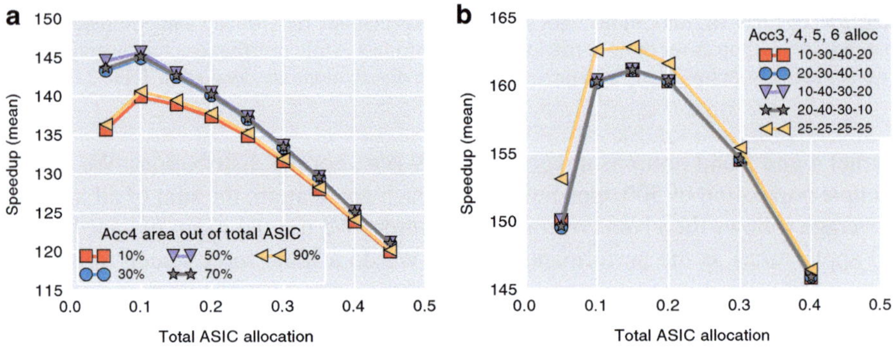

Fig. 2.10 Speedup of a system composed of dim silicon and ASICs. We show configurations with tow accelerators (**a**) and four accelerators (**b**). The X-axis is the total allocation to ASIC kernels, relative to the die area budget. The legend labels show the allocation of a kernel relative to the total ASIC allocation. (**a**) ASIC kernels of 4 and 5. (**b**) ASIC kernels of 3, 4, 5, and 6

a general-purpose workload leads to limited overall performance benefit for the workload as a whole. However, accelerators for library-call kernels tend to be more beneficial in the overall speedup of the workload due to their higher overall coverage. We plot potential allocations of two, three, and four accelerators for library-call kernels in Fig. 2.10. As the number of accelerators increases, the best achievable performance increases as well, from 140 to 155. With a given number of accelerators, the system achieves the best performance when area allocations to each accelerator are evenly distributed. The total area allocation to all hardware accelerators is limited to less than 20 %.

Dim Silicon with Accelerators(RL and ASIC) on General-Purpose Workload

In the previous two subsections, we have shown the benefits of accompanying conventional but dim cores with RL and ASIC accelerators, respectively. Both of them exhibit performance improvement over a baseline system composed of dim cores only. However, it is not necessary to achieve a better performance by combining both types of U-cores. As shown in Fig. 2.11, the best performance is achieved with RL-only system organization. This counter-intuitive result comes from two reasons:

- In Lumos, the accelerator performance scales proportionally to the area it is allocated. In addition, we have a conservative assumption on the performance ratio between ASIC and its counterpart RL implementation (see section "Sensitivity of ASIC Performance Ratio" for the impact of alternative performance ratios). With these assumptions, RL will be more powerful than the ASIC implementation as long as the RL allocation is adequately larger than the ASIC. Moreover, a large RL allocation is justified by the high utilization achievable across multiple kernels. Consequently, the system ends up with an RL-only configuration.

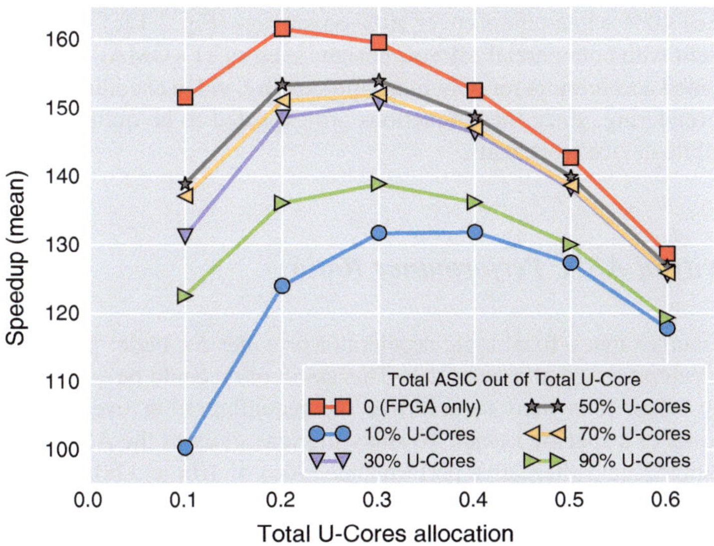

Fig. 2.11 Speedup of a system composed of dim silicon, RL, and ASICs. The *X*-axis is the total allocation to U-cores, including both RL and ASIC accelerators, relative to the die area budget. Legend labels indicate the total allocation of ASIC accelerators, relative to the total U-cores allocation. We assume an evenly distributed allocation among ASIC accelerators, since it shows the best performance in the previous analysis

- With a general-purpose workload, the average coverage of a kernel is small, due to either small coverage among applications (library-call kernels) or rare presence (application-specific kernels). This exaggerates the cost of a single ASIC accelerator that is only helpful for a specific kernel. As a result, the RL implementation is more favorable due to its versatility across kernels.

Benefit of ASIC Accelerators

Although RL works better with general-purpose workloads, an ASIC accelerator becomes beneficial when its targeted kernel is common enough across applications in a workload. To capture this scenario, we add one special kernel to the set of ten kernels we have used in previous analyses. The coverage of the special kernel is fixed across all applications, as well as the total coverage of all other kernels. We generate workloads by varying the coverage of the special kernel and all other kernels.

Figure 2.12 shows results of 10 and 30 % coverage of all other kernels. One of the most significant observations is that the dedicated ASIC accelerator is not beneficial at all until its targeted kernel covers more than the total of all other kernels within an application, e.g., in the case where the special kernel has an average coverage of 40 % while others cover 10 % on average (Fig. 2.12c). This observation is consistent with commercial MPSoC designs such as TI's OMAP4470 [19], which has dedicated accelerators for only image processing, video encoding/decoding, and graphics rendering, since these functions are expected to be quite common in the target mobile device workloads.

Sensitivity of ASIC Performance Ratio

Lumos assumes that a fixed logic accelerator provides 5× better performance than the RL accelerator with the same size. This assumption could be quite conservative, especially when it is hard to amortize the reconfiguration overhead over RL's running time of a specific kernel. To study the sensitivity of the ASIC performance ratio, we investigate alternative performance ratios of 10× and 50× using a similar workload setup in the previous subsection.

In Fig. 2.13a, the coverage of the special kernel is fixed at 20 %, while the total coverage of all other kernels varies from 10 to 40 %. When the performance of the dedicated ASIC accelerator is merely 5× better than RL, the system ends up with zero allocations to the dedicated accelerator, unless the coverage of its targeting kernel (20 %) is larger than the total coverage of all other kernels (10 %). However, when the performance of the dedicated accelerator boosts to 50× better than RL, the dedicated accelerator will always hold its place with 10 % allocation out of all u-cores area. When the performance ratio is 10×, the dedicated accelerator

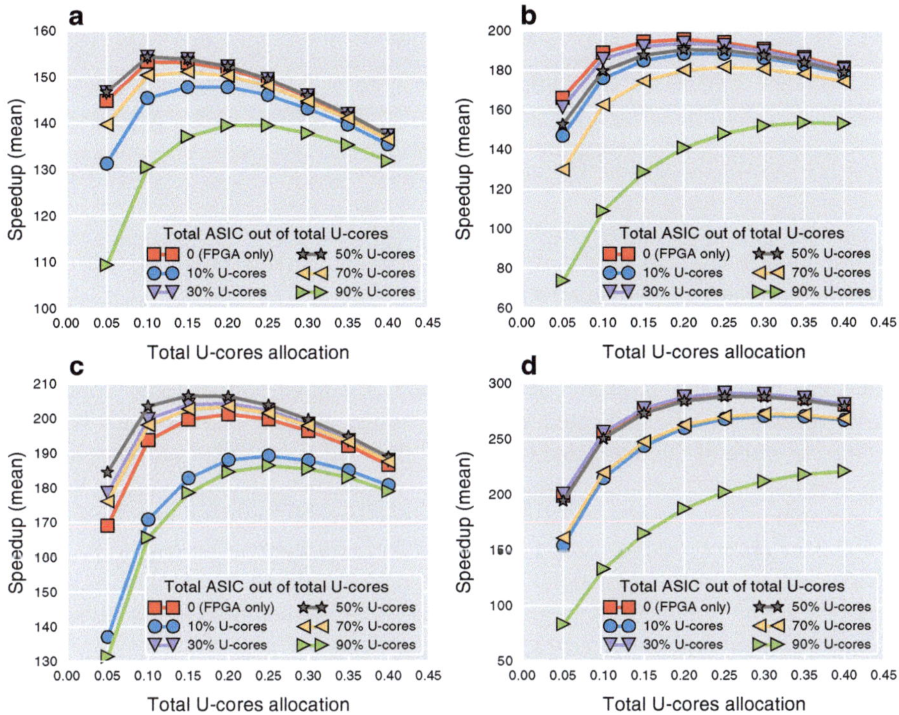

Fig. 2.12 To study the benefit of a dedicated ASIC accelerator, we add one special kernel to the set of ten kernels used before. We vary the total coverage of all other kernels at 10 % (*left plots*) and 30 % (*right plots*), while setting the coverage of the special kernel at 20 % (*upper plots*) and 40 % (*lower plots*). The results show that the dedicated ASIC accelerator in only beneficial when its targeted kernel has a dominant coverage with applications, e.g., in **c**. (**a**) The special kernel covers 20 %, other kernels cover 10 %. (**b**) The special kernel covers 20 %, other kernels cover 30 %. (**c**) The special kernel covers 40 %, other kernels cover 10 %. (**d**) The special kernel covers 40 %, other kernels cover 30 %

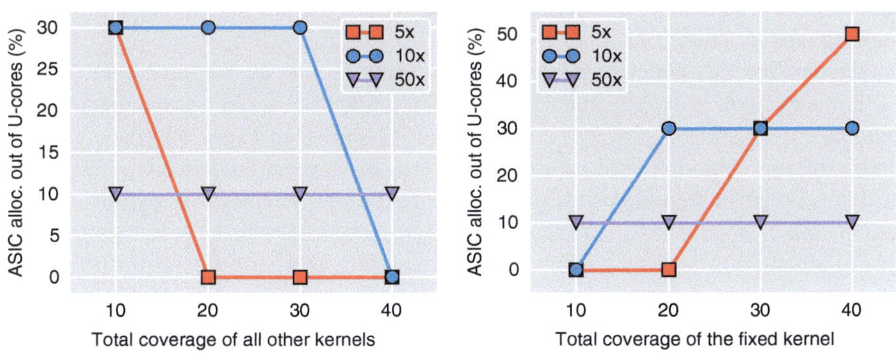

Fig. 2.13 Sensitivity study on ASIC performance ratio. (**a**) 20 % coverage of the fixed kernel. (**b**) 20 % coverage of all the other kernels

Table 2.7 Characteristics of an O3 core at 45 nm, normalized to the in-order core at the same technology node

Perf.	Power	Area
2.2×	3.5×	3.46×

is beneficial unless the total coverage of all other kernels is as large as 40 %, overwhelming the special kernel's coverage of 20 %.

In Fig. 2.13b, the total coverage of all other kernels is fixed at 20 %, while the special kernel's coverage varies from 10 % through 40 %. It is similar that the dedicated accelerator is preferred when there is a huge gap between either: (1) the performance of the dedicated and the reconfigurable accelerator (e.g., 50×), or (2) the coverage of the dedicated kernel and the total coverage of all other kernels (e.g., 2× larger).

In summary, the benefit of ASIC accelerators depends on its relative throughput to RL accelerators on the same kernel. With a large throughput gap (e.g., 50×), ASIC accelerators are beneficial even when the targeted kernel has a limited presence across applications (e.g., as low as 10 %). Otherwise, RL is preferable, as long as reconfiguration overhead can be neglected.

Alternative Serial Cores

Although massive parallelism has been observed in several computing domains, such as high-performance computing, there are many applications with a limited parallel ratio. In this case, adding a "beefy" out-of-order (O3) core is more beneficial, especially when dim cores get diminishing returns on throughput. We extract the performance of the O3 core from SPEC2006 scores of a Core i7 processor and calculate its power and area using McPAT, which are normalized and summarized in Table 2.7.

We assume the same inter-technology scaling factors as we have used for in-order cores in all previous analyses. We also assume the serial core is gated off in parallel mode to save power for dim cores. As shown in Fig. 2.14, even with an embarrassingly parallel application, the die area investment on dedicated O3 core is still beneficial. In the case of ideal parallelism, the performance loss by introducing an O3 core is around 14 % at 45 nm. The performance loss reduces to less than 1 % at 16 nm, due to diminishing performance returns from a larger number of throughput cores, and lower percentage area impact of one O3 core.

Alternatively, the serial code can be executed by the best core selected from several out-of-order candidates, as proposed in [15]. In this work, Najaf et al. observed a 10 % performance improvement by selecting from two alternative out-of-order cores, compared to an optimal-in-average out-of-order core. We model

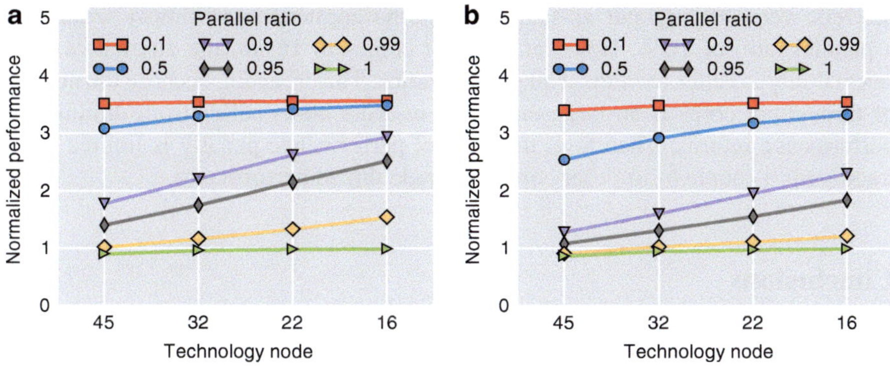

Fig. 2.14 Relative performance by using O3 core to run the serial part of an application. System budget is large in (**a**) and small in (**b**), as characterized in Table 2.5. Y-axis is the performance normalized to the system at the same technology node, which only includes in-order cores and uses one of the in-order cores as the serial core. (**a**) Large chip with area of 200 mm^2. (**b**) Small chip with area of 107 mm^2

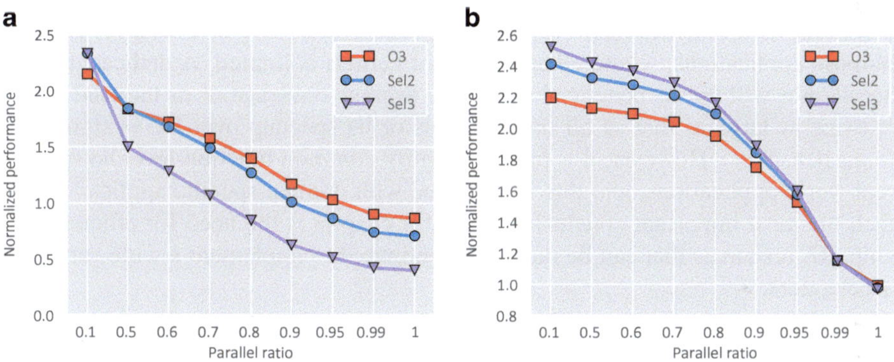

Fig. 2.15 Performance comparison among systems implementing the serial core as an out-of-order (O3) core, core-selectability of two O3 cores (Sel2), and core-selectability of three O3 cores (Sel3). The system budget is small from Table 2.5. Y-axis is the performance normalized to the system at the corresponding technology node, which only includes in-order cores and uses one of in-order cores as the serial core. (**a**) 45 nm. (**b**) 16 nm

core-selectability with assumptions of 10 % better performance but twice the area as an out-of-order core. To study the potential limits, we model an alternative organization of core-selectability with a total of three O3 cores to be selected, and assume another 5 % performance boost by introducing more selections (a total of 15 % over one O3 core "group"). We use the small system budget summarized in Table 2.5. As shown in Fig. 2.15a, the original core-selectability (Sel2) shows less benefit as long as the parallel ratio is larger than 50 % at 45 nm. At 16 nm (Fig. 2.15b), the original core-selectability outperforms the conventional core with almost all parallel ratios. The core-selectability in the alternative organization (Sel3) delivers worse throughput at 45 nm unless the application is almost serial with a parallel ratio of 10 %. However, when it comes to 16 nm, the alternative core-selectability is better for almost all parallel ratios. This is, again, because the number of throughput cores is so large that more in-order cores suffer from diminishing performance returns. Therefore, the parallel performance penalty is limited when trading off a couple of in-order cores for single thread performance.

Conclusions

In this chapter, we describe *Lumos*, a framework for exploring the performance of heterogeneous systems operating at near-threshold (e.g., with dim cores). We show that dim cores manage to provide a moderate speedup up to 2× over conventional CMP architecture (suffering from dark silicon issue with further technology scaling). However, the poor per-core frequency of dim cores leads to diminishing returns in throughput, creating an opportunity for more efficient hardware accelerators such as RL and ASIC. Lumos models the general-purpose workload via a statistical approach. We show that RL is more favorable for general-purpose applications, where commonality of kernels is limited. A dedicated ASIC accelerator will not be beneficial unless the average coverage of its targeted kernel is twice as large as that of all other kernels or its speedup over RL is significant (e.g., 10×–50×). However, it is hard to identify common function-level hotspots in real-world applications, and this is even true with domain-specific applications. In fact, the most important conclusion from this chapter is the need for efficient, on-chip, RL resources that can be rapidly reconfigured to implement a wide variety of accelerators.

References

1. C. Bienia, S. Kumar, J. Singh, K. Li, The PARSEC benchmark suite: characterization and architectural implications, in *International Conference on Parallel Architectures and Compilation Techniques, PACT '08* (2008)
2. S. Borkar, A.A. Chien, The future of microprocessors. Commun. ACM **54**(5) (2011). doi:http://doi.acm.org/10.1145/1941487.1941507
3. J.A. Butts, G.S. Sohi, A static power model for architects, in *International Symposium on Microarchitecture, MICRO '00* (2000)
4. B.H. Calhoun, S. Khanna, R. Mann, J. Wang, Sub-threshold circuit design with shrinking CMOS devices, in *International Symposium on Circuits and Systems, ISCAS '09* (2009)
5. E.S. Chung, P.A. Milder, J.C. Hoe, K. Mai, Single-chip heterogeneous computing: does the future include custom logic, FPGAs, and GPGPUs? in *International Symposium on Microarchitecture, MICRO '10* (2010). doi:http://dx.doi.org/10.1109/MICRO.2010.36
6. H. Esmaeilzadeh, E. Blem, R. St Amant, K. Sankaralingam, D. Burger, Dark silicon and the end of multicore scaling, in *International Symposium on Computer Architecture, ISCA '11* (2011)
7. D. Fick, R.G. Dreslinski, B. Giridhar, G. Kim, S. Seo, M. Fojtik, S. Satpathy, Y. Lee, D. Kim, N. Liu, M. Wieckowski, G.K. Chen, T.N. Mudge, D. Sylvester, D. Blaauw, Centip3De:

a 3930DMIPS/W configurable near-threshold 3D stacked system with 64 ARM Cortex-M3 cores, in *IEEE International Solid-State Circuits Conference, ISSCC '12* (2012)
8. V. Govindaraju, C.H. Ho, K. Sankaralingam Dynamically specialized datapaths for energy efficient computing, in *International Symposium on High Performance Computer Architecture, HPCA '11* (2011)
9. R. Hameed, W. Qadeer, M. Wachs, O. Azizi, A. Solomatnikov, B.C. Lee, S. Richardson, C. Kozyrakis, M. Horowitz, Understanding sources of inefficiency in general-purpose chips. in *International Symposium on Computer Architecture, ISCA '10* (2010)
10. M.D. Hill, M.R. Marty, Amdahl's law in the multicore era. Computer **41**(7) (2008). doi:10.1109/MC.2008.209
11. W. Huang, K. Rajamani, M.R. Stan, K. Skadron Scaling with design constraints: predicting the future of big chips. IEEE Micro **31**(4) (2011). doi:10.1109/MM.2011.42
12. Intel® Core™2 Quad Processor Q9550S (2007). http://ark.intel.com/products/40815
13. Intel® Xeon® Processor W5590 (2009). http://ark.intel.com/products/41643
14. S. Li, J.H. Ahn, R.D. Strong, J.B. Brockman, D.M. Tullsen, N.P. Jouppi, McPAT: an integrated power, area, and timing modeling framework for multicore and manycore architectures. in *International Symposium on Microarchitecture, MICRO '09* (2009)
15. H.H. Najaf-abadi, N.K. Choudhary, E. Rotenberg (2009) Core-selectability in chip multiprocessors, in *International Conference on Parallel Architectures and Compilation Techniques, PACT '09*. doi:10.1109/PACT.2009.44
16. Nangate FreePDK45 Generic Open Cell Library (2008). http://www.si2.org/openeda.si2.org/projects/nangatelib
17. R. Narayanan, B. Özışıyılmaz, J. Zambreno, G. Memik, A. Choudhary Minebench: a benchmark suite for data mining workloads, in *IEEE International Symposium on Workload Characterization, IISWC '06* (2006)
18. N. Nethercote, J. Seward Valgrind: a framework for heavyweight dynamic binary instrumentation, in *Proceedings of ACM SIGPLAN Conference on Programming Language Design and Implementation, PLDI '07* (2007)
19. OMAP4470 Mobile Application Processor (2011). http://focus.ti.com/pdfs/wtbu/OMAP4470_07-05-v2.pdf
20. Oracle SPARC T4 Processor (2011). http://www.oracle.com/us/products/servers-storage/servers/sparc-enterprise/t-series/sparc-t4-processor-ds-497205.pdf
21. J.M. Rabaey, A. Chandrakasan, B. Nikolić, *Digital Integrated Circuits: A Design Perspective*, 2nd edn. (Prentice-Hall, Englewood Cliffs, NJ, 2003)
22. S. Sinha, G. Yeric, V. Chandra, B. Cline, Y. Cao, Exploring sub-20 nm FinFET design with predictive technology models, in *Proceedings of the ACM/IEEE Design Automation Conference* (2012)
23. M.B. Taylor, Is dark silicon useful?: harnessing the four horsemen of the coming dark silicon apocalypse, in *Design Automation Conference, DAC '12* (2012)
24. A.C.N. Tseng, D. Brooks, Analytical latency-throughput model of future power constrained multicore processors, in *Workshop on Energy-Efficient Design, WEED '12* (2012)
25. S.K. Venkata, I. Ahn, D. Jeon, A. Gupta, C. Louie, S. Garcia, S. Belongie, M.B. Taylor, SD-VBS: the San Diego vision benchmark suite, in *IEEE International Symposium on Workload Characterization, IISWC '09* (2009)
26. G. Venkatesh, J. Sampson, N. Goulding, S. Garcia, V. Bryksin, J. Lugo-Martinez, S. Swanson, M.B. Taylor, Conservation cores: reducing the energy of mature computations, in *International Conference on Architectural Support for Programming Languages and Operating Systems, ASPLOS '10* (2010)
27. L. Wu, M.A. Kim, Acceleration targets: a study of popular benchmark suites, in *The First Dark Silicon Workshop, DaSi '12* (2012)
28. T. Zidenberg, I. Keslassy, U. Weiser. MultiAmdahl: how should I divide my heterogeneous chip? IEEE Comput. Archit. Lett. **11**(2), 65–68 (2012)

Chapter 3
The SiLago Solution: Architecture and Design Methods for a Heterogeneous Dark Silicon Aware Coarse Grain Reconfigurable Fabric

Ahmed Hemani, Nasim Farahini, Syed M.A.H. Jafri, Hassan Sohofi, Shuo Li, and Kolin Paul

Introduction

The design complexity and the low power demands of modern applications are outstripping the improvements in performance offered by the technology scaling. Rabaey characterized these phenomena as Shannon beats Moore in his Hotchips tutorial on wireless communication applications [1]. Rabaey's observation is true for other application domains as well and we call this gap as the architecture efficacy gap. To bridge this ever widening gap, computer architects have responded with a series of innovations during the last 40 years. This includes virtual memory [2, 3], multi-banking [3], pipelining [4], instruction level parallelism [5], thread level parallelism [5], multi/many processor [6], network-on-chip (NoC) [7], branch prediction [8], and many innovative design methods [9]. One of the main goals of these innovations is to provide greater computational, silicon, and engineering efficiencies.

As we move further into the nano-scale regime, the dark silicon era is imminent, where the power considerations will allow only a fraction of a chip to be powered [10], resulting in exponentially smaller percentage of the chip that can be active with each processor generation [11]. To understand this problem, consider Table 3.1. The table shows how transistor properties change with each process generation as a function of *S*—the scaling factor before and after the 2005, the year that demarcates the Dennard and the post-Dennard scaling eras. In the Dennard scaling era, it was possible to simultaneously scale the threshold and the supply voltages. In this era,

A. Hemani (✉) • N. Farahini • S.M.A. H. Jafri • H. Sohofi • S. Li
Royal Institute of Technology (KTH), Kista, Sweden
e-mail: hemani@kth.se; farahini@kth.se; Jafri@kth.se; sohofi@kth.se; shuol@kth.se

K. Paul
Indian Institute of Technology (IIT), New Delhi, India
e-mail: kolin.paul@gmail.com

A.M. Rahmani et al. (eds.), *The Dark Side of Silicon*,
DOI 10.1007/978-3-319-31596-6_3

Table 3.1 Origin of the dark silicon era

Transistor property	Dennard scaling era	Post Dennard scaling era
ΔDensity	S^2	S^2
ΔFrequency	≈S	≈S
ΔCapacitance	1/S	1/S
$\Delta V_{DD}{}^2$	$1/S^2$	≈1
Power = $\Delta QFCV_{DD}{}^2$	1	S^2

the transistor properties were governed by *Dennard's scaling* [12], which implies that power consumption was proportional to the area of a transistor. In the post-Dennard scaling era, the threshold or supply voltage could no longer be easily scaled without causing either exponential increase in leakage or transistor delay [13]. The table shows that as the transistors increase by S^2, their frequency increases by S, and their capacitance Q decreases by $1/S$. The Dennard and the post-Dennard scaling eras differ in supply voltage V_{DD} scaling; during Dennard scaling, V_{DD} scaled down by $1/S$, but in the post-Dennard scaling era, V_{DD} remains almost constant because the threshold voltage Vt cannot be scaled down easily. As a result, the change in power consumption $\Delta P = \Delta QFCV_{DD}{}^2$ during Dennard scaling era was constant, whereas in the post-Dennard era it increases by S^2 for the same die size and clock frequency. This increase in power for the next generation devices will be sufficient to burn them [11].

In future, since the dark silicon constraint will prevent all transistors from being turned on simultaneously, it is imperative that we use the transistors that can be turned on with highest possible silicon and computational efficiencies. It is well established [1] that hardware implementations compared to software implementations have significantly higher silicon and computational efficiencies. However, the conventional wisdom and some evidence [1] also point to the fact that custom hardware implementation comes at the cost of steep engineering and manufacturing costs. Motivated by these arguments, the VLSI design community has settled for an architectural style that customizes at two levels to achieve improved use of limited transistors that can be turned on. The first level that we call as the mild level of customization relies on heterogeneity in processors [14]. The second level, which we call as the medium level of customization, relies on mapping power and performance critical parts of the functionality to custom hardware designs called accelerators. This architectural style based on two levels customization is known as accelerator-rich heterogeneous multi-processors [15–17]. To tackle the steep engineering cost, the VLSI design community has settled on platform based design around popular general purpose processors and their associated interconnect and peripheral systems, dominated by ARM [18]. These platforms are built around pre-designed and pre-verified IPs at RTL that enables rapid composition of SOCs and integration of custom designed accelerators. In spite of these judicious measures and limiting customizations to timid levels to address the engineering costs, the cost of delivering a manufacturable design has risen to levels that are stifling innovation; it takes 45 MUSDs to make a state-of-the-art SOC and another five MUSDs to manufacture it.

We argue that this state-of-the-art approach for architecture and design methodology has four limitations. A more detailed analysis of these four limitations is presented in the section "State-of-the-Art Review of the Techniques to Deal with Dark Silicon Constraints." Overcoming these four limitations is the key contribution of this chapter and forms the basis for a dark silicon aware architectural template and design methodology:

1. The customization in the accelerator-rich heterogeneous multi-processors is software-centric, i.e., the bulk of functionality is implemented in the inefficient software style and only a small fraction that is power and performance critical is moved to hardware accelerators. In contrast, we propose an approach, where by default functionality is mapped to hardware on a coarse grain reconfigurable fabric for computation and storage and only flexibility critical and control intensive functionality is mapped to small simple processors that we call as flexilators.
2. The customization is compute centric, i.e., the use of heterogeneous processors and accelerators both target customizing, at mild to medium levels, only the computation. The approach proposed in this chapter in contrast adopts a more aggressive approach to customizing the entire architecture that includes computation, storage, interconnect, control, and address generation. This customization can be done at compile/build time but also at runtime.
3. The customization is static and decided at design or build/compile time. In today's multi-application scenarios, it is impossible to know the optimal degree of parallelism and voltage frequency operating point of an application in advance. This depends on the runtime conditions, load and deadlines, available energy, etc. The dark silicon aware architecture proposed in this chapter not only enables complete and aggressive customization, but it also enables dynamic runtime customization in terms of degree of parallelism and the voltage frequency operating point deployed for each application depending on the runtime conditions.
4. The granularity of standard cell as atomic physical design object to compose designs of complexity of 100 s of millions of gates is not scalable. This mismatch impedes the progress of synthesis tools beyond RTL[1] and results in large verification cost because manual refinement of abstract system models down to RTL lacks the correct by construction guarantee of the synthesis tools. Additionally, the large gap between abstract system models and the standard cells based physical design results in unpredictability in cost metrics (area, energy, and performance), resulting in costly synthesis and verification iterations further aggravating the engineering cost. These limitations strongly impact the ability to customize and exploit the efficient hardware implementation style that is essential to efficiently use the limited number of transistors that can be turned on in the dark silicon era.

[1]Even after two decades of research, HLS is still not mainstream and even if it becomes, it effectively synthesizes and optimizes only an algorithm at a time in isolation and algorithm is too small a unit in comparison to the complex multi-application functionality that SOCs host.

The method proposed in this chapter not only proposes a more complete and aggressive customization of the SOC architecture but also proposes a design methodology that makes such customization orders of magnitude more efficient compared to the state-of-the-art standard cell based EDA flows. The key idea behind this methodology is to raise the level of abstraction of physical design target from Boolean level standard cells to micro-architecture level SiLago blocks. This is in compliance with Moore's EDA law [19]. The SiLago method [20] uses a structured grid based layout scheme to compose a DRC and timing clean manufacturable design by abutting the pre-hardened SiLago blocks. The micro-architecture level physical design target reduces the abstraction gap between abstract system models and the physical design target to improve and enable true application-level synthesis [21, 22]. This goes beyond high-level synthesis in taking a hierarchy of algorithms that implement a modem or a codec class of functionality, explore the design space, and generate solutions in terms of implementation of functions (FSM + Datapaths) as building blocks as opposed to micro-architectural blocks used by HLS. It does global optimization across algorithms, accept global system-level area, energy and latency (in cycles) constraints, and automatically synthesize the inter-algorithm interconnect and the global control, necessary to orchestrate the execution. In contrast, HLS explores design space and generates solutions in terms of micro-architectural blocks. Some HLS tools superficially handle hierarchy of algorithms but synthesize one algorithm at a time, expect the user to budget constraints for each algorithm, refine inter-algorithm interconnect and global control.

State-of-the-Art Review and Problem Analysis

In this section, we review the state-of-the-art in techniques that have been developed to deal with the dark silicon constraints. This is followed by analyzing the problems with the proposed methods. This analysis then becomes the basis for proposing the SiLago solution, composed of overcoming both the architectural limitations and the design methodology limitations.

State-of-the-Art Review of the Techniques to Deal with Dark Silicon Constraints

The phenomena of dark silicon was most likely first observed by Taylor in 2005, when he attempted to prove the scalability of the Raw Tiled Multicore processor [15, 23]. The term *dark silicon* was probably coined by ARM CTO Mike Muller. Since then extensive work has been carried out to tackle the dark silicon.

In the dark silicon era, heterogeneity is considered as one of the most potent techniques used to harness the on-chip transistors (for improving energy/power efficiency) [24]. The existing research, that tackles the dark silicon using heterogeneity, can broadly be classified into four domains: (a) *Platform identification and customization*. The research in this domain attempts to determine the platform (FPGA, Processor, CGRA, or a hybrid) that is the most suitable for the dark silicon era. The work in this domain has revealed that the accelerator-rich designs, i.e., computing platforms containing a multitude of application-specific and general purpose accelerators, are particularly useful to combat the utilization wall [15, 25]; (b) *Additional constraint identification*. The works in this domain determine the additional criteria to customize (micro-) architectural features and the number and type of various components for dark silicon era [26, 27]; (c) *Power management* (i.e., Dynamic Voltage and Frequency Scaling (DVFS) and power off/sleep modes). The research in this domain deals with the architecture and algorithms to scale operating point such that the power consumption kept below the budget [10, 28–31]; (d) *Near-Threshold Computing*, i.e., operating the components at a very low voltage to power-on more cores [32]. In this we will explore these four domains, followed by what lacks in the existing research and the main contributions of this chapter.

Platform identification is one of the key considerations in designing a chip for the dark silicon era. While general propose processors provide ease of implementation, they lack the efficiency of specialized hardware. It has been shown that general purpose processor can be 3–6 orders less efficient (in terms of performance/power ratio) in some cases [33]. On the other hand, in the era of platforms hosting multiple applications, the specialized hardware is neither optimal nor desirable (since making a new chip is costly). Reconfigurable hardware (in particular the coarse-grained reconfigurable architectures) and accelerator-rich designs are considered to be the platforms ideally suited for the dark silicon era. These designs embed various specialized coarse-grained accelerators alongside the conventional processors. It is predicted that the number of accelerators per chip will be close to 1500 by 2022, according to an ITRS prediction [34]. Taylor et al. [15] showed that the utilization wall can be overcome by increasing the number accelerators. Cong et al. [25] presented an architecture to effectively manage a pool of accelerators to keep the power consumption within the required limits. Lyons et al. [16] presented a shared memory scheme for the accelerator-rich designs. ARM big.LITTLE is another example of a heterogeneous multi-core integrating high performance out-of-order cores with low power. This architecture enables optimization strategies which reduce the system power consumption through balancing the power-performance workload [18]

Adding dark silicon as an additional constraint to determine the optimal number (and type) of various micro-architectural components has also yielded promising results. Jason et al. [26] proposed a new design metric and a stochastic optimization algorithm for efficient chip design in dark silicon era. Turakhia et al. [27] designed a technique to determine the optimal number of cores of each type for a platform. The main goal of this work was to maximize the performance within a power budget. Mitra [35] proposed a heterogeneous multi-core architecture and the runtime management strategies to improve the energy efficiency.

Power management is a very frequently used technique that scales the voltage/frequency operating point to keep the power consumption within the limits. Jafri et al. [29, 36] presented an energy aware reconfigurable platform that uses existing resources to synchronize communication between resources operating at different frequencies. The main goal of their technique was to keep the power below threshold. Muthukaruppan et al. [10] introduced a hierarchical power management framework for asymmetric multi-cores. This work relies on control theory and coordinates multiple controllers in a chip for achieving optimal power-performance efficiency. Bokhari et al. [37] proposed a multi-layer NoC architecture that uses DVFS technique by optimizing each layer for a particular voltage frequency range to reduce the NoC power consumption. Matsutani et al. [38] proposed a runtime power-gating technique to reduce the leakage power of the NoC in CMPs. Liu et al. [39] introduced a dynamic thread mapping scheme that maximizes the performance under power constraints for heterogeneous many cores. Pricopi et al. [40] developed a software-based modeling technique to estimate the performance-power workload of the asymmetric core architecture and evaluated the technique on ARM big little.

Near threshold voltage (NTV) operation was presented by Intel and tested on various processors [32]. It has potential to improve the energy efficiency by an order of magnitude by allowing to keep the supply voltage very low. Even though it has a great potential, it is a circuit level technique and will not be discussed further in this chapter.

The related work reveals that in most of the works, the decision about heterogeneity (i.e., the number and type of computing and memory elements) is taken at design time. Therefore, the existing platforms are only suitable for a specific class of application. When a new class is instantiated, the important performance metrics (i.e., energy, power, and area) suffer badly. Efficient support of a new application class requires complicated chip redesign which itself is very costly [41]. The only decisions taken at run/compile-time are the voltage and frequency scaling (in form of DVFS). Even in DVFS, the decisions about the synchronization hardware and the buffers are taken at the design time.

Problem Analysis

Heterogeneity in the form of processors with different performance and power footprint like ARM big.LITTLE and accelerator-rich architectures [25] are the two mainstream strategies to address the dark silicon constraints. The rationale behind this approach is that because we cannot turn on all transistors, we need an implementation style that enables more functions to be realized from the transistors that can be turned on simultaneously. In other words, we need greater silicon and computational efficiencies. Powerful general purpose VLIW or superscalar processors with deep pipelines, sophisticated logic for branch prediction, control, data, and structural hazard avoidance, etc., are known to waste a significant amount of energy in overheads; Dally et al. [42] report up to 90 % of energy wasted in overhead. Simpler processors have smaller overhead and require fewer transistors and can implement functions that do not need the performance of powerful processors with greater efficiency. This is the justification for the heterogeneity in processors. Another track that the VLSI community has pursued to achieve greater silicon and computational efficiency has been to use accelerators to speed up inner loops or kernels whose performance demands cannot be met by the host processor. Accelerators are custom hardware designs, whose silicon and computational efficiencies are known to be 3–5 orders more than that of general purpose RISC processors [33]. Even if accelerators implement specific functions and we need different set of transistors for each of them, this is justified by the dark silicon constraint that we anyway cannot turn on all the transistors, so why not use them for efficient custom hardware accelerators. Heterogeneity and accelerators are two different approaches but the underlying rationale is common—customize the hardware to improve computational and silicon efficiencies. These two categories of customization, heterogeneity and accelerator-rich architectures, are well justified, but we argue that they along with their state-of-the-art design methodology that are used to design them are deficient in four respects:

Partial Computation-Centric Customization

The two customizations, processor heterogeneity and accelerators, are partial and focus only on improving the efficiency of the computation—the logic and arithmetic operations. They do not customize the storage, interconnect, control, and address generation that we argue have a far greater impact on the overall power efficiency. What is required is the ability to customize all aspects to match the functionality in terms of its datapath, storage and control requirements, its power and performance requirements, and also the runtime conditions. Already in 130 nm technology 51 % of power was consumed in interconnect and as Kecklar et al. [72] have shown recently the efficiency of wire is scaling half as much as that of logic. As a result, architectures that enable short local wires and minimize or eliminate the need to transport data/address will be more compatible with the technology trend and enable more efficient usage of logic and wires that can be powered on in the dark silicon era. Memory is another key architectural component that dominates all three cost metrics—area, energy, and latency of the overall system. For this reason, it is important to be able to customize the storage architecture to the needs of the application, rather than have a large bulky centralized storage or a distributed storage that induces long wires. The control in state-of-the-art accelerator-rich heterogeneous multi-processor architectures is often centralized and based on deeply pipelined fetch-decode-execute sequencers, whose inefficiency has been well documented [42]. These control architectures, while general, add significant bulk in terms of program storage, complex logic to avoid hazards, and long wires to arithmetic, interconnect, and storage.

Inefficient Software-Centric Implementation Style

The two categories of customizations adopted by state-of-the-art solutions, the processor heterogeneity and use of accelerators, are both software-centric. Software-centricity implies that by default functionality is implemented as software and only power and performance critical parts are moved to hardware accelerators or alternatively, the functionality is mapped to the processor category with the least power consumption that is able to meet the performance requirement. This implies that the bulk of functionality that is not mapped to accelerator, though not performance critical, is implemented in an implementation style that is known to be significantly less efficient.

Figure 3.1 isan example of an industrial software-centric heterogeneous multi-processor accelerator-rich platform [14]. This SOC is a typical ARM platform, where an ARM processor is a system controller and serves to host general purpose applications. Compute intensive applications are mapped to domain-specific processor for greater efficiency; video applications are mapped to Trimedia and audio applications are mapped to the Audio DSP. Do note that, while Trimedia is quite capable of also hosting the audio applications but for the typical use case when the end user is only listening to music, it is more energy efficient to run MP3 type applications on smaller and more efficient audio DSP to exploit the heterogeneity. In spite of being domain specific, these processors often need accelerators to meet the performance demands of latest generation of applications. Additionally, there are hardware macros for the modems. The customization of such SOCs is only in terms of the computation, the heterogeneous processors have dedicated DSP and SIMD instructions and data width and some accelerators for these processors. But the storage and interconnect are general purpose, both within the processors and across the SOC. This generality, designed to increase the reusability, results in bloated designs that have a large area, energy, and performance overhead. To

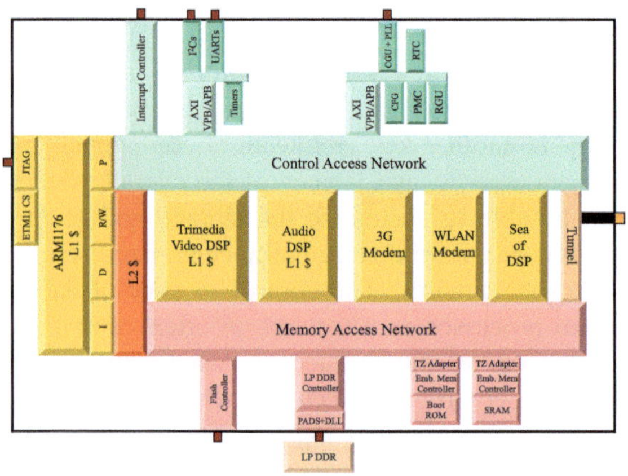

Key Limitations:

1. Bulk of the functionality implemented in inefficient software style
2. The heterogeneous processors and accelerators only customize the computation while the storage and interconnect are general purpose
3. The large and inefficient general purpose resources (processors and interconnect) are non-deterministically shared in temporal sense.
   a. This further degrades their ineffiiency and makes it harder to dimension them sacrificing the composability of such systems
   b. The voltage frequency operating point of these large time-shared resources has to be set for the worst case application requirement
4. The partial architectural customisation is design time static with little or no possibility of compile/build time or run time customisation

Fig. 3.1 An industrial accelerator-rich heterogeneous multi-processor SOC design

justify and amortize the cost of these expensive silicons, they are shared in a time multiplexed manner that further degrades the efficiency and makes it hard to dimension such systems because the execution patterns of the hosted applications are non-deterministic. Such designs are neither composable, nor predictable.

Lack of Dynamic Runtime Customization

Typical SOCs today host multiple applications, whose communication and concurrency pattern is non-deterministic. As a result, it is hard to decide at compile/build time the optimal degree of parallelism and the voltage frequency operating point at which an application should be executed. There are three sub-problems with the present day accelerator-rich heterogeneous SOCs like the one shown in Fig. 3.1: (a) the degree of arithmetic level parallelism in the architecture is a static design time decision in the form of number of issues in VLIW or superscalar and often beyond the control of the end user; this is the case with the ARM1176, Trimedia, and Audio DSP processor in Fig. 3.1. In case of accelerators, the degree of parallelism is again a static design time decision and at best a compile time decision. (b) Multi-processors are suitable for task or thread level parallelism but not suitable for collaborating to provide a higher degree of arithmetic level parallelism. Another factor is that increasing arithmetic parallelism does not help unless one can also increase the bandwidth to the storage. (c) The voltage island boundaries in the SOC are static and because processors typically host multiple applications, the voltage frequency operating point is adjusted to the worst case, the average improvement in power efficiency is lower [43] and finally, unless all applications sharing a voltage island are idle, it cannot be powered down.

Large Engineering Cost for Customization

The conventional wisdom in the VLSI design community has been that doing custom hardware design is expensive both in engineering and manufacturing terms and should be avoided. This is partly the motivation for the software-centric platform based design despite of inferior silicon and computational efficiencies. The rationale has been to use pre-designed and pre-verified IPs and platforms to rapidly compose a SOC. New hardware logic design is restricted to relatively small accelerators. In spite of adopting this methodology, the engineering cost for a complex SOC is 45 MUSDs in addition to five MUSDs in manufacturing cost [41]. In light of these numbers, proposing complete customization to make SOCs dark silicon friendly requires a radical rethink and solution; in this chapter, we propose such a solution. Before we elaborate this solution, we analyze the reasons why the engineering cost of SOCs with heterogeneity and accelerators has ballooned to such unacceptable levels.

Figure 3.2 shows the typical SOC design flow that is used for multi-processor heterogeneous accelerator-rich industrial SOCs of the type shown in Fig. 3.1. The flow is divided into two parts: the system design and logic implementation.

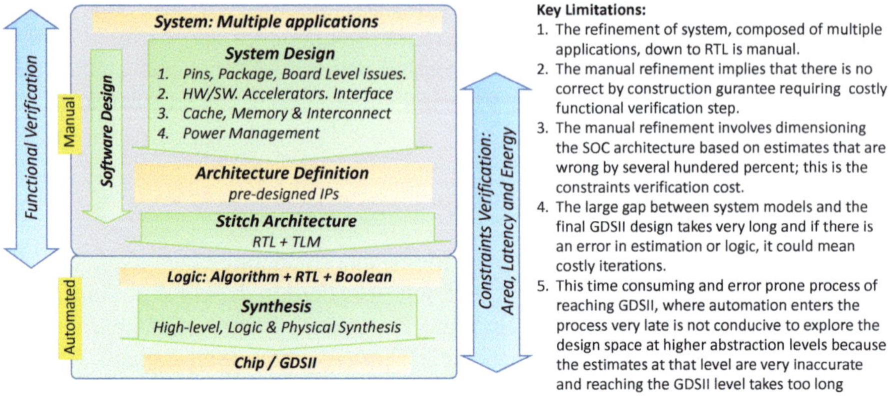

Fig. 3.2 State-of-the-art SOC design flow and its limitations

The system design that involves architecting the SOC in terms of IPs, identification of accelerators, etc., is manual. The output of this phase is dominantly RTL with trivially small parts that are algorithmic. This manual refinement lacks correct by construction guarantee and requires expensive functional verification. The second part of the design flow is logic implementation that is largely automated relying primarily on RTL and physical syntheses with some use of HLS. The correct-by-construction guarantee of these syntheses tools obviates the need for functional verification at gate and circuit levels. There is another verification cost that stems from the large gap between the abstract system-level models and the GDSII design. This verification cost is shown as the constraints verification cost. This makes predicting the cost metrics from system-level very difficult and is often off the mark by several 100 % even for small designs as quantified in the section "SiLago Design Methodology." This can result in costly iterations requiring re-synthesis and in worse-case re-verification.

These two verification components have made SOC engineering a very expensive proposition and unless this challenge is addressed, the objective of complete customization to achieve significantly more functionality per unit current and area to deal with dark silicon constraints cannot be fulfilled.

To address the problems identified in this section, we propose a solution that has two parts. The first part, elaborated in the section "The SiLago Platform," deals with the architectural template that is hardware centric and enables complete and dynamic customization to overcome problems analyzed in sections "Partial Computation-Centric Customization," "In-Efficient Software-Centric Implementation Style," and "Lack of Dynamic Runtime Customization." The second part elaborated in the section "Automating the Complete Customization" uses the architectural template described in the section "The SiLago Platform" as the basis for a design methodology that enables automation of the customization to overcome the problems analyzed above in the section "Large Engineering Cost for Customization".

The SiLago Platform

In this section, we present the SiLago platform and how it address the three architectural problems with the state-of-the-art accelerator-rich software-centric heterogeneous multi-processor SOC designs in their ability to deal with the dark silicon constraints. The SiLago platform is a generic concept; however, in this chapter, we present a specific design that targets signal processing applications like modems and codecs. It can host multiple such applications concurrently. The SiLago platform we present in this chapter has two coarse grain reconfigurable fabrics, one to support streaming signal processing computation, control and address generation and the second to support streaming scratchpad storage with local address generation. The two fabrics enable hardware style implementation of streaming data parallel portions of signal processing sub-systems. The efficiency achieved is marginally lower than ASIC, as validated in the section "Computational and Silicon Efficiencies of the SiLago Platform." Complementing the hardware style implementation resource are flexilators that implement control and flexibility critical adaptive signal processing loops; flexilators play the same role in the hardware centric SiLago platform as the accelerators play in the software-centric state-of-the-art SOCs. The key feature of the SiLago platform is its ability to cluster resources from the coarse grain reconfigurable and flexilator fabrics to dynamically carve out private execution partitions (PREXes) to fully customize computation, storage, interconnect, address generation and control to achieve ASIC like efficiency. The platform also supports a fine grain power management that enables dynamically deciding the clock and voltage domain boundaries and adjusts the voltage frequency operating point based on runtime conditions. Finally, the SiLago platform has RISC based system controller to control, configure, monitor, instantiate, and terminate applications. The system controller monitors the available resources, load and deadlines, and can dynamically deploy applications with varying degree of parallelism to further enhance the power efficiency that can be achieved by dynamic voltage frequency scaling alone. SiLago platform is a design template, the actual dimensions of the fabric and its components are decided by the application-level synthesis phase that is described in the section "Sub-System/Application Level Synthesis Using Sylva."

We next describe the different components of the SiLago platform and how they contribute to SiLago platform being dark silicon aware.

Dynamically Reconfigurable Resource Array

Dynamically reconfigurable resource array (DRRA) is a coarse grain reconfigurable fabric that targets implementing streaming data parallel signal processing functions. It is composed of DRRA cells that are organized in two rows as shown in Fig. 3.3.

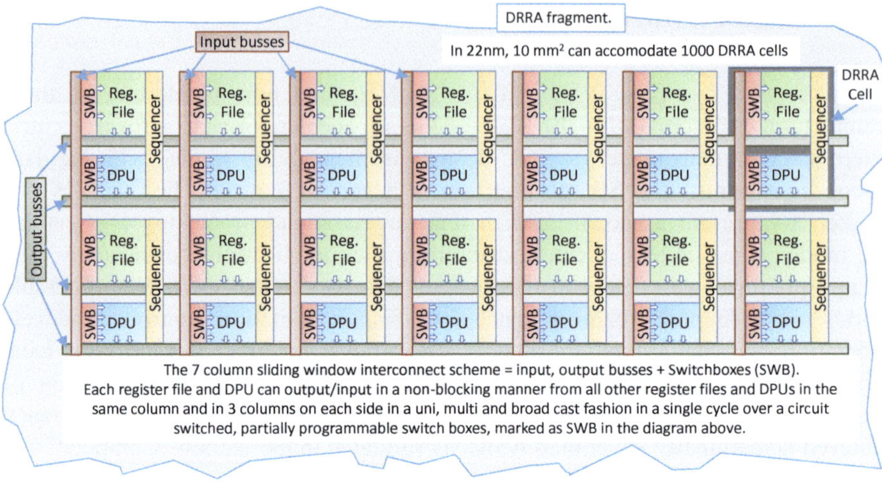

Fig. 3.3 Dynamically reconfigurable resource array (DRRA)

Each DRRA cell is composed of four components: (a) register file (Reg. File), (b) sequencer, (c) datapath unit (DPU), and (d) two switch boxes (SWBs) as shown in Fig. 3.3.

Register File The DRRA *register file* has two read and two write ports and by default has 32 16-bit words. The two ports are designed to deal with the typical complex number read and write required in signal processing applications but in general could be any two read/write operands. Each port is equipped with its own address generation unit (AGU) that enables streams of data with spatial and temporal programmability. Spatial programming implies a variety of addressing modes like linear, circular buffer, bit-reverse, etc. [44–46]. Temporal programming allows to insert a programmable delay before the stream starts, between the consecutive reads/writes, and at the end of the stream. AGUs are hardware configurable finite state machines (FSMs) and an instance of customization at micro-architectural level. The AGUs not only provide more efficient address generation but they also save on the cost of address transportation and make the configware size more compact [47]. This contributes to extracting more function per mm^2 and per joule to make SiLago dark silicon friendly. Additionally, DRRA cells equipped with specific custom addressing modes can be instantiated by the SiLago design flow if required as long as they have the same interface and physical footprint as the standard/default register file in the SiLago fabric. AGUs can deal with two dimensional affine address functions. To deal with non-affine functions, a unit called runtime address constraints computation unit (RACCU) that is part of sequencer is used. Details of RACCU are presented in [48].

DPU The DRRA *DPU* has four inputs that would typically correspond to two complex numbers but in general represents four arbitrary operands, some of which may not be used in some DPU modes. It supports typical DSP operations like MAC, optionally with an adder in front to deal with symmetric FIRs, half butterfly, sum of difference, difference of sum, programmable truncation, rounding, etc. The DPU offers a significant freedom in design time customization. Like register file, the DPUs with the same interface and physical footprint can be instantiated by the syntheses tools. Some DPUs would be tailored to implement typical DSP arithmetic used in inner-modem (i.e., FFT, FIR, etc.), while others could be tailored for bit level processing for outer modems (i.e., scrambler, interleaver, etc.) [48, 49].

Sequencer The DRRA *sequencer* is principally a configuration unit but it also serves as a simple sequencer that is primarily designed to handle control of compile time static signal processing functionalities. Unlike the traditional processor sequencer, the DRRA sequencer has a small local store of 32 words. The fetch-decode and execute path is single cycle, i.e., the sequencer itself does not induce any pipeline stages. The principal modus operandi of the sequencer is to setup a vector operation by (a) configuring register file(s) to source and sink a stream of data in right spatio-temporal pattern, (b) connecting the DPU(s) and register file(s) input/output ports by programming the switchboxes, and (c) putting the DPU into the right mode. Each such vector operation is followed by potentially evaluating some condition and repeating the same vector operation with different address range and temporal constraints for register file(s) [48] and different DPU mode or a completely different vector operation. This change in vector operations or the constraints of vector operation is decided by loop and/or branch instructions. DRRA is designed to principally implement vector operations with minimal branching and loops. A functionality that is dominated by loops and branches is instead mapped to flexilators (see section "Flexilators and System Controller") that are tightly coupled to the sequencers. The DRRA sequencers can also be chained into a hierarchy of FSMs. This capability is motivated by the need to emulate the way the control is organized in ASICs. For a modem, there is typically a modem level FSM that interacts with the MAC layer and RF/Analog front end and also controls and configures the Rx/Tx state machines. The Rx/Tx state machines in turn controls the individual signal processing transforms. To enable organizing such a hierarchy of parallel distributed FSMs, DRRA sequencers can be triggered by neighboring sequencers and in turn trigger them in a sliding window fashion. Though this is a design time decision, the present implementation can trigger sequencer in the same column and sequencers in three columns on each side. Details of this capability can be found in [50].

Sliding Window Interconnect The DRRA *intra-regional interconnect* scheme consists of a sliding window nearest neighbor connectivity [51, 52]. Each DPU and register file outputs to a bus (bundle of wires) that straddles three columns on each side. Thus including the self-column, it forms a seven-column window. Since each column has a similar seven-column window, the overall scheme forms an overlapping, seven column sliding windows. These output busses, horizontal in their orientation, are intersected by input busses, vertical in their orientation, as shown in Fig. 3.3. At the intersection of each input and output bus sits switchboxes, four per column, two for the two register files and two for the two DPUs. These switchboxes (multiplexors) can be programmed to select the input source from any register file or DPU in the same column or from the three columns on each side and connect to the inputs of their respective sink—the register file or DPU. These switchboxes provide a non-blocking, uni-, multi-, or broadcast capability to all DPUs and register files within the sliding column window. We elaborate in the section "Private Execution Partitions: Complete and Dynamic Hardware Centric Custom Implementation for a Predictable and Composable System" how the interconnect scheme is utilized to create complete customization to make DRRA a critical component in making the SiLago platform dark silicon aware.

Distributed Memory Architecture

DRRA is primarily a computational fabric where register files provide a high bandwidth, low latency (same cycle) source, and sink of data to/from the DPUs within the sliding window span. However, signal processing applications need larger scratchpad memory to hold frames/buffers/packets besides the off chip DRAM based main memory. Further, the ability of DRRA to provide arbitrary degree of parallelism to implement CDFGs would be wasted if it were not possible to provide a matching parallel access to a large scratchpad memory. The distributed memory architecture (DiMArch) is a fabric that fulfills these two objectives: (a) it provides a large scratchpad memory to the DRRA fabric and (b) it provides a matching parallelism in storage access to the computational parallelism provided by the DRRA. DiMArch, shown in Fig. 3.4, like DRRA, is made up of regular cells that are realized as SiLago blocks. Each DiMArch cell has at its core an SRAM bank, whose size is a design time decision but dimensioned to occupy physically the width of one or two DRRA columns. In the present implementation, we have SRAM banks of 2 KB size. Each bank has spatio-temporal programming capability for address generation and data streaming similar to that for the register files. These banks are glued together by a circuit switched NOC, whose switches can also be programmed by simple hardwired but configurable FSMs that enable the fabric of SRAMs to be programmed in spatio-temporal manner to make the cluster of SRAM banks look like one large SRAM; e.g., for first n cycles send the data south from east, for next $n + 1$ to $n + m$ cycles send data south from north, and so on. The reason for choosing the circuit switched network, for the data transfer, is that it is deterministic and has low overhead (when the traffic patterns are deterministic).

A packet switched NOC is used to configure the SRAM banks and the circuit switched data NOCs. The packet switched NOC employs simple dimension order routing with buffer less switches. The motivation for using the packet switched NOC is that it allows to conveniently configure arbitrary node in the network. The reason for choosing the low overhead buffer less switches is that for most of the DSP applications, the configuration will occur only infrequently. Therefore, due to scarce traffic, the network will mostly be free from congestion. For more details on the DiMArch fabric, see [53, 54].

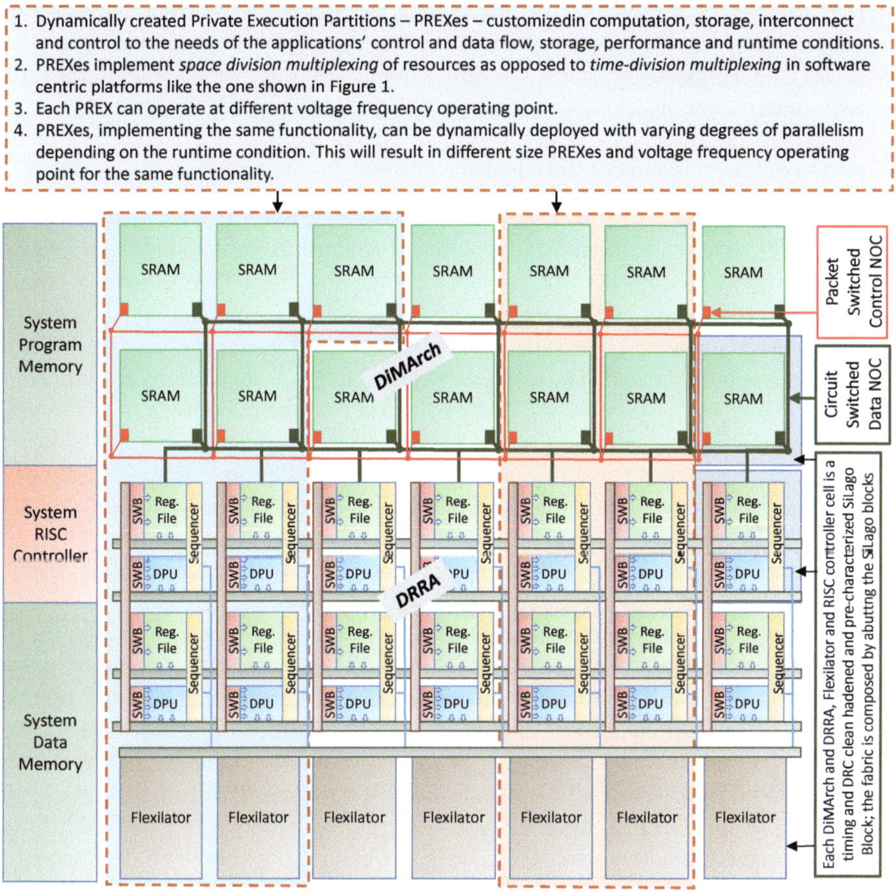

Fig. 3.4 SiLago platform instance with DRRA, DiMArch, flexilators, RISC system controller, and two private execution partitions (PREXes)

Dynamic Customization of Parallelism and Voltage Frequency Scaling

DVFS has become the mainstay of almost all modern SOCs to reduce dynamic power consumption. Additional power management features like power gating to turn off idle resources to reduce static leakage current is also a mainstream power management practice. The typical DVFS and power-gating implementation in the software-centric heterogeneous multi-processor SOCs like the one shown in Fig. 3.1 has two critical limitations:

1. Voltage island boundaries are design time static and these islands are few and thereby large. For instance, the design in Fig. 3.1 has only four voltage islands

[14]. Because these islands are large, they are shared by multiple applications with varying voltage frequency needs forcing the power management controller to set the voltage frequency operating point for the worst case application requirement, thereby running other applications at sub-optimal points.
2. Unless all applications sharing a voltage island are idle, the island cannot be turned off causing unnecessary leakage current.

These limitations make such designs incompatible with the dark silicon constraints and to overcome them SiLago platform enables dynamic creation of private execution partitions like the ones shown in Fig. 3.4. Each PREX hosts a single application, whose voltage frequency point can be decided considering the needs of that application alone and when the application is done these resources (the SiLago blocks) can be power gated. The efficiency can be further improved by dynamically deciding the degree of parallelism. Depending on runtime conditions, if additional resources are available, a more parallel version of the application can be deployed to reduce the number of cycles required to enable use of a lower voltage frequency operating point. We call this dynamic parallelism and voltage frequency scaling (*DPVFS*). Effectively, this ability allows the SiLago platform to transform spatial slack in the form availability of more resources into a temporal slack that can be exploited to lower the voltage frequency operating point.

This approach of combining parallelism and voltage frequency scaling has been adopted by other researchers like [29–31, 55–57]. However, the focus of these work is on task level parallelism across multi-processors that we have argued before are not scalable with dark silicon constraints. Further, these works also have static voltage islands that restrict their flexibility. In contrast, the SiLago platform enables fine-grained data level parallelism within an application and also task level parallelism in the form multiple private execution partitions (PREX) and allows voltage frequency island boundaries to be decided at runtime based on the degree of parallelism that decides the size of the PREXes.

Fine Grain SiLago Power Management Infrastructure We next explain the key power management infrastructure of the SiLago blocks (DiMArch, DRRA, flexilators, and System Control RISC are SiLago blocks). Each SiLago block is equipped with *Voltage Control Unit (VCU)* and *Clock Generation Unit (CGU)* [36, 58]. Each VCU contains a PMOS power switch and the necessary logic to drive it. The power switch selects one of the global supply voltages as the local supply voltage for the SiLago blocks. This allows quantized voltage regulation. The clock generation unit depending on the selected clock generates the clock of the desired frequency. We have chosen globally ratiochronous locally synchronous (GRLS) [59] design style to synchronize data transfer among different clock domains and implement DVFS in the SiLago platform. The main motivation for choosing GRLS is that it promises higher performance compared to GALS systems because it avoids the round trip delays of handshake [59, 60] and higher efficiency compared to mesochronous systems (requiring all modules to operate at same frequency) [59, 61]. For this work, we have chosen two operating voltages, due to the library limitations. Our previous investigation [58] revealed that up to four supply levels can be efficiently integrated on the chip (the flexibility/cost for providing more levels is not justifiable). The proposed approach results in limited loss of flexibility compared to a solution allowing continuous choice of supply voltages. However, the latter approach would have unacceptably high overhead for the fine grain power management we have the SiLago platform.

Dynamic Boundaries of Voltage Islands To be able to create arbitrary sized PREXes and place them anywhere in the fabric and to enable safe data transfer between PREXes operating at different voltage frequency points, we need two infrastructural enhancements. The first is the ability to set supply voltage of each SiLago block individually, thus by setting the same supply voltage and frequency of all the SiLago blocks that goes into a PREX, we are able to create a PREX that operates at a certain voltage frequency point. The second infrastructural need is to be able to safely transfer data between two arbitrarily sized and place PREXes, in practice this means synchronization and a regulation algorithm and a buffer to absorb a bounded burst; this has been detailed in [62]. There are two options for providing this second infrastructural need. The first option is to design and associate special SiLago blocks that deals with this need (synchronization, data regulation, and buffer) and instantiate them along with each SiLago block. This solution is clearly not scalable because the need for these special SiLago blocks will only be at the boundaries and would thus result in unacceptable overhead as most of these special purpose SiLago blocks will rarely be used. The second is to enhance each SiLago block with synchronization capability and enable it to use the existing buffer and controller in the SiLago blocks to provide the data regulation and buffer needs. This second alternative results in each SiLago block becoming slightly more expensive but is scalable because these blocks can be used for both isolation between two PREXes or their normal SiLago block functionality. As a result, SiLago blocks between two PREXes are dynamically configured to provide this second infrastructural need of isolation. These dynamically created SiLago blocks are called dynamically reconfigured isolation cells (DRICs). Since, any SiLago block can be configured to temporarily play the role of DRIC, boundaries and positions of PREXes can be arbitrary and dynamically drawn.

A simple motivational example of PDVFS and DRICs is annotated in Fig. 3.5. Note that agile and efficient reconfiguration enables the runtime management system to continuously monitor the load and available resources and load/deadlines to dynamically deploy different versions of applications varying in degrees of parallelism and thereby requiring different voltage/frequency operating points. An efficient compression and reconfiguration infrastructure has been developed that supports agile and efficient deployment of solutions in varying parallelism [63, 64]. This is described next.

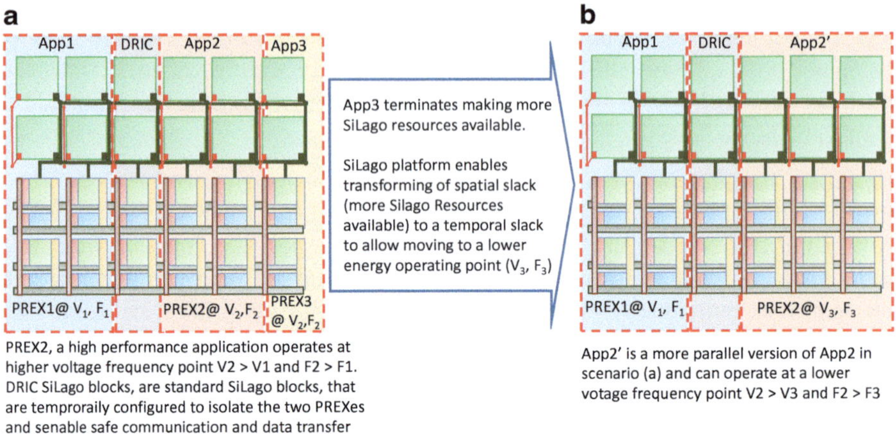

Fig. 3.5 Motivational example for PDVFS and use of DRICs. (**a**) PREX2, a high performance application operates at higher voltage frequency point V2 > V1 and F2 > F1. DRIC SiLago blocks are standard SiLago blocks that are temporarily configured to isolate the two PREXes and enable safe communication and data transfer. (**b**) App2′ is a more parallel version of App2 in scenario (a) and can operate at a lower voltage frequency point V2 > V3 and F2 > F3

The Global Interconnect and Reconfiguration Infrastructure in the SiLago Platform

The SiLago platform has two levels of interconnect: intra-region and inter-region. Examples of regions in the Silago Platform shown in Fig. 3.4 are DRRA, DiMArch, flexilators, and system controller, etc. The intra-region interconnect scheme for DRRA, DiMArch has been discussed above and also shown in Fig. 3.4. However, there is also a global inter-region interconnect scheme based on two NOCs, one high bandwidth circuit switched data NOC and the other is a packet switched control NOC for control, supervision, and configuration. Note that DiMArch also has a different version of these NOCs. The inter-region NOCs connect to the region at one or more network interface (*NI*) points, depending on the region and its dimensions. For instance, in a large SiLago platform instance, multiple network interface points will be instantiated for DRRA and DiMArch. However, for regions like system controller, memory controller (shown in Fig. 3.7), etc., a single connect network interface point suffices. The regional network interface(s) is then responsible for further distribution of the data/control using the intra-region interconnect.

The reconfiguration infrastructure relies on both the inter-region/global data and control NOCs. The data NOC with its higher bandwidth is used to distribute the configware for DRRA sequencer memories and the control NOC is used for power management, communication with PREX on start, stop, and done status.

Depending on the application's reconfiguration requirements, SiLago platform has a range of five options varying in their agility of reconfiguration and overhead. These reconfigurable, reconfiguration options are detailed in [36, 55]. These reconfiguration options differ from related research [65–67] in that we not only offer these five options but we can combine time to further add flexibility to cater to different needs. The most serial solution relies on system controller decompressing the configware and distributing to the sequencer memories and the most parallel solution relies on decompressed configware stored in DiMArch banks to load the sequencer memories in parallel and in fewer cycles; this obviously needs larger DiMArch dimension.

Compared to FPGAs, the reconfiguration in the SiLago platform is orders of magnitude more agile and efficient. This stems from the coarse grain micro-architecture level resources, rather than the Boolean level resources in FPGAs. For instance, for a WLAN transmit fragment, an FPGA will require a couple of milliseconds in the best of the case when the configware is on chip. The same WLAN functionality, when mapped to the SiLago platform will require in the slowest of the five options, 5000 cycles (25 μs @ 200 MHz) and only 60 cycles (300 ns @ 200 MHz) in the fastest option [36, 55].

Flexilators and System Controller

While the DRRA and DiMArch form the core of the SiLago platform to support fully customized hardware style implementation to make it a dark silicon aware platform, it will be incomplete without the support of resources to implement some part of functionalities in software. The SiLago platform is a generic concept, the specific design we present in this chapter focuses on signal processing applications and this has strongly influenced the design decisions regarding the fabric regions—the DRRA, DiMArch, flexilators, etc., and the SiLago blocks that populate these regions. In signal processing applications, we distinguish between three broad categories of sub-functionalities in signal processing systems as illustrated in Fig. 3.6.

The streaming functions are compute and storage intensive data parallel signal processing transforms. These categories of functions almost always have nearest

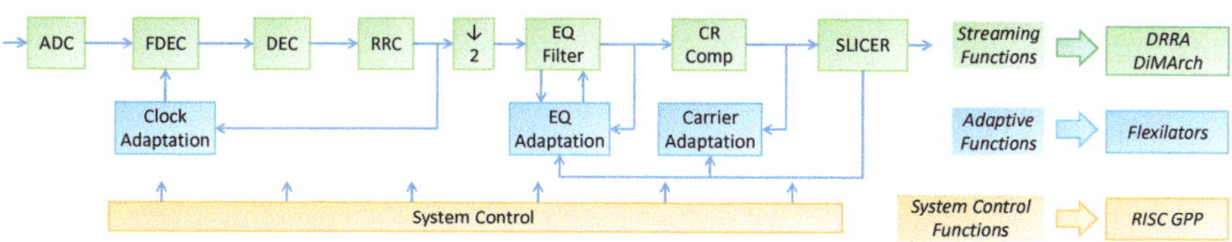

Fig. 3.6 Three categories of functions in signal processing systems and how they are mapped to different regions in the SiLago platform

neighbor connectivity, as shown with a real industrial example in Fig. 3.6, that matches very well with the nearest neighbor interconnect scheme of the DRRA DiMArch fabrics. These functionalities also exploit the rich address generation capability of these two fabrics. Do note that DRRA and DiMArch are not purely arithmetic and storage fabrics as other CGRA fabrics tend to be. They do have a rich parallel distributed control to handle the necessary FSMs and their hierarchy necessary for control of streaming functions.

Signal processing sub-systems also have functions that are adaptive and control intensive. Examples of these functions in the radio domain are frequency tracking, DC-offset cancelation, clock and carrier adaptation, AGC control, etc. These functions are characterized by small but latency sensitive control loops whose bounds, unlike streaming functions, are not compile time static and are control intensive. Another key characteristic of these adaptive functions is that they take inputs from multiple transforms and can potentially send their results, often some form of correction or adjustment of parameters to multiple streaming transforms as shown in Fig. 3.6. For these reasons, the SiLago platform hosts them on small simple processors called flexilators that implement typically single functions and have connectivity to reach sequencers of multiple columns. Flexilators today have 16 bit arithmetic and have no pipeline in terms of fetch-decode-execute to minimize the overhead. Only the multiplication operation has a two stage pipeline. Flexilators have 128 word memory that is sufficient for most adaptive functions and are tightly coupled with the sequencers of the DRRA. The data transfer of adaptive functions does not require heavy bandwidth but is latency sensitive. To meet these requirements, we have a low bandwidth circuit switched interconnect scheme that enables each flexilator to reach arbitrary column. In practice, the mapping has a high degree of spatial locality, even if it is not a nearest neighbor type of connectivity.

The third category of functionality is system control that controls, configures, and monitors operations and resources in different regions. System control is realized by vanilla RISC processor and in our experiments, we use the public domain LEON processor. Triggered by external stimuli, the system controller creates and instantiates PREXes for applications and sets their voltage frequency operating points. It monitors the available resources and factors in the load and deadlines to decide the position of the PREX in the fabric and degree of parallelism and its voltage frequency operating point. It is also responsible for controlling the data transfer between external sources/sinks and the SiLago platform. Infrastructural resources like memory controllers, PLLs, voltage regulators, etc., are not shown in Fig. 3.4 to avoid cluttering.

Private Execution Partitions: Complete and Dynamic Hardware Centric Custom Implementation for a Predictable and Composable System

Private execution partitions (PREXes) is key design concept in the SiLago platform that brings together many individual features of the SiLago platform to enable complete and dynamic customization to closely match the functional requirements of applications and also factor in the runtime conditions to enable deploying an optimal implementation. The key to achieving complete customization in the SiLago platform and a hardware centric implementation style is the parallel distributed micro-architecture level resources and an interconnect scheme that allows clustering them to create arbitrarily complex Control and Data Flow Graphs (CDFGs) in varying degrees of parallelism. Critically, these CDFGs include arbitrarily parallel access to large scratchpad storage. We next enumerate the various customization possibilities that exist in the SiLago platform:

1. The sliding window interconnect scheme of DRRA enables chaining of DPUs to create arbitrary dataflow graphs. This is akin to standard cells in ASICs and LUTs in FPGA being chained to implement arbitrary Boolean level data flow graphs. Note that the interconnect scheme enables uni-, multi-, and broadcasting capability that can be implemented by single cycle parallel distributed configuration.
2. Not only DPUs, but the register files can also be chained to implement larger register files. The addressing modes of register file can be customized in terms of addressing modes, range, and their temporal rhythm.
3. The DPUs and register files can be clustered to create arbitrarily complex, partially overlapping data flow graphs with hierarchical parallel distributed control in the form of sequencers [50] and AGUs in register files.
4. The DiMArch allows clustering of SRAM banks by partitioning the DiMArch fabric into parallel streaming sources and sinks of data to and from DRRA. This enables the SiLago platform to customize the storage and the access to storage according to the needs of the application and thereby address one of the critical limitations of state-of-the-art SOCs. Additionally, by having local AGUs, the SiLago platform eliminates the cost of address transportation in addition to using efficient lightweight configurable hardware FSMs [48]. By deploying small SRAM banks and just as many as needed, we significantly reduce the power consumption to enable the meager dark silicon era power supply to be fully utilized. A special high bandwidth interconnect exists between DRRA register files and the DiMArch fabric. Multiple parallel paths can potentially exist between a DiMArch partition and the corresponding DRRA partition as shown in Fig. 3.4. In addition, multiple PREXes can exist simultaneously.
5. By combining the clusters of DRRA, DiMArch, and flexilators, SiLago platform enables dynamic creation of private execution partitions (PREXes) that are ASIC like macros for implementation of complete modem/codec sub-systems with silicon and computational efficiency [42] that is modestly lower than ASIC.

This goes significantly beyond the timid static software-centric customization of heterogeneous processors and use of accelerators. This efficiency is achieved because of complete customization and hardware like parallel distributed storage, control and data flow, and efficient local address generation.

6. Another key advantage that PREXes offer is their predictability and composability. Predictability is a result of SiLago blocks being hardened timing and DRC clean GDSII macros that have been characterized. As we elaborate in the section "SiLago Physical Design Platform Overview," the SiLago blocks are micro-architecture level equivalents of the standard cells in the traditional EDA flows. Moreover, with a grid based structured layout scheme that enables composition of the SiLago fabric by abutment, not only micro-architectural operations but also the wires are pre-characterized. Effectively, this makes SiLago a micro-architecture level physical design platform, i.e., as soon as a design is refined to micro-architecture level, the cost metrics are perfectly predictable.

   PREXes also make the SiLago platform composable because of them being *private*, i.e., they are not shared. Once a PREX is created, the energy and performance guarantees are not violated when new application PREXes are instantiated, because PREXes do not share resources. In essence, PREXes make the SiLago platform a space division multiplexing platform instead of time division multiplexing that is the basis for the state-of-the-art accelerator-rich software-centric heterogeneous multi-processor platforms.

Differentiating SiLago Platform with Other Implementation Styles

The key difference that the SiLago platform has with the three categories of implementation platforms is as follows:

1. ASICs (Standard cells) and FPGAs have a sea of parallel distributed resources but they are at Boolean level and require first clustering them to create synthetic micro-architecture level blocks. And then clustering these blocks into a solution that reflects the functionality. It is well known that FPGAs devote a significant percentage of their resource in creating such clusters. ASICs are in some sense the golden standard for silicon and computational efficiency but their drawback is that these decisions are design time and many customizations require compile and runtime decisions.
2. The SiLago fabric as is visually depicted in Figs. 3.3 and 3.4 can be easily mistaken for homogeneous many cores like Tilera and GPUs. The key difference is that the innards of the DRRA cell are exposed and can be fused in a combinatorial sense with the resources in the sliding window span. Furthermore, the DRRA cell boundary is principally the reach of the local sequencer in its ability to configure the register files and switchboxes and provide the local control. The same applies to DiMArch cells. As a result, the DRRA/DiMArch cell boundaries are not similar to a typical processor boundary; it is very porous

and is designed to cluster resources to create complex micro-architecture level structures. In contrast, processors boundaries and interconnect are designed to parallelize at larger granularity of tasks and not at micro-architecture level operations as is done in the SiLago platform. Fusing a DPU in one processor with DPU in another to create a complex data flow graph in a homogeneous multi-core system will incur large latency and energy penalty in addition to tools not supporting such fusion. *These homogeneous multi-core platforms rely on temporal iteration to realize variety and size in functionality; in contrast, the SiLago platform relies on spatial combination of micro-architectural resources to realize variety and size in functionality. This is also the fundamental difference between software and hardware implementation styles.*

3. Other coarse grain reconfigurable architectures (CGRAs) [15, 23, 68, 69] to some extent allow DPUs and register files to be clustered to create small and simple dataflow graphs but lack resources to enable implementation of an arbitrary parallel distributed control flow and large parallel distributed and partitionable storage like DiMArch. These aspects are centralized in these CGRAs by the processor and CGRAs are instantiated as reconfigurable accelerators.

Automating the Complete Customization

In the previous section, we presented the SiLago platform as the architectural basis for complete, dynamic hardware centric customization to make SiLago based design dark silicon constraint compatible by enabling more functionality and higher performance to be squeezed out of the limited power that can be supplied and dissipated. As stated earlier in sections "Introduction" and "Lack of Dynamic Runtime Customization," the large engineering cost associated with hardware implementation is part of the reason why the VLSI design community has moved towards software-centric solutions. The SiLago platform not only enables complete hardware centric customization, but it also enables automation of such customization. This automation eliminates both functional and constraint verification related costs and also enables true and efficient application-level synthesis to eliminate the large engineering costs associated with present day standard cell based SOC design flows as analyzed in the section "Lack of Dynamic Runtime Customization" and shown in Fig. 3.2.

SiLago Physical Design Platform Overview

The SiLago design platform is not just architectural as elaborated in the section "The SiLago Platform" but it is also physical as we elaborate in this section. The platform is divided into regions that provide specific types of functionalities as shown in Fig. 3.7 that corresponds to the functionalities shown earlier in Fig. 3.4. Additional functionalities shown in Fig. 3.7 are infrastructural functionalities on the right.

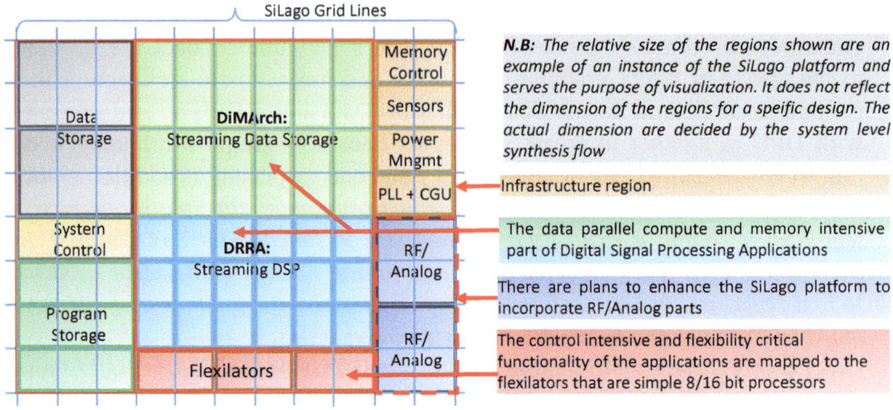

Fig. 3.7 SiLago platform example

SiLago stands for Silicon Large Grain Objects; the use of word silicon signifies that they are hardened, i.e., their physical design is done. This is in contrast to IPs that are typically soft and at register transfer or gate level. The SiLago blocks are not just hardened but they are hardened to occupy one and exactly fit one or more contiguous grid cells as shown in Fig. 3.7. SiLago grid is a virtual object that imposes a physical design discipline to simplify composing a design instance based on SiLago platform. The pins of the SiLago blocks, both functional and infrastructural (clock, reset, power, etc.), are brought to the periphery of the SiLago block and thus at the grid lines at right positions and right metal layers to enable composing a SiLago fabric instance by abutment.

The SiLago blocks are 3–4 orders of magnitude larger than the standard cells and replace standard cells as the atomic physical design objects. In the SiLago design flow, a design is composed by the synthesis tools in terms of SiLago blocks rather than standard cells. SiLago blocks, in the interest of remaining foundry compatible, are designed and hardened using standard ASIC design flows and realized in terms of standard cells. They are characterized with post-layout sign-off quality data and thus have measurement level accuracy of cost metrics for area, average energy, and latency in cycles. Measurement level means equivalent to measuring on chip (silicon) and this confidence stems from the sign-off quality design data and analysis tools. This is another reason that gives credence to the use of "silicon" in SiLago. SiLago blocks export measurement level accurate characterized micro-architectural level operations to high-level and system-level syntheses tools. This is analogous to standard cells and characterized technology library exporting Boolean level operations to logic and physical syntheses tools. *Effectively, the SiLago platform raises the abstraction/granularity level of the physical design target to micro-architecture level compared to the Boolean level of the standard cell based physical design. The use of word platform is justified because it exports characterized micro-architectural operations to higher levels of synthesis like algorithmic, application, and SOC/System levels.*

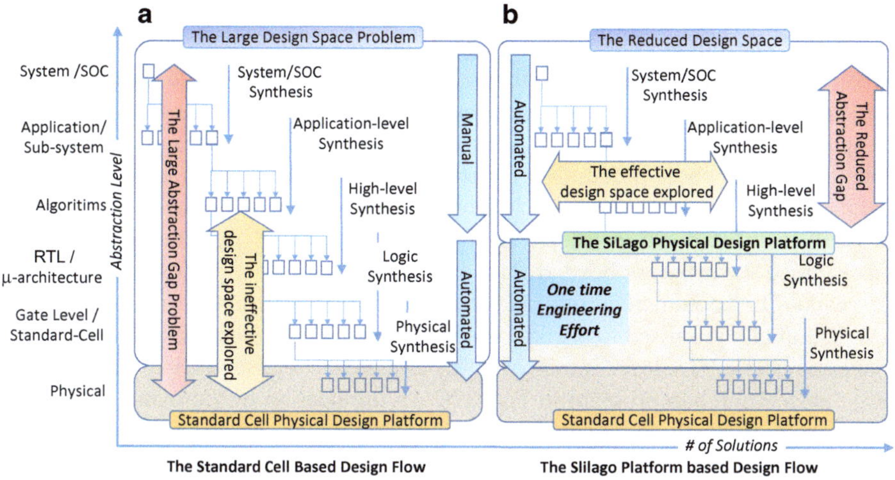

Fig. 3.8 Standard cell based design flow vs. SiLago platform based design flow. (**a**) The standard cell based design flow. (**b**) The SiLago platform based design flow

SiLago Platform Based Design Flow vs. Standard Cell Based Design Flow

With the brief introduction to the SiLago physical design platform concepts, we next discuss how it solves the problems faced by the standard cell based design flow. The arguments we present are visualized in Fig. 3.8 that shows the design space in the form of an inverted tree, where the leaf nodes are the final design solutions and the root and intermediate nodes are the abstraction level at which the design process starts. This implies, as is shown in Fig. 3.8, higher the abstraction level at which the design process starts, greater the number of solutions that can be reached. The highest abstraction level is that of system composed of multiple applications like WLAN, JPEG MPEG, etc., followed by application-level; applications are composed of a hierarchy of algorithms and then follows the well-known abstraction levels of algorithm, RTL/logic, and physical.

Analysis of Problems with the Standard Cell Based Design Flow First consider the standard cell based design flow shown Fig. 3.8a. The design flow is top-down, successively refining the design down to physical level, where the building blocks are standard cells and wires. The cost metrics of a design are known with certainty only when the physical design is done that decides the dimension/choice and position of each standard cell and wires connecting them. All intermediate design levels above physical design are soft and require the syntheses (manual or automatic), to make decisions based on estimates of the final cost metrics at physical design level. The accuracy of these estimates exponentially degrades as the abstraction gap and the complexity of the design increases. Our experiments show that even for modest application-level complexity of JPEG, wireless LAN fragments, starting at relatively low algorithmic abstraction level, the cost metrics predicted by the commercial synthesis tools is off by an average of 300 %. Not only are these estimates wrong by a large margin, it takes a long time to reach the physical design level to know the margin by which the estimates were wrong. This makes the whole process very inefficient because the wrong estimates forces iterations that are very expensive, it takes more than 22 h to synthesize a fragment of a JPEG codec down to the physical level starting from algorithmic-level. This has stymied the progress beyond RTL synthesis; high-level synthesis is even after 25 years of first commercial introduction in mid-1990s, a fringe design method that has perennially promised to become mainstream but hasn't become yet. As a result, synthesis above RTL level is today largely a manual exercise. This is shown as the large abstraction gap problem in Fig. 3.8. An implication of the manual design is that it warrants costly functional verification step because manual refinement unlike automatic synthesis cannot provide correct by construction guarantee.

Not only is the abstraction gap large in the standard cell based design flow, the design space to be searched is also large. If there are A applications in a System, L algorithms in each application, M micro-architecture level operations in each algorithm, S standard cells for each micro-architectural operations, and P physical design options for each standard cell, the total number of solutions is of the order of $O((((P)^S)^M)^L)^A)$. The problem of effectively searching this large 100–1000 million gate designs is compounded by the large synthesis time it takes to reach the physical design stage. This reduces the effective design space that the standard cell based tooling can search to a thin vertical design space shown in Fig. 3.8b. This design space, at lower abstraction levels, is relatively less effective because the granularity of objects and their impact on cost metrics is small compared to the granularity of design objects and their impact on design decisions at higher level.

SiLago platform Based Design Flow Solves the Problem The SiLago platform based design flow shown in Fig. 3.8b corrects this problem by raising the abstraction/granularity of the physical design platform to micro-architecture level. This reduces the abstraction gap and also the design space that should be searched. Additionally, the design space that is searched is at higher level where the design objects have a larger granularity and a greater impact on the design cost metrics. Note that in the SiLago design flow, the design space at micro-architecture level and below is fixed and decided down to the layout (GDSII) level as a onetime engineering effort. This is similar to the standard cell designs remaining fixed in terms of transistors and their layout no matter which functionality is composed in terms of these standard cells. Besides the difference in granularity and abstraction, the SiLago platform introduces additional physical design discipline in the form of a structured grid based layout scheme, where the inter-SiLago wires, functional and infrastructural are also fixed.

These factors enable the synthesis tools to know the implication of their (synthesis) decisions on cost metrics (average energy, latency in cycles, and

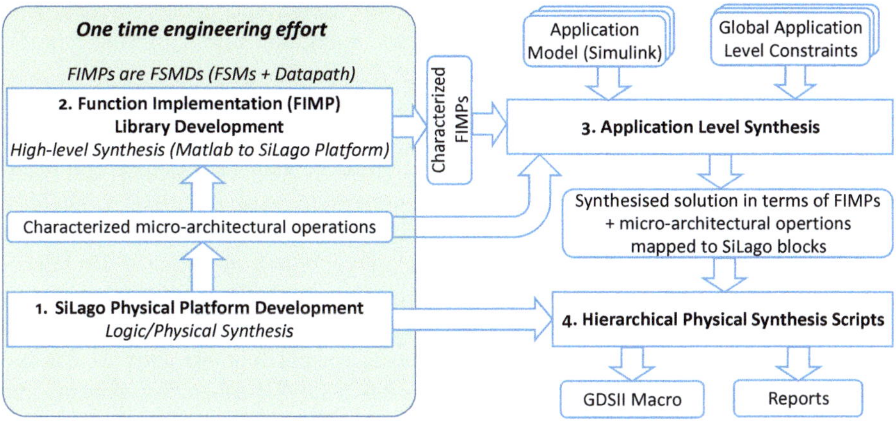

Fig. 3.9 The SiLago design flow

area) with measurement level accuracy. *This largely eliminates the cost metrics constraints verification problem (Fig. 3.2). This improved accuracy and 3–4 orders improvement in synthesis efficiency—due to reduced abstraction gap and design space—enables more efficient and effective synthesis from algorithmic, application/sub-system and SOC levels. This automated synthesis guarantees correct by construction and eliminates the large functional verification problem.* Finally, eliminating the large engineering costs—the functional and constraints verification—the SiLago platform enables not only total customization but also makes customization engineering efficient.

The SiLago Design Flow: Automating the Customization

The SiLago design flow has four steps as shown in Fig. 3.9. The first two are onetime engineering effort, while the next two are performed for every application that needs to be synthesized. In the step 1, the SiLago Physical Platform is designed, verified, implemented, and characterized. This is akin to standard cell library development phase in the conventional standard cell based EDA flows. It can also be compared to the class library development in an objected oriented program development framework. The net result is of this step is the export of characterized micro-architecture level operations to higher levels of syntheses (algorithmic-level and application-level) tools. Another outcome of this step are hierarchical physical synthesis scripts that are used in step 4 that is elaborated later in this section.

In step 2, a library of function implementations (FIMPs) in the form of FSMDs (FSM + Datapath) is developed. These FIMPs are the building blocks that the SiLago application-level synthesis tool uses to synthesize applications and explore design space. FIMPs are composed of one or more contiguous SiLago blocks. This step is elaborated in the section “Function Implementation Library Using AlgoSil High-Level Synthesis.”

In step 3, the application-level synthesis tool takes the application models in Simulink and explores design space in terms of FIMPs varying in their architecture and degree of parallelism for the same function to generate a globally optimized solution to meet the global application-level constraints and automatic generation of global interconnect and control to orchestrate the execution of FIMPs using micro-architectural operations. The solutions are generated in terms of the micro-architectural resources available in the SiLago platform. This step is elaborated in the section “Sub-system/Application Level Synthesis Using Sylva.”

Finally, in step 4, the hierarchical synthesis scripts that are part of SiLago physical design platform development composes a GDSII macro by abutting the selected SiLago blocks to generate the final GDSII macro and synthesis result reports.

SiLago Physical Design Platform Development

In this onetime engineering effort, SiLago blocks of different types are hardened and made timing and DRC clean and ready for composition of fabric by abutment. The types of SiLago blocks depend on the region, but even within the region, SiLago blocks could have different types depending on the neighboring SiLago blocks they need to abut with, and in some cases, SiLago blocks of same region will differ also in their internal functionality, for instance, having different computation modes and addressing modes, etc. The SiLago blocks are designed at RTL and the standard ASIC design flow is used to harden them and characterize their cost metrics with post-layout sign-off quality design data. This decides the size of the SiLago blocks and the number of grid cells they take and their orientation. The size of the grid cell is a function of these physical synthesis results to minimize the white space and represents the overlap and iteration with the architectural design phase.

Closely related to the physical design of the individual SiLago blocks are physical design rules for composing the SiLago fabric instances of arbitrary dimensions. Some parts of these rules are global and concerns how the clock, reset and power grid are to be implemented, dimensioned, and absorbed within each metal layer to enable composition by abutment. Besides these global infrastructural nets, the position and dimensioning of the global functional interconnect related nets are also similarly designed. Lastly, physical design of intra-region nets is designed and their pins brought to the periphery at right position and metal layer to enable composition by abutment. As a result, each SiLago block has nets on its periphery. These nets allow a SiLago block it to connect with either some permitted SiLago blocks from the same region or with the Silago blocks of a neighboring region directly by abutment.

The Engineering Effort Concern Many readers would be justifiably concerned that by raising the abstraction of physical design target to micro-architectural level, we are shifting the engineering effort to the development of SiLago physical design platform. We argue that this effort is scalable because of the regular tiled nature of the DRRA, DiMArch, and flexilator fabrics that requires a finite number of SiLago block types to be designed, verified, implemented, and characterized; once a SiLago block of a certain type is hardened, it can be reused without repeating any further RTL/logic synthesis, verification and characterization effort. The inter-SiLago block pattern is also not arbitrary but repetitive and restricted to a few patterns. For instance, the sliding window interconnect scheme and the restrictions on the topology of neighboring regions restrict the combinations of neighboring SiLago blocks. The system controller and the infrastructural region are made up of pre-designed IPs, the effort is in adapting them as SiLago blocks to comply with the physical design discipline of the SiLago platform that aligns the rim of these SiLago blocks to the grid lines and brings out the interconnect and functional wires at right positions and right metal levels to enable composition by abutment. *The main engineering effort is in developing the fabric architecture, interconnect and physical design, and composition scheme.*

Characterization Besides architectural and physical design and verification at RTL, the other major effort in the SiLago platform development is that of characterization. Once again, this is based on the standard EDA design flows. The design is synthesized down to physical level and all the RC parameters are back annotated from the post-layout sign-off quality design data. Such designs are then comprehensively simulated to extract the average energy values for the micro-architectural operations. The latency in cycles is accurately predicted by the static timing analysis region. Area is also accurately known as a result of the physical synthesis. As energy characterization relies on large simulation for designs that are sensitive to input data pattern, the SiLago blocks are designed to have modest complexity—10–100 k gates to enable efficient and accurate characterization. For SRAM banks, accurate data is available from the SRAM generator.

The SiLago blocks are encased in their own power rings. This provides a high degree of isolation for SiLago operations. This ensures that an operation in a SiLago block consumes energy that is independent of the activity in other SiLago blocks. Concurrent operations within a SiLago block do couple to a certain extent but their coupling is modest and can be modeled as there are a small finite number of combinations due to the simplicity and modest complexity of the SiLago blocks.

The end result of the physical platform design activity is (1) characterized micro-architectural operations, (2) physically designed hardened Silago Blocks that align with SiLago grids and connect to neighboring SiLago blocks by abutment, and (3) scripts to compose a SiLago fabrics of and arbitrary dimension according to the application requirements. The number and type of SiLago blocks to be instantiated in a specific design is automatically decided by the high-level synthesis and application-level synthesis tools that also generate the configware and software as explained next.

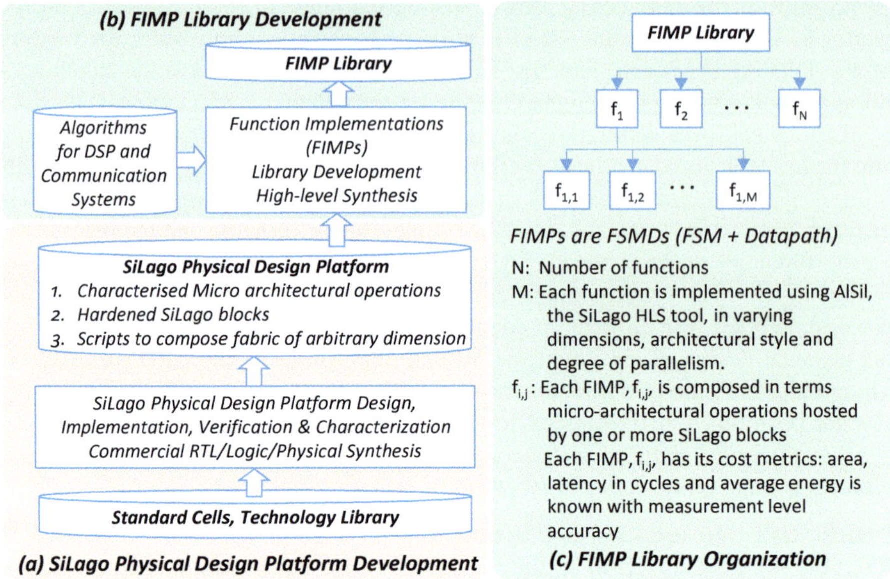

Fig. 3.10 SiLago physical design platform and FIMP library development. (**a**) SiLago physical design platform development. (**b**) FIMP library development. (**c**) FIMP library organization

Function Implementation Library Using AlgoSil High-Level Synthesis

AlgoSil, Algorithm to Silicon, is a high-level synthesis tool that synthesizes algorithms/functions in Matlab in terms of micro-architectural operations supported by the SiLago blocks. These synthesized solutions are in the form of FSMs + Datapaths—FSMDs and called FIMPs for Function Implementations (FIMPs). The role of AlgoSil in the SiLago design flow is to generate a library of FIMPs. Each function is implemented in many different architectural styles and degrees of parallelism and thereby varying in their cost metrics as shown in Fig. 3.10c. These variations create the design space at higher levels of abstraction that the application-level synthesis (section "Sub-system/Application Level Synthesis Using Sylva") explores. This library has functions that correspond to Matlab's signal processing and communication toolboxes. The application model in Simulink, shown in Figs. 3.9 and 3.11, is composed in terms of the functions in the AlgoSil created FIMP library.

AlgoSil accepts the algorithms or the functions modeled in Matlab as input. A rich set of pragmas that are symbolic and parametric allow the library developer to guide the synthesis process towards a desired architecture. These pragmas guide AlgoSil in terms of allocation and binding but the scheduling and synchronization of the vector operations is done automatically by AlgoSil. Allocation is often a constraint and expressing it as a pragma does not dilute the automation more than what other high-level synthesis tools do. Binding, we believe is directly

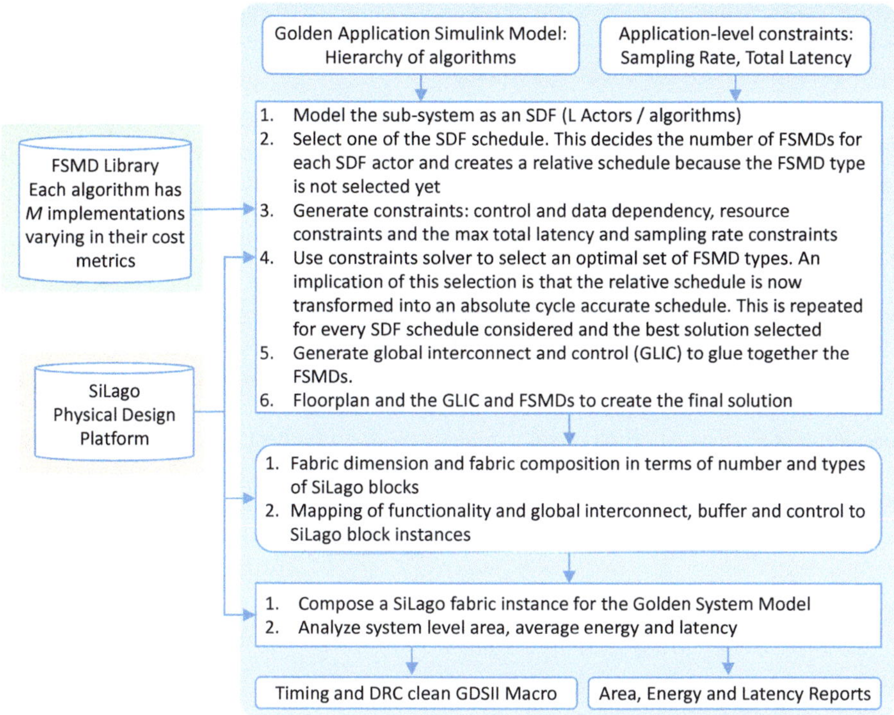

Fig. 3.11 Sylva: sub-system/application-level synthesis based on the SiLago platform

associated with allocation for a target like DRRA-DiMArch and for DSP like functions with their regular structured data flow, it is very visible to the library developer. By making the pragmas symbolic and parametric, the same Matlab code can be used for different dimensions and different degrees of parallelism. This feature is used to sweep through the parameters that control the degree of parallelism to generate a range of solutions when doing design space exploration during application-level synthesis. This is similar to high-level synthesis having access to micro-architectural blocks like adders and multipliers in varying degrees of latency and costs.

The scheme of pragmas exists mainly for data parallel DSP functions whose architecture shows a high degree of regularity, e.g., FIR, FFT, Matrix Multiplication, etc. But, not all DSP algorithms have the regularity and the desired architecture may not be obvious to guide AlgoSil with pragmas. In such case, AlgoSil has the ability to also work without the guidance of the pragmas. In short, the role AlgoSil, a high-level synthesis tool, is to develop a library that provides the design space for the application-level synthesis described next.

Differentiation AlgoSil differs from conventional HLS tools in three key respects. (1) Conventional HLS tools generate a singles FSM to control the entire datapath and storage. AlgoSil in contrast generates parallel distributed controllers. Some of these controllers are AGUs in register files and DiMArch cells, while others control parallel functional computation in different DRRA cells. (2) Conventional HLS tools do not deal with address and address constraints generation as separate functionality but fuses them into the main computation flow. AlgoSil in contrast treats them as separate functionalities and maps them to separate resources with their own control. See [48] for more details on address and address constraints generation in the SiLago platform. (3) Conventional HLS tools need to estimate the impact of RTL/Logic and physical synthesis on the final cost metrics when it makes synthesis decisions. For AlgoSil, micro-architecture is physical, so as soon as it has a solution at micro-architectural level, it knows the cost metrics with measurement level accuracy making the HLS process dramatically more efficient and accurate.

This efficiency and accuracy of high-level synthesis is vital in the SiLago flow because it enables efficiently generating a rich library of FIMPs for the application-level synthesis tool to use with the confidence of knowing the cost metrics accurately. It is the application-level synthesis that provides a quantum improvement in design productivity of complete customization and thereby helps combat the dark silicon problem.

Sub-System/Application-Level Synthesis Using Sylva

An application or sub-system level is defined as a hierarchy of algorithms and constraints are specified for the entire application and not for individual algorithms as is done in typical high-level synthesis (HLS) tools. The category of applications that we focus with the current SiLago platform are modems and codecs. For such signal processing applications, the typical constraints are max total latency and sampling rate and objective of optimization is to meet these constraints while minimizing the area and energy of the entire sub-systems not individual algorithms. Application-level synthesis generates its solutions and explores the design space in terms of FIMPs that implements the individual algorithms. In contrast, HLS generates its solutions and explores the design space in terms of micro-architectural blocks. If there are M algorithms in an application and each algorithm has L FIMPs available in the FIMP library (section "Function Implementation Library Using AlgoSil High-Level Synthesis" and Fig. 3.10c) the task of application-level synthesis is to search the design space of L^M solutions and find a solution that meets the max total latency and sampling rate constraint while minimizing the *total* area and energy. As stated in the previous section "Function Implementation Library Using AlgoSil High-Level Synthesis," the L implementations differ in their degree of parallelism/architecture and its implications on the cost metrics (average energy, area, and latency). Additionally, the application-level synthesis is expected to automatically glue together the chosen FIMPs with global interconnect that implements the data transfer among the FIMPs and global control that orchestrates their execution. This global glue logic is generated using SiLago blocks. Note that some of the functions may be implemented in software on flexilators and Sylva knows how to interconnect in both data and control sense the FIMPs implemented on flexilators and DRRA and both of them to the scratchpad memory in the DiMArch.

Sylva—(Sub-system level Architectural Synthesis)—is the application/sub-system level synthesis that is used as part of the SiLago design flow as shown in Fig. 3.11 and presented in detail in [21, 22]. This flow is an end-to-end fully automated design flow from Simulink to GDSII that is 3–4 orders of magnitude more efficient compared to application-level synthesis done with commercial HLS tools and significant manual refinement as elaborated below under the heading "differentiation." By being fully automated, the correct by construction guarantee eliminates the functional verification. Raising the physical design platform to micro-architectural level guarantees with measurement level accurate cost metrics eliminates the constraints verification problem. Besides eliminating the two most expensive engineering costs, the SiLago design flow also eliminates the task of architecting the application in terms of FIMPs. Collectively, these improvements in terms of dramatically improving the engineering efficiency but also in terms of automating customization and use of hardware centric design style that enables more functionality per unit current drawn and area used compared to the software-centric implementation style. This is what makes the SiLago platform dark silicon friendly. This objective cannot be achieved just by architecture or design method, they both need to holistically address the problem of dark silicon in a synchronized manner.

Differentiation Many commercial HLS tool flows claim application-level synthesis. However, the only superficial similarity they have is the *claim* that like Sylva (Fig. 3.11) they can handle a hierarchy of functions. In reality, these HLS tools still deal with one algorithm at a time and interconnect between these algorithms and the global control necessary to orchestrate the entire solution is manually refined. These tools also require the end user to partition the global constraints into individual algorithmic-level constraints. In effect, the hierarchy that these HLS tools handle is a manually refined version with interconnect and global control and often re-writing of the code in a tool specific style to get best results. This manual refinement breaks the correct by construction guarantee forcing the expensive verification step. In contrast, Sylva is able to deal with the hierarchy without manual refinement and does global optimization for the whole hierarchy and not individual algorithms. Finally, by using FIMPs as the basis for design space exploration and having micro-architecture level physical design platform, Sylva is orders of magnitude more efficient and accurate as we substantiate in the section "SiLago Design Methodology" and explores design space more effectively at higher abstraction level of FSMDs rather than micro-architecture and standard cells as is done by HLS tools.

Experimental Results

In this section, we quantify the benefits and the claims of the SiLago platform and design methodology in the next two sub-sections "Computational and Silicon Efficiencies of the SiLago Platform" and "SiLago Design Methodology."

Computational and Silicon Efficiencies of the SiLago Platform

In this section, we quantify and compare computational and silicon efficiencies of implementing signal processing functionalities in different implementation styles and compare them to SiLago platform. These results validate the claim that SiLago provides significantly improved computational and silicon efficiencies and achieve results, especially computational efficiency that is comparable to ASICs. This superior efficiency comes from hardware centric implementation style and complete and dynamic customization. The end result of these benefits is that it enables more functionality to be powered by the same amount of current and in less silicon. While dynamic power consumption is still the most commonly used metric, it is worth pointing out that silicon efficiency also indirectly contributes to SiLago platform being dark silicon aware because larger silicon implies more transistors and unless they are turned off they will leak; both sub-threshold and gate oxide leakage currents are proportional to the total width of transistors, i.e., proportional to area of the design.

Complete Customization of the SiLago Platform

In this sub-section, we report results of experiment done to quantify and validate the claim that complete customization makes the SiLago platform nearly as efficient as the ASIC and thus dark silicon friendly compared to the more general purpose and software-centric solutions like FPGAs, GPUs, GPPs (General Purpose Processors), etc. Figure 3.12 compares the results of two typical signal processing transforms and their computational silicon efficiencies on different platforms. As expected, the i7, a GPP, has the lowest silicon and computational efficiencies. FPGAs, in spite of implementing in hardware style, are less efficient compared to GPUs because FPGAs lack support for streaming data parallel applications that one typically needs. FPGAs are a classic case of compute centric optimization, DSP related arithmetic was added to FPGAs but not tightly coupled DSP related interconnect and storage. In contrast, GPUs were built from grounds up to support vector operations and thus achieves better results compared to FPGAs. This also validates our thesis that not just compute but a complete optimization is required to achieve superior silicon and computational efficiencies. SiLago outperforms GPUs because it provides the infrastructure to customize the entire datapath for the two

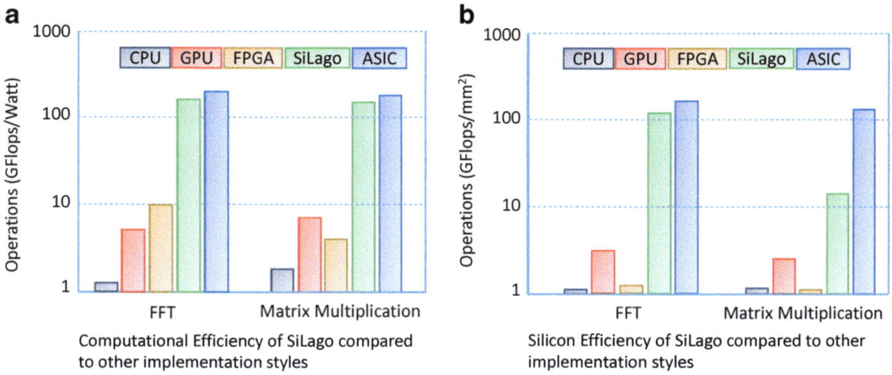

Fig. 3.12 Comparison of the energy and silicon efficiency of different computing platforms. (**a**) Computational efficiency of SiLago compared to other implementation styles. (**b**) Silicon efficiency of SiLago compared to other implementation styles

applications. In GPU, both applications FFT and Matrix multiplication would run on exact same datapath, interconnect, and storage; only the program code would differ. In contrast, in the SiLago platform, all aspects—computation, interconnect, storage, control, and address generation—are fully customized for the two applications and the implementation is parallel and distributed. This is the reason why the SiLago implementation is more than an order of magnitude better than the GPU.

Compared to ASIC, the SiLago platform has modestly lower (~20 %) computational efficiency. This again is attributed to the fact that like ASIC, the SiLago platform enables complete customization, albeit at micro-architecture level. The silicon efficiency of the SiLago platform compared to ASIC has a greater difference compared to difference in computational efficiency. There are two reasons for this. The first is that in the SiLago platform, each SiLago block has the same size and resources, irrespective of the application being mapped to it. This under-utilization is what explains the greater difference. The second reason is that the results shown in Fig. 3.12 do not include the cost of large memory which would be the same for both ASIC and SiLago. If this large constant factor is included for both ASIC and SiLago, the difference would be substantially less.

The numbers for the CPU i7 core, GPU GTX255, and FPGA LX760 are based on the experiments reported in [24]. For ASIC, we have used Cadence CtoS high-level synthesis tool to generate the hardware and compared the results with SiLago results; both ASIC and SiLago results are based on simulation of post-layout design data. *NB: the numbers shown for SiLago are conservatively scaled for floating point operations in DPUs and 32 bit data for fair comparison because SiLago has 16 bit datapath.*

The SiLago design flow offers two alternatives to address the larger difference in area, compared to power consumption, between ASIC and SiLago platform. The first alternative is to automatically eliminate the unused features and modes of the SiLago design instance created by the application-level synthesis flow (section

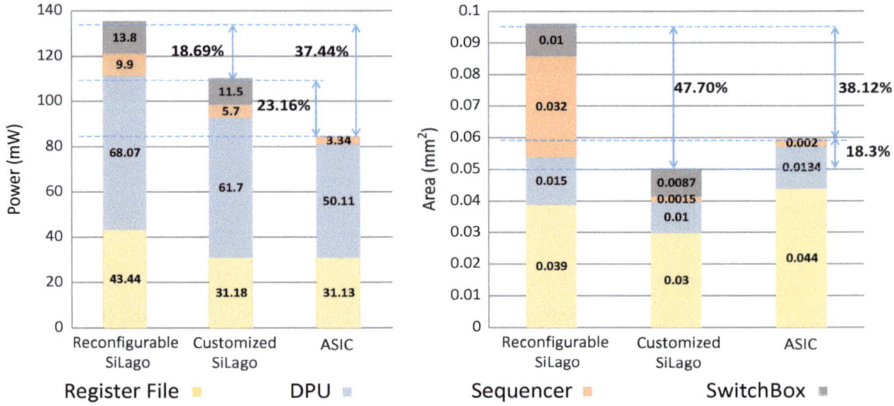

Fig. 3.13 Industrial case study showing the results of folding two applications to the same set of SiLago resources and removing unused SiLago features and comparison to ASIC [70]

"Sub-System/Application Level Synthesis Using Sylva") as a post processing step. The second alternative is to fold multiple applications/transforms onto the same SiLago resources, if their uses case is such that only one of them can be active at a time. This often happens in radio applications, where the terminal device is working with one standard but not another (4G or 3G but not both at the same time). We report results of an industrial example to demonstrate the results that can be achieved when these two alternatives are applied [70] and shown in Fig. 3.13. In effect, we report results of three cases: Fully reconfigurable SiLago platform with both applications mapped to the same SiLago resources, partially reconfigurable SiLago platform that is able to deal with requirements of the two applications, again folded to the same SiLago resources and finally ASIC implementation. We report both area and power consumption for the three cases in Fig. 3.13. As can be seen, the fully reconfigurable SiLago platform solution is 37 % off ASIC power consumption but when unused resources are eliminated, the difference is only 23 %. For the area, the results are bit surprising because while the fully reconfigurable SiLago solution is worse off by 38 % but when the unused resources, a large part of sequencer memory, is eliminated the SiLago solution is 18 % better than ASIC. This happens because, in an ASIC flow, unless a significant engineering effort is invested, it is not possible to map two ASIC macros onto the same set of standard cells.

Later in the sub-section "The Overhead Incurred by SiLago's Application Level Synthesis," we also report an average of overhead incurred by the SiLago method compared to ASIC for three substantial applications: WLAN, JPEG, and LTE and validate that these overheads are modest and comparable to the loss in design quality when the VLSI design community switched to standard cell based design automation from design of full custom macros based on the Carver-Mead methodology.

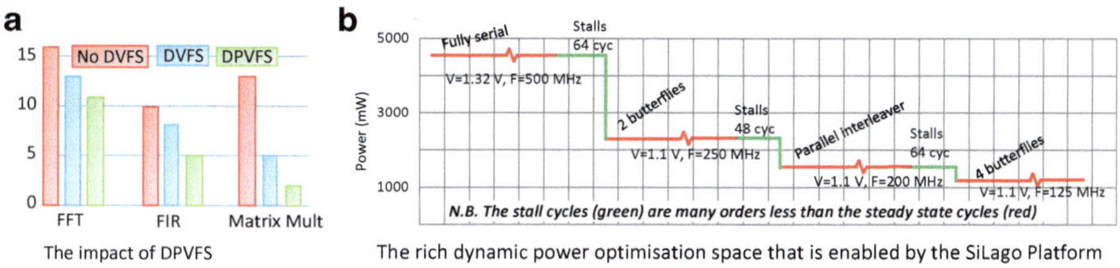

Fig. 3.14 Energy savings by using DPVFS. (**a**) The impacts of DPVFS. (**b**) The rich dynamic power optimization space that is enabled by the SiLago platform

Dynamic Parallelism and Voltage Frequency Scaling

SiLago's hardware centric complete customization brings the computational efficiency close to ASICs and is significantly better than the software-centric implementations and FPGAs. This claim was quantified and validated in the previous section. In this section, we quantify the benefits of dynamic parallelism and voltage frequency scaling (DPVFS). Figure 3.14a shows implementation of three typical signal processing transforms on the SiLago platform to compare three power management regimes. The No DVFS case operates at the highest voltage operating point 1.32 V, 500 MHz and obviously has the highest power consumption. The DVFS scheme assumes static degree of parallelism and operates at 1.1 V, 400 MHz as the lowest operating point at which it can meet the deadline. The DPVFS scheme case exploits the extra parallelism that is possible to lower the frequency, 200 MHz at 1.1 V, but still meet the deadline because the extra parallelism takes fewer cycles. This added flexibility gives the SiLago platform an added elasticity to adjust to the runtime conditions and exploit it to lower the overall power consumption. As we noted earlier that compared to ASICs, the SiLago platform uses more area. Some of this added area is due to the extra SiLago resources that are not used all the time but it could be used to dynamically adjust the parallelism to lower the power consumption.

We observe that ASICs lack this ability to exploit the runtime slack that might become available non-deterministically. *As a result, depending on the runtime conditions, it is possible that the SiLago based implementation, in spite of its modestly inferior computational efficiency, could still result in overall lower energy consumption compared to ASIC because of its ability to exploit the DPVFS that the ASIC implementations cannot.*

Figure 3.14b further validates the DPVFS potential of the SiLago platform. It shows, how a WLAN applications (a Tx fragment as a representative example) can operate at different voltage frequency operating points by exploiting variations if parallelism of its component transforms as annotated on the trace of execution shown in Fig. 3.14b. Starting from a fully serial solution that requires operating at a higher voltage frequency point, the runtime gradually deploys more parallelism in the component transforms (algorithms) in the WLAN Tx chain to achieve successively lower voltage frequency operating point and thereby lower the overall energy consumption. Note that the stall cycles are few and compared to the steady state shown by red lines it is three or more orders of magnitude less. The stall cycles are easily handled by gently easing to the new operating point by operating at a slightly higher frequency for the next OFDM symbol to make up for the lost cycles to stall and then easing into the steady state frequency.

The DPVFS vs DVFS scheme can also be compared in a more abstract sense by saying that DVFS responds to a temporal slack to lower/raise the voltage frequency operating point, whereas DPVFS has the ability to respond also to a spatial slack (SiLago resources) and transform it into an equivalent temporal slack to adjust the operating point time.

DPVFS Overheads The DPVFS scheme has moderate overheads. The SiLago blocks with added fine grain power management infrastructure and synchronization and regulation algorithm adds to 12 % in area. Besides this per SiLago block increase there is an overhead of DRIC cells that depends on the number and size of PREXes (Private Execution Partitions). For the case of JPEG and WLAN applications, the number of DRIC cells needed was five, whereas the total number of SiLago blocks that are used is 45. The power consumption numbers reported above factor in the power consumption cost of the DRIC cells.

SiLago Design Methodology

In the previous sub-section, we have quantified and validated the claim that the SiLago platform enables complete, dynamic hardware centric customization. In this section, we quantify the claim that it also enables automation of customization and dramatically reduces the engineering cost compared to the traditional standard cell based SOC design flow. Specially, the SiLago design flow brings three benefits that address the corresponding problems analyzed with traditional SOC design flow in the section "Lack of Dynamic Runtime Customization" and illustrated in Fig. 3.2. These are: (1) Enables more accurate and efficient system-level design space exploration. (2) Eliminates the constraints (on cost metrics area, average energy, and latency in cycles) and functional verification, and (3) Reduces the synthesis time from application-level to GDSII. In this sub-section, we quantify and validate these claims using the following experimental setup. The SiLago fabric that we used for our experiments is based on a SiLago platform composed of DRRA, DiMArch, RISC system controller, system data, and memory regions as shown in Fig. 3.7. The fabric was implemented using commercial EDA SOC/ASIC design flow from high-level synthesis down to physical synthesis in 40 nm CMOS process and constrained by 200 MHz clock. This was designed, implemented, and characterized by observing the design disciplines to ensure the space invariance and linearity of micro-architectural operations [20]. The size of the grid cell was $185 \times 185\ \mu m^2$ and SiLago blocks of different regions occupy number of grid cells as shown in Fig. 3.7.

To validate the benefit of the SiLago design flow, compared to the commercial SOC design flow, we selected three system-level functionalities: (a) the JPEG encoder, designed for resolution 1920×1080 and frame rate of 60 Hz; (b) the LTE uplink transmitter that has one bit input, BPSK constellation mapping, 64 sub-carriers, two users and output width of 16 bits; and (c) the WLAN 802.11a transmitter also has one bit input and uses 64 QAM constellation mapping and an output width of eight bits. To analyze the efficacy of the SiLago platform enabled design space exploration, we used the application-level synthesis tool proposed in [21] and briefly described in the section "Sub-System/Application Level Synthesis Using Sylva."

The SiLago flow was able to generate a large number of solutions in fairly moderate time, e.g., 89 JPEG solutions in ~4 min as shown in Fig. 3.15a. Not only are the solutions generated rapidly down to GDSII level but also their accuracy in predicting the energy consumed is very high as quantified in Fig. 3.15b. The accuracy was measured by simulating these applications at Simulink level and generating a trace of pre-characterized micro-architecture level operations and adding their energy cost metrics. This prediction was compared to simulating the gate level designs annotated with post-layout data; this was a very significant effort but necessary to quantify the accuracy. We have not reported the accuracy of estimates for area and latency in cycles because these estimates were perfect. In contrast, the standard cell based commercial tools take extra-ordinarily long time to reach GDSII level, e.g., it takes 30 h of syntheses time to produce a single solution down to GDSII level (Fig. 3.15a). The situation is further aggravated, when the accuracy of predicted energy is considered (Fig. 3.15b). This large inaccuracy in prediction will result in costly iterations. This inefficiency in long synthesis time and inaccuracy is a direct result of the large abstraction gap and design problems analyzed in the section "SiLago Platform Based Design Flow vs. Standard Cell

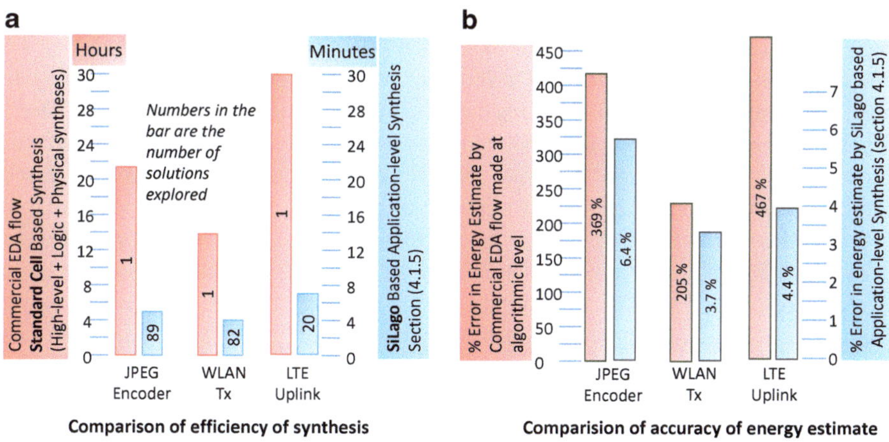

Fig. 3.15 Improvements of system-level design space exploration and synthesis time. (**a**) Comparison of efficiency of synthesis. (**b**) Comparison of accuracy of energy estimate

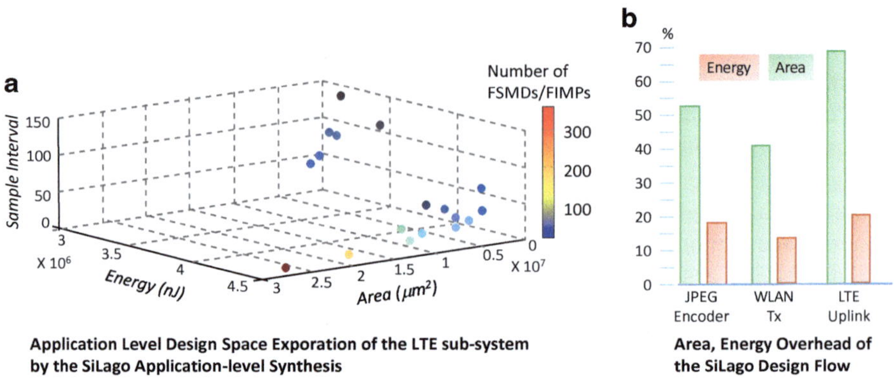

Fig. 3.16 Overhead of the SiLago design flow and the application-level DSE of the LTE sub-system. (**a**) Application-level design space exploration of the LTE sub-system by the SiLago application-level synthesis. (**b**) Area, energy overhead of the SiLago design flow

Based Design Flow" and illustrated in Fig. 3.8. Because of this large synthesis time and inaccurate prediction, we generated a single solution for the commercial flow; additional results would not have add any further information. While it is true that high-level synthesis can explore design space at micro-architecture level, the results reported are crude estimates as validated by numbers shown in Fig. 3.15b. Moreover, for fair comparison, it is essential to compare the application-level to GDSII synthesis time and accuracy as is achieved by the SiLago flow.

To further substantiate the DSE prowess of the SiLago flow, we show design points that it explores for the LTE sub-systems in Fig. 3.16a. The three dimensional design space is spanned by sampling interval, area, and energy. The solutions vary in number and types of FIMPs used [21]; the type information is not visible in the plot. All solutions meet the max total latency constraint. The blue colored dots are the most serial solutions with less energy consumption and medium to large sampling interval (low sampling rate). On the other hand, yellow and red dots are the solutions that use many FIMPs and thus are more parallel solutions that can handle fast sampling (small sampling interval) but consume more energy and area. We emphasize that the SiLago flow explores the designs space in terms FIMPs as shown in Fig. 3.8b as more effective DSE. In contrast, high-level synthesis-based DSE does not explore all the points shown in Fig. 3.16a, but the user guides the tool to one of the points as the starting point and then the tool explores the region around it by changing number and type of micro-architectural blocks and arrangement of standard cells, resulting in small variations in the generated designs and as we quantified in Fig. 3.15, the synthesis time is unreasonably long to reach GDSII and the accuracy of prediction is unacceptably poor.

While the improvement in synthesis time and accuracy of prediction are significant benefits in themselves, an even bigger benefit is the elimination of functional verification because of correct by construction guarantee of end-to-end synthesis from application-level down to GDSII. These benefits directly derive from the core SiLago idea of raising the abstraction of physical design target to micro-architecture level and the grid based structured layout scheme based on composition by abutment.

The Overhead Incurred by SiLago's Application-Level Synthesis

The use of SiLago flow results incurs some loss in design quality, which is defined as the percentage of degradation in area and energy cost metrics for a specific sampling rate compared to an ASIC implementation. For area, the loss in design quality results from (a) grid based regular design and its overheads and (b) not all resources in a SiLago block being utilized, depending on the application. Increasingly, many applications are dominated by fixed memory cost and for such applications, the area overhead is significantly less compared to logic dominated designs. We also note that lower utilization of the resources is naturally compatible with the dark silicon constraint. The loss in design quality for energy is much less because the number of logical operations done by SiLago and commercial standard cell based flows produced solutions is identical. The degradation in design quality shown in Fig. 3.16b is comparable to the degradation of standard cells based automated designs compared to full-custom [71].

The Efficiency of FIMP Library Development

To enable the application-level design space exploration that has been quantified above, a rich FIMP library is required. Such a library provides many FIMP options for the same function that varies in architecture and/or degree of parallelism reflected in variations in cost metrics (section "Function Implementation Library Using AlgoSil High-Level Synthesis" and Fig. 3.10). In the SiLago flow, we use a HLS tool proposed by [72] to generate the FIMP library. Fig. 3.17a reports the total synthesis time required to generate 15 FFTs from algorithmic-level down to physical level for the two competing design flows. For the SiLago flow, the only time required is for high-level synthesis to refine the FFT function in terms of the pre-characterized SiLago operations, fully customized (control, data, storage, and address) for architecture and varying degrees of parallelism. Since micro-architecture is the level of abstraction of the physical design in the SiLago flow, as soon as a design is refined down to micro-architecture level, the design is largely done and its cost metrics know with post-layout accuracy. In contrast, for the standard cell based commercial EDA flow, implementing 15 FFTs down to layout level is a significant effort, spread across all three syntheses phases; the final cost metric is not known until the physical design is reached. The comparison is quantified in Fig. 3.17a that has averaged the synthesis time required for the 15 FFTs. As in the case of application-level synthesis, the accuracy of estimates is almost perfect for the SiLago method, whereas in the case of standard cell based flow, it is unacceptably poor as shown in Fig. 3.17b. Figure 3.17c, d quantifies the moderate area and energy overheads resulting from under-utilization of some

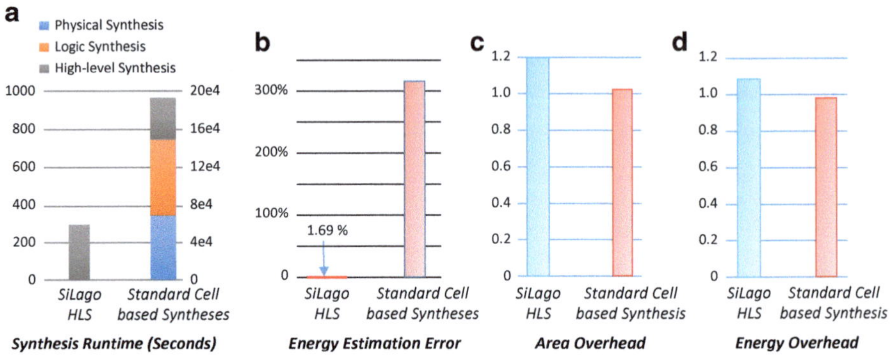

Fig. 3.17 Quantifying efficiency of function implementation (FIMP) library generation. (**a**) Synthesis runtime (seconds). (**b**) Energy estimation error. (**c**) Area overhead. (**d**) Energy overhead

SiLago block resources. The efficiency of SiLago HLS is also vital when the SiLago platform is ported to a new technology node. As long as the SiLago platform exports the same micro-architectural operations, the FIMP configware and shape in terms of the SiLago blocks will remain the same, only the cost metrics will change to reflect the new technology node. If instead, standard cell was the physical design target, the cost generating or re-generating the FIMP library would be unacceptably high.

The long synthesis time and error in estimation are symptoms of the malaise that stems from the large abstraction gap and large design space problems (Fig. 3.8) that has impeded the progress in standard cell based synthesis beyond RTL and justifies our thesis that with increasing complexity, it is time to increase the abstraction and granularity of the physical design objects from Boolean level standard cells to micro-architecture level blocks like SiLago blocks. Equally important is to introduce new physical design discipline that goes beyond the standard pitch of standard cells for the complexity of design we need to deal with 30 years after standard cells were introduced. Our proposal is to introduce the grid based structured layout scheme that enables composition of design by abutment. Collectively these measures, enable efficient and accurate synthesis from higher abstraction with correct-by-construction guarantee. Furthermore, they eliminate functional and constraints verification.

Conclusion and Future Work

In this chapter, we have offered an alternative to the state-of-the-art software-centric SOC implementation style that relies on mapping bulk of the functionality to processors. Realizing that customization is a key to cope with the dark silicon constraints, the VLSI design community has promoted two kinds of customizations. The first is to use heterogeneous processors that are more efficient for different kinds of functionalities and/or capacity. The second customization is to map a small fraction of the total functionality that is power and performance critical to custom designed hardware units called accelerators; the remaining functionality is mapped to processors. We have argued that this strategy is not scalable because the bulk of functionality remains in software that is known to be significantly less efficient compared to hardware. The small degree of customization that is done targets mainly computation while interconnect, control, and access to storage remain general purpose. Besides the customization being partial, the cost of designing such SOCs that are called accelerator-rich heterogeneous SOCs is very high; 45 MUSDs out of a total cost of 50 MUSDs. We have analyzed the reason for this high cost as primarily being functional and constraints verification. These costs in turn stem from manual system-level architecting and refinement down to register transfer level that requires detailed functional verification of billion gate SOCs at RTL that all the applications and their use cases work. The constraints verification problem stems from the large abstraction gap between system-level and the Boolean level standard cell based physical design. Because of this large abstraction gap, it is hard to accurately know the cost metrics at system-level resulting in costly iterations as it takes long time to reach the physical design level. The dogma in the VLSI Design community is that use of general purpose processors and platform based design will lead to reduction in engineering cost. Custom hardware design implies large engineering cost and since the engineering cost even with limited customization in the form of accelerators is already very high, any suggestion of increasing customization goes against the conventional wisdom.

It is in this background, we have proposed the SiLago platform that offers complete customization and is hardware centric which implies that by default the functionality is mapped to hardware implementation style and only flexibility critical and control intensive functionality is implemented in software on simple processors called flexilators. The interconnect scheme allows the SiLago resources to be clustered to tailor the computation, storage, control, interconnect, and address generation to dynamically create private execution partitions tailored to the functional and non-functional constraints of the application. Fine grain power management allows dynamically drawing the boundaries of voltage frequency islands and PREXes can operate at a voltage frequency operating point tailored to its needs and when an application is done, the fine grain power management allows the resources to power gated. Additionally, when more resources are available, the runtime management system can deploy a more parallel version of PREX and further lower the voltage frequency operating point or vice versa if there is a shortage of resources. This is called dynamic parallel voltage frequency scaling; it allows spatial slack to be transformed into temporal slack. Efficient and agile reconfiguration option allows the instantiation of applications or switch to applications with different degree of parallelism with minimal overhead. Collectively these features contribute to make SiLago platform dark silicon aware and enable more functionality per unit current drawn and area.

The SiLago platform not only enables complete hardware centric dynamic customization but equally critically it also enables automates the complete customization to dramatically lower the engineering cost of producing a fully customized dark silicon friendly SiLago based SOC. This solution is based on raising the abstraction of the physical design target to micro-architecture level from the present day Boolean level standard cells. The SiLago platform is conceptualized as a fabric divided into regions that specialize in certain functionalities like DSP computation, scratchpad memory, system control, infrastructure, etc. Each region is populated by a region specific SiLago blocks that occupies one or more contiguous SiLago grid cells. SiLago blocks are hardened—designed down to physical level—and characterized with post-layout sign-off quality data to know the cost metrics of micro-architectural operations of SiLago blocks with measurement level accuracy. The hardened SiLago blocks absorb not just the functionality but also local and global functional and infrastructural nets. Pins of SiLago blocks are brought out at right place and metal layers to enable composition by abutment. This SiLago physical design target exports measurement level accurate cost metrics of micro-architecture level operations and a structured grid based physical design scheme to high-level and application-level synthesis tools. These tools work more effectively and efficiently because of the accurate knowledge of cost metrics of both micro-architecture level operations and interconnect. The efficiency comes from the fact that to produce a solution, a design needs to be refined down to micro-architecture level and not to the Boolean level. The design space searched is also more effective because it is at higher abstraction level where the granularity of design decisions and their impact is higher compared to the lower levels.

In conclusion, the SiLago platform offers solution that improves silicon, computational, and engineering efficiencies in a way that enables creating SOC solutions that are dark silicon friendly—delivers more functionality for unit current drawn, area used, and design hours expended. The big unsolved problem that remains is to lower the manufacturing cost. We are working on it and have reasons to believe that the SiLago method can be the basis for also dramatically lowering the manufacturing cost as well.

References

1. J.M. Rabaey, Silicon architectures for wireless systems—part 1. Presented at the Tutorial HotChips, 2001, Memorial Auditorium, Stanford University, Stanford
2. H.G. Cragon, *Memory Systems and Pipelined Processors* (Jones and Bartlett Publishers, Sudbury, 1996)
3. J. Barth, D. Anand, J. Dreibelbis, E. Nelson, A 300 MHz multi-banked eDRAM macro featuring GND sense, bit-line twisting and direct reference cell write, in *2002 IEEE International Solid-State Circuits Conference (ISSCC 2002)*. Digest of Technical Papers, vol. 1 (2002), San Francisco, pp. 156–157
4. N. Zea, J. Sartori, B. Ahrens, R. Kumar, Optimal power/performance pipelining for error resilient processors, in *2010 IEEE International Conference on Computer Design (ICCD)* (2010), Amsterdam, pp. 356–363
5. K. Sankaralingam, R. Nagarajan, L. Haiming, K. Changkyu, H. Jaehyuk, D. Burger et al., Exploiting ILP, TLP, and DLP with the polymorphous trips architecture. *Micro, IEEE,* vol. 23 (2003), pp. 46–51.

6. C. Ranger, R. Raghuraman, A. Penmetsa, G. Bradski, C. Kozyrakis, Evaluating MapReduce for multi-core and multiprocessor systems, in *IEEE 13th International Symposium on High Performance Computer Architecture, 2007 (HPCA 2007)* (2007), Phoenicx, Arizona, pp. 13–24
7. S. Kumar, A. Jantsch, J.P. Soininen, M. Forsell, M. Millberg, J. Oberg et al., A network on chip architecture and design methodology, in *Proceedings. IEEE Computer Society Annual Symposium on VLSI, 2002* (2002), pp. 105–112
8. T.-Y. Yeh, Y.N. Patt, Alternative implementations of two-level adaptive branch prediction. Presented at the proceedings of the 19th annual international symposium on computer architecture, Queensland, Australia, 1992
9. A. Jantsch, P. Ellervee, J. Oberg, A. Hemani, H. Tenhunen, A software oriented approach to hardware/software codesign. *Proceedings of the Poster Session of CC*, 1994
10. T.S. Muthukaruppan, M. Pricopi, V. Venkataramani, T. Mitra, S. Vishin, Hierarchical power management for asymmetric multi-core in dark silicon era, in *2013 50th ACM/EDAC/IEEE Design Automation Conference (DAC)* (2013), pp. 1–9
11. M.B. Taylor, Is dark silicon useful? Harnessing the four horsemen of the coming dark silicon apocalypse, in *2012 49th ACM/EDAC/IEEE Design Automation Conference (DAC)* (2012), pp. 1131–1136
12. R.H. Dennard, V.L. Rideout, E. Bassous, A.R. LeBlanc, Design of ion-implanted MOSFET's with very small physical dimensions. IEEE J. Solid State Circuits **9**(5), 256–268 (1974)
13. M. Shafique, S. Garg, J. Henkel, D. Marculescu. The EDA challenges in the dark silicon era, in *2014 51st ACM/EDAC/IEEE Design Automation Conference (DAC)* (2014), pp. 1–6
14. A. Hemani, P. Klapproth, Trends in SOC architectures, in *Radio Design in Nanometer Technologies*, ed. by M. Ismail, D.D.E.L. Gonzalez (Springer, The Netherlands, 2006), pp. 59–81
15. G. Venkatesh, J. Sampson, N. Goulding, S. Garcia, V. Bryksin, J. Lugo-Martinez et al., Conservation cores: reducing the energy of mature computations. SIGPLAN Not. **45**, 205–218 (2010)
16. M.J. Lyons, M. Hempstead, G.-Y. Wei, D. Brooks, The accelerator store: a shared memory framework for accelerator-based systems. ACM Trans. Archit. Code Optim. **8**, 1–22 (2012)
17. J. Cong, B. Xiao, Optimization of interconnects between accelerators and shared memories in dark silicon. Presented at the Proceedings of the International Conference on Computer-Aided Design, San Jose, California, 2013
18. Z. Yuhao, V.J. Reddi, High-performance and energy-efficient mobile web browsing on big/little systems, in *2013 IEEE 19th International Symposium on High Performance Computer Architecture (HPCA 2013)* (2013), pp. 13–24
19. A. Hemani, Charting the EDA roadmap. IEEE Circuits Devices Mag. **20**, 5–10 (2004)
20. N. Farahini, A. Hemani, H. Sohofi, S. Li, Physical design aware system level synthesis of hardware. Presented at the 2015 International Conference on Embedded Computer Systems: Architectures, Modeling, and Simulation (SAMOS XV) (2015)
21. L. Shuo, N. Farahini, A. Hemani, K. Rosvall, I. Sander, System level synthesis of hardware for DSP applications using pre-characterized function implementations, in *2013 International Conference on Hardware/Software Codesign and System Synthesis (CODES + ISSS)* (2013), pp. 1–10
22. L. Shuo, A. Hemani, Global control and storage synthesis for a system level synthesis approach, in *IEEE 21st Annual International Symposium on Field-Programmable Custom Computing Machines (FCCM 2013)* (2013), pp. 239–239
23. M.B. Taylor, Tiled microprocessors. Ph.D. Thesis, Massachusetts Institute of Technology, 2007
24. E.S. Chung, P.A. Milder, J.C. Hoe, M. Ken, Single-chip heterogeneous computing: does the future include custom logic, FPGAs, and GPGPUs? in *2010 43rd Annual IEEE/ACM International Symposium on Microarchitecture (MICRO)* (2010), pp. 225–236

25. J. Cong, M.A. Ghodrat, M. Gill, B. Grigorian, G. Reinman, Architecture support for accelerator-rich CMPs, in *2012 49th ACM/EDAC/IEEE Design Automation Conference (DAC)* (2012), pp. 843–849
26. J. Allred, S. Roy, K. Chakraborty, Designing for dark silicon: a methodological perspective on energy efficient systems. Presented at the proceedings of the 2012 ACM/IEEE international symposium on low power electronics and design, Redondo Beach, California, USA, 2012
27. Y. Turakhia, B. Raghunathan, S. Garg, D. Marculescu, HaDeS: architectural synthesis for heterogeneous dark silicon chip multi-processors, in *2013 50th ACM/EDAC/IEEE Design Automation Conference (DAC)* (2013), pp. 1–7
28. R. Cochran, C. Hankendi, A.K. Coskun, S. Reda, Pack & cap: adaptive DVFS and thread packing under power caps. Presented at the proceedings of the 44th annual IEEE/ACM international symposium on microarchitecture, Porto Alegre, Brazil, 2011
29. S.M.A.H. Jafri, O. Bag, A. Hemani, N. Farahini, K. Paul, J. Plosila et al., Energy-aware coarse-grained reconfigurable architectures using dynamically reconfigurable isolation cells, in *2013 14th International Symposium on Quality Electronic Design (ISQED)* (2013), pp. 104–111
30. S.M.A.H. Jafri, S.J. Piestrak, K. Paul, A. Hemani, J. Plosila, H. Tenhunen, Energy-aware fault-tolerant CGRAs addressing application with different reliability needs, in *2013 Euromicro Conference on Digital System Design (DSD)* (2013), pp. 525–534
31. S.M.A.H. Jafri, M.A. Tajammul, A. Hemani, K. Paul, J. Plosila, H. Tenhunen, Energy-aware-task-parallelism for efficient dynamic voltage, and frequency scaling, in CGRAs, in *2013 International Conference on Embedded Computer Systems: Architectures, Modeling, and Simulation (SAMOS XIII)* (2013), pp. 104–112
32. H. Kaul, M. Anders, S. Hsu, A. Agarwal, R. Krishnamurthy, S. Borkar. Near-threshold voltage (NTV) design—opportunities and challenges, in *2012 49th ACM/EDAC/IEEE Design Automation Conference (DAC)* (2012), pp. 1149–1154
33. P. Schaumont, I. Verbauwhede, Domain-specific codesign for embedded security. Computer **36**, 68–74 (2003)
34. Itrs 2007 system drivers (2007), online available: http://www.itrs.net/
35. T. Mitra, Energy-efficient computing with heterogeneous multi-cores, in *2014 14th International Symposium on Integrated Circuits (ISIC)* (2014), pp. 63–66
36. S.M.A.H. Jafri, O. Ozbag, N. Farahini, K. Paul, A. Hemani, J. Plosila et al., Architecture and implementation of dynamic parallelism, voltage and frequency scaling (PVFS) on CGRAs. J. Emerg. Technol. Comput. Syst. **11**, 1–29 (2015)
37. H. Bokhari, H. Javaid, M. Shafique, J. Henkel, S. Parameswaran, darkNoC: designing energy-efficient network-on-chip with multi-Vt cells for dark silicon, in *2014 51st ACM/EDAC/IEEE Design Automation Conference (DAC)* (2014), pp. 1–6
38. H. Matsutani, M. Koibuchi, D. Ikebuchi, K. Usami, H. Nakamura, H. Amano, Ultra fine-grained run-time power gating of on-chip routers for CMPs, in *2010 Fourth ACM/IEEE International Symposium on Networks-on-Chip (NOCS)* (2010), pp. 61–68
39. L. Guangshuo, P. Jinpyo, D. Marculescu, Dynamic thread mapping for high-performance, power-efficient heterogeneous many-core systems, in *2013 IEEE 31st International Conference on Computer Design (ICCD)* (2013), pp. 54–61
40. M. Pricopi, T.S. Muthukaruppan, V. Venkataramani, T. Mitra, S. Vishin, Power-performance modeling on asymmetric multi-cores, in *2013 International Conference on Compilers, Architecture and Synthesis for Embedded Systems (CASES)* (2013), pp. 1–10
41. W.J. Dally, C. Malachowsky, S.W. Keckler, 21st century digital design tools. Presented at the Proceedings of the 50th Annual Design Automation Conference, Austin, Texas, 2013
42. W.J. Dally, J. Balfour, D. Black-Shaffer, J. Chen, R.C. Harting, V. Parikh et al., Efficient embedded computing. Computer **41**, 27–32 (2008)
43. S. Herbert, D. Marculescu, Variation-aware dynamic voltage/frequency scaling, in *IEEE 15th International Symposium on High Performance Computer Architecture, 2009 (HPCA 2009)* (2009), pp. 301–312

44. M.A. Shami, A. Hemani, Address generation scheme for a coarse grain reconfigurable architecture, in *2011 IEEE International Conference on Application-Specific Systems, Architectures and Processors (ASAP)* (2011), pp. 17–24
45. N. Farahini, A. Hemani, K. Paul, Distributed runtime computation of constraints for multiple inner loops, in *2013 Euromicro Conference on Digital System Design (DSD)* (2013)
46. M.A. Shami, A. Hemani, Classification of massively parallel computer architectures, in *2012 IEEE 26th International Parallel and Distributed Processing Symposium Workshops & PhD Forum (IPDPSW)* (2012), pp. 344–351
47. N. Farahini, A. Hemani, Atomic stream computation unit based on micro-thread level parallelism, in *2015 IEEE 26th International Conference on Application-specific Systems, Architectures and Processors (ASAP)* (2015), pp. 25–29
48. N. Farahini, A. Hemani, H. Sohofi, S.M.A.H. Jafri, M.A. Tajammul, K. Paul, Parallel distributed scalable runtime address generation scheme for a coarse grain reconfigurable computation and storage fabric. Microprocess. Microsyst. **38**, 788–802 (2014)
49. M.A. Shami, A. Hemani, Morphable DPU: smart and efficient data path for signal processing applications, in *IEEE Workshop on Signal Processing Systems, 2009 (SiPS 2009)* (2009), pp. 167–172
50. M.A. Shami, A. Hemani, Control scheme for a CGRA, in *2010 22nd International Symposium on Computer Architecture and High Performance Computing (SBAC-PAD)* (2010), pp. 17–24
51. M.A. Shami, A. Hemani, An improved self-reconfigurable interconnection scheme for a coarse grain reconfigurable architecture, in *NORCHIP, 2010* (2010), pp. 1–6
52. M. A. Shami, A. Hemani, Partially reconfigurable interconnection network for dynamically reprogrammable resource array, in *IEEE 8th International Conference on ASIC, 2009. ASICON'09* (2009), pp. 122–125
53. M.A. Tajammul, M.A. Shami, A. Hemani, S. Moorthi, NoC based distributed partitionable memory system for a coarse grain reconfigurable architecture, in *2011 24th International Conference on VLSI Design (VLSI Design)* (2011), pp. 232–237
54. T. Sato, H. Watanabe, K. Shiba, Implementation of dynamically reconfigurable processor DAPDNA-2, in *2005 IEEE VLSI-TSA International Symposium on VLSI Design, Automation and Test, 2005 (VLSI-TSA-DAT)* (2005), pp. 323–324
55. S.M.A.H. Jafri, A. Hemani, K. Paul, J. Plosila, H. Tenhunen. Compact generic intermediate representation (CGIR) to enable late binding in coarse grained reconfigurable architectures, in *2011 International Conference on Field-Programmable Technology (FPT)* (2011)
56. C. Ykman-Couvreur, E. Brockmeyer, V. Nollet, T. Marescaux, F. Catthoor, H. Corporaal. Design-time application exploration for MP-SoC customized run-time management, in *Proceedings of the 2005 International Symposium on System-on-Chip, 2005* (2005), pp. 66–69
57. C. Ykman-Couvreur, V. Nollet, T. Marescaux, E. Brockmeyer, F. Catthoor, H. Corporaal. Pareto-based application specification for MP-SoC customized run-time management, in *International Conference on Embedded Computer Systems: Architectures, Modeling and Simulation, 2006 (IC-SAMOS 2006)* (2006), pp. 78–84.
58. J.-M. Chabloz, A. Hemani, Power management architecture in McNoC, in *Scalable Multi-core Architectures*, ed. by D. Soudris, A. Jantsch (Springer, New York, 2012), pp. 55–80
59. J.M. Chabloz, A. Hemani, Distributed DVFS using rationally-related frequencies and discrete voltage levels, in *2010 ACM/IEEE International Symposium on Low-Power Electronics and Design (ISLPED)* (2010), pp. 247–252
60. P. Teehan, M. Greenstreet, G. Lemieux, A survey and taxonomy of GALS design styles. IEEE Des. Test Comput. **24**, 418–428 (2007)
61. I. Loi, F. Angiolini, L. Benini, Developing mesochronous synchronizers to enable 3D NoCs, in *Design, Automation and Test in Europe, 2008 (DATE '08)* (2008), pp. 1414–1419
62. J.-M. Chabloz, A. Hemani, Low-latency maximal-throughput communication interfaces for rationally-related clock domains. IEEE Trans. Very Large Scale Integr. VLSI Syst. **22**(3), 641–654 (2013)

63. S.M.A.H. Jafri, G. Serrano, M. Daneshtalab, N. Abbas, A. Hemani, K. Paul et al., TransPar: transformation based dynamic parallelism for low power CGRAs, in *2014 24th International Conference on Field Programmable Logic and Applications (FPL)* (2014), pp. 1–8
64. S.M.A.H. Jafri, G. Serrano, J. Iqbal, M. Daneshtalab, A. Hemani, K. Paul et al., RuRot: run-time rotatable-expandable partitions for efficient mapping in CGRAs, in *2014 International Conference on Embedded Computer Systems: Architectures, Modeling, and Simulation (SAMOS XIV)* (2014), pp. 233–241
65. M. Huebner, M. Ullmann, F. Weissel, J. Becker, Real-time configuration code decompression for dynamic FPGA self-reconfiguration, in *Proceedings of the 18th International Parallel and Distributed Processing Symposium, 2004* (2004), p. 138
66. V. Tunbunheng, M. Suzuki, H. Amano, RoMultiC: fast and simple configuration data multicasting scheme for coarse grain reconfigurable devices, in *Proceedings of the 2005 IEEE International Conference on Field-Programmable Technology* (2005), pp. 129–136
67. H. Amano, Y. Hasegawa, S. Tsutsumi, T. Nakamura, T. Nishimura, V. Tanbunheng et al.., MuCCRA chips: configurable dynamically-reconfigurable processors, in *Solid-State Circuits Conference, 2007. ASSCC '07. IEEE Asian* (2007), pp. 384–387
68. Heysters, P.M.; Smit, G.J.M.. Mapping of DSP algorithms on the MONTIUM architecture, in *International Parallel and Distributed Processing Symposium, 22-26 April 2003* (2003), p. 6
69. F.-J. Veredas, M. Scheppler, W. Moffat, Bingfeng Mei. Custom implementation of the coarse-grained reconfigurable ADRES architecture for multimedia purposes, in *International Conference on Field Programmable Logic and Applications, 24–26 August 2005* (2005), pp. 106–111
70. N. Farahini, S. Li, M.A. Tajammul, M.A. Shami, G. Chen, A. Hemani, W. Ye, 39.9 GOPs/watt multi-mode CGRA accelerator for a multi-standard basestation, in *IEEE International Symposium on Circuits and Systems (ISCAS), 19–23 May 2013* (2013), pp. 1448–1451
71. W.J. Daily, A. Chang, The role of custom design in ASIC chips, in *Proceedings 2000—Design Automation Conference, 2000* (2000), pp. 643–647
72. S.W. Keckler, W.J. Dally, B. Khailany, M. Garland, D. Glasco, GPUs and the future of parallel computing. IEEE Micro **31**, 7–17 (2011)

Chapter 4
Heterogeneous Dark Silicon Chip Multi-Processors: Design and Run-Time Management

Siddharth Garg, Yatish Turakhia, and Diana Marculescu

Introduction

Dark silicon chips are expected to be *heterogeneous* in nature, consisting of, for example, a multitude of dedicated hardware accelerators to assist the on-chip cores [5, 6] or heterogeneous chip multi-processors (CMPs) consisting of different types of general-purpose cores.

However, in addition to heterogeneity that is designed into the chip, manufacturing process variations further introduce differences in the power and frequency of even identical cores on a chip. Indeed, process variations have typically been thought of as a major design concern for CMOS ICs. The magnitude of process variations is increasing with technology scaling and with decreasing supply voltages, transistors are more susceptible to variations in their process parameters. Dark silicon chips, on the other hand, offer a new opportunity to *exploit* process variations. In particular, since we are allowed to *pick and choose* which cores on the chip to turn on, we can potentially harness process variations to our benefit by picking the subset of cores that best fit the application characteristics. We refer to this intuitive idea as Cherry-picking.

This chapter deals with both intentional micro-architectural heterogeneity in dark silicon CMPs and the heterogeneity introduced by process variations. In this context,

---

S. Garg (✉)
New York University, New York, NY, USA
e-mail: sg175@nyu.edu

Y. Turakhia
Stanford University, Stanford, CA, USA
e-mail: yatisht@stanford.edu

D. Marculescu
Carnegie Mellon University, Pittsburgh, PA, USA
e-mail: dianam@cmu.edu

A.M. Rahmani et al. (eds.), *The Dark Side of Silicon*,
DOI 10.1007/978-3-319-31596-6_4

we answer the following questions: (1) what will future heterogeneous dark silicon CMPs look like (and how can we design them); and (2) how can we exploit on-chip heterogeneity in run-time scheduling of multi-threaded applications on such CMPs.

Cherry-Picking: Exploiting Process Variation Induced Heterogeneity

Figure 4.1 shows an overview of the Cherry-picking idea using an example of a 16-core homogeneous CMP with 4 redundant cores (25 % dark silicon). Multi-threaded applications are mapped on to the CMP by choosing the optimal *subset* of cores, i.e., cherry-picking cores, that maximize application performance under a power budget. The unmapped cores are left dark. The 4 extra cores provide an

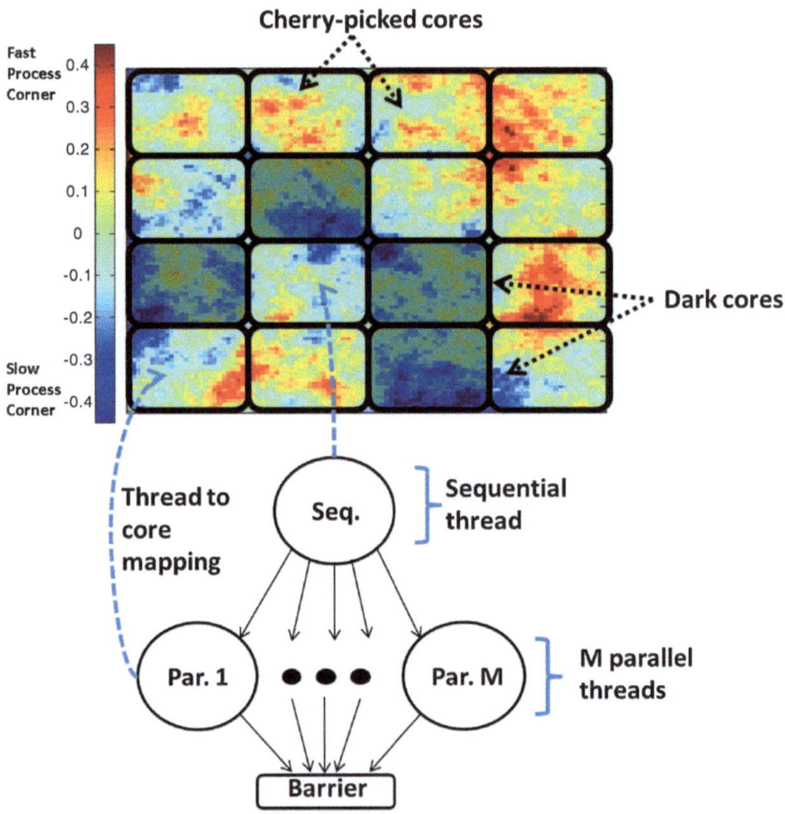

Fig. 4.1 A homogeneous CMP with N cores overlaid on top of a process variation map. Multi-threaded applications with M parallel threads ($M \leq N$) are mapped to the CMP by cherry-picking the best cores within a power budget. The remaining cores are dark

additional set of options for the scheduler which would not have been available if the CMP had only 16 cores, i.e., no dark silicon. In fact, as we will see in the experimental results, increasing the number of dark cores provides even greater performance benefits. We validated these ideas on multi-threaded applications from the SPLASH-2 and PARSEC benchmark suites and perform detailed simulations using a modified version of the Sniper multi-core simulation infrastructure [4].

Process Variation Modeling

Our platform is tiled homogeneous CMPs consisting of N_{cores} identical tiles. Each tile consists of a processing core, a private L1 cache, a section of the shared L2 cache, and an on-chip router to communicate with neighboring cores. As a result of manufacturing process variations, the power dissipation and maximum frequency of each core will be different, depending on the statistics of the process variation distribution. We start by discussing the process variation model used in this part of the work and the resulting core frequency and power distributions.

To model process variation, the chip surface is modeled as a fine grid of dimensions $N_{chip} \times N_{chip}$. Let $p_{ij}(i,j \in [1, N_{chip}])$ represent the value of the process parameter at grid cell (i,j), for example, the channel gate length of the gates that correspond to that grid cell. In the presence of process variations, p_{ij} can be modeled as a Gaussian random variable with mean μ_p and standard deviation σ_p. In addition, the process parameters at two different grid points are correlated with a correlation coefficient, $\rho_{ij,kl}$, that reduces with increasing distance. Based on the experimentally validated model proposed by Xiong et al. [15], we express the spatial correlation between two grid points as

$$\rho_{ij,kl} = e^{-\alpha\sqrt{(i-k)^2+(j-l)^2}} \; \forall i,j,k,l \in [1, N_{chip}], \tag{4.1}$$

where the value of α determines how quickly the spatial correlations die out.

In [3], the authors have shown that frequency of a digital circuit under the impact of process variations can be accurately modeled as the worst-case delay of N_{cp} identical critical paths. Assuming that all the gates in a critical path lie entirely within a grid cell and critical paths are uniformly distributed over the core area, we can write the maximum frequency of core $i(i \in [1, N_{core}])$ as [8]:

$$f_i^{MAX} = K' \min_{k,l \in S_{CP,i}} \left(\frac{1}{p_{kl}}\right), \tag{4.2}$$

where K' is a technology dependent constant and $S_{CP,i}$ represents the set of N_{CP} grid cells in core i that contains critical paths.

The power consumption of core i depends on its dynamic and leakage power components. Although the direct impact of process variations on dynamic power consumption is negligible, the leakage power consumption depends exponentially on process parameters such as effective gate length. The total power consumption of core i is written as

$$P_i = \sum_{k,l \in S_i} \left(C_{kl}^{sw} V_{DD}^2 f_i + V_{DD} I_{kl}^S e^{-K'' p_{kl}} \right) \tag{4.3}$$

$$= P_D f_i + P_{L,i} \tag{4.4}$$

where S_i represents the set of all the grid cells in core i, C_{kl}^{sw} represents the average switched capacitance of a grid cell, V_{DD} is the chip supply voltage, I_{kl}^S is the nominal sub-threshold leakage current for the gates in a grid cell, and K'' is a technology dependent constant. The values of the technology dependent constants can be derived by curve fitting circuit simulation outputs.

Performance Modeling for Multi-Threaded Applications

We focus on multi-threaded applications from the scientific computing domain, such as those found in the SPLASH-2 and PARSEC benchmark suites. These applications consist of two phases of execution—a sequential phase, which consists of a single thread of execution; and a parallel phase in which multiple threads process data in parallel. The parallel threads of execution in a parallel phase typically synchronize on a barrier, in other words, all threads must finish execution before the application can proceed to the next phase. Therefore, the latency of a parallel phase is dominated by the worst-case execution latency across all parallel threads, and the total execution latency is the sum of the latencies of the sequential and parallel phases. Based on this observation, we model the execution time of an application, E, as

$$E = \frac{W_{seq}}{f_{seq}} + \frac{W_{par}}{M \times \min_{i \in [1,M]}(f_{par,i})}, \tag{4.5}$$

where W_{seq} is the percentage of program execution spent in the sequential phase and W_{par} is the percentage of program execution spent in the parallel phase for an implementation with only one parallel thread. M is the number of parallel threads in the application. Furthermore, f_{seq} and $f_{par,i}$ refer to the frequency at which the sequential and the ith parallel thread are executed, respectively. These values depend on the scheduling of threads to cores in the CMP system, *and* the frequency assigned to each core. Determining the optimal scheduling of threads to cores and the core frequency values is the primary goal of this work. To this end, we assume that each application is pre-characterized in terms of its W_{seq} and W_{par} values (Fig. 4.2).

Figure 4.8 shows simulation data that validates the proposed model on a 16-core CMP. Here, the sequential thread is mapped to Core 0 and the parallel threads

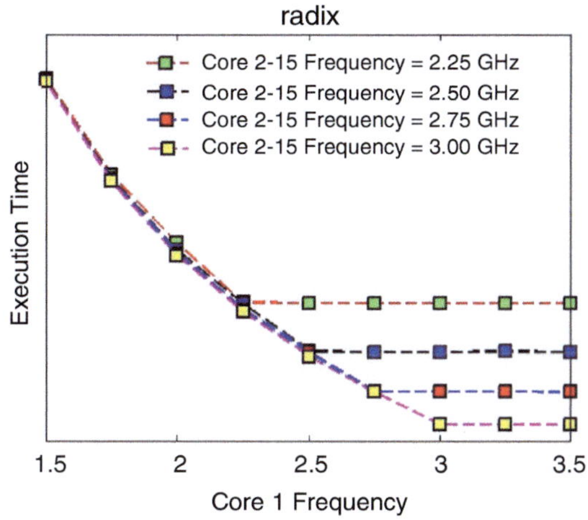

Fig. 4.2 Execution time for different core frequency values for the radix benchmark. The sequential thread is mapped to Core 0 and the parallel threads are mapped to Cores 1–15

are mapped to Cores 1–15. The frequency of Cores 2–15 is kept constant and the frequency of Core 1 is varied. We observe that when Core 1 is slower than Cores 2–15, it dominates the application execution time. However, when the frequency of Cores 2–15 is slower than Core 1, the application execution time no longer changes with increasing Core 1 frequency. Additional data validating the proposed model will be presented in Section "Experimental Results" where we discuss our experimental results.

Variation-Aware Core Selection and Scheduling

As mentioned before, dark silicon chips consist of more cores than can be simultaneously powered on at any given point in time because of a thermal design power (TDP) constraint. Assume that the TDP constraint is given by a designer specified value P_{max}. We will now propose an online optimization strategy to optimally determine for a homogeneous CMP affected by process variations: (a) *which* cores are turned on and which cores are left dark, (b) which core to run the sequential thread and parallel threads on, and (c) what frequency to run every core at. The goal of the online optimization is to maximize performance (minimize application execution latency) under the TDP constraint.

Let $x_{seq}(x_{seq} \in [1, N_{core}])$ represent the core on which the sequential thread executes and, correspondingly, $x_{par,i}(x_{par,i} \in [1, N_{core}])$ be the core on which the ith parallel thread executes. Furthermore, $f_i(i \in [1, N_{cores}])$ represents the frequency at which each core in the CMP executes. To maximize performance within a power budget, we need to solve the following optimization problem:

$$\min_{x,f,M} \frac{W_{seq}}{f_{x_{seq}}} + \frac{W_{par}}{M \times \min_{i \in [1,M]}(f_{x_{par,i}})}, \tag{4.6}$$

subject to:

$$\sum_{i\in[1,M]} P_{x_{par,i}} = \sum_{i\in[1,M]} P_D f_{x_{par,i}} + P_{L,x_{par,i}} \leq P_{max} \tag{4.7}$$

$$f_i \leq f_i^{MAX} \; \forall i \in [1, N_{cores}] \tag{4.8}$$

$$x_{seq} \neq x_{par,i} \; \forall i \in [1, N_{cores}] \tag{4.9}$$

$$x_{par,i} \neq x_{par,j} \; \forall i,j \in [1, N_{cores}], i \neq j \tag{4.10}$$

$$x_{seq} \in [1, N_{cores}] \tag{4.11}$$

$$x_{par,i} \in [1, N_{cores}] \tag{4.12}$$

Equation (4.7) represents the dark silicon constraint, i.e., the TDP dissipation during the parallel phase must be less than the maximum power budget. In this formulation, we have assumed for mathematical simplicity that the cores that are off do not dissipate any power (perfect power gating), although a more realistic scenario in which off cores dissipate a fraction of leakage power is modeled in our experimental results. Note that in the dark silicon constraint, the leakage power dissipation differs from one core to another due to process variations. Equation (4.8) represents the process variation constrained maximum operating frequency of each core. If required, cores can be assigned frequencies lower than their respective maximum frequency constraint to conserve power and meet the maximum power budget. Equations (4.9) and (4.10) ensure that only one thread is mapped to each core. The optimization formulation presented here represents a mixed-integer non-linear program (MINLP). We note that Teodorescu et al. [12] have previously proposed a simpler integer linear programming (ILP) based variation-aware scheduling solution. However, their formulation only addresses multi-programmed and not multi-threaded workloads and does not take into account dark silicon constraints, i.e., they assume that the number of threads is always equal to the number of cores.

We now propose a polynomial time solution to the optimization problem formulated above. Assume that x^* and f^* are the optimal solutions for this problem.

Lemma 1. *The optimal solution satisfies one of two mutually exclusive conditions. If*

$$\sum_{i\in[1,M]} P_D f^{MAX}_{x*_{par,i}} + P_{L,x^*_{par,i}} \leq P_{max}$$

then

$$f^*_{x*_{par,i}} = f^{MAX}_{x*_{par,i}}.$$

Else,

$$f^*_{x*_{par,i}} = f^*_{x*_{par,j}} = f_{min}$$

and

$$f_{min} < f^{MAX}_{x*_{par,i}} \forall i \in [1, M].$$

Proof. The if condition merely states that if there exists an optimal scheduling policy in which each parallel core runs at its maximum frequency while meeting the dark silicon TDP constraint, there is no benefit in scaling down the frequency of any core. On the other hand, if such a solution does not exist, then the operating frequencies must be scaled down to reduce power dissipation. Let $f_{min} = \min_{i\in[1,M]}(f_{x_{par,i}})$. From Eq. (4.8) and given the monotonically decreasing relation between frequency and power dissipation, it is easy to show that there is no benefit for any core to operate a frequency higher than f_{min}. Therefore, $f^*_{x*_{par,i}} = f_{min} \forall i \in [1, M]$.

Now, assume that the sequential thread is scheduled for execution on core i, and the *slowest* parallel thread is scheduled on core j. Let the set $Q^{i,j}$ represent the set of cores on the chip (not including i and j) with frequency *greater* than f_j^{max}. The elements of the set Q^{ij} are ordered in ascending order of their leakage power dissipation. With the help of this definition, we can now compute the optimal scheduling and frequency assignment for the remaining $M - 1$ parallel threads.

Lemma 2. *If $|Q^{ij}| \geq M - 1$, then the optimal mapping of the remaining $M - 1$ threads can be written as:*

$$x_{par,k} = Q^{ij}_{k-1} \forall k \in [2, M].$$

The optimal frequency assignment is determined based on the following conditions. If

$$P_{max} - MP_D f_j^{max} - L_j - \sum_{k=1}^{M-1} L_{Q_k^{ij}} \geq 0$$

then

$$f_{x_{par,k}} = f^{MAX}_{x_{par,k}} \forall k \in [1, M].$$

Else,

$$f_{x_{par,k}} = \theta_{ij} f^{MAX}_{x_{par,j}} \forall k \in [1, M].$$

where

$$\theta_{ij} = \frac{P_{max} - L_j - \sum_{k=1}^{M-1} L_{Q_k^{ij}}}{MP_D}.$$

Proof. The if condition corresponds to the case in which the dark silicon power budget can be met while still running a subset of the cores at their respective maximum frequencies. The else condition corresponds to the case in which the core frequencies are scaled to meet the power budget. Based on Lemma 1, all parallel threads must execute at the same frequency. We make use of this fact to determine the scaling factor θ that decides how much each core is slowed down to meet the power budget while maximizing performance.

Finally, note that Lemma 2 provides the optimal scheduling and frequency assignment *given* the two cores on which the sequential and slowest parallel threads run. There are $O(N_{core}^2)$ such pairs in a CMP with N_{core} cores. These pairs can be exhaustively searched (in polynomial time) to determine the best overall scheduling and frequency assignment that maximizes performance within the dark silicon TDP budget.

Algorithm 1 is a formal description of the proposed scheduling and frequency assignment algorithm.

Experimental Methodology

We evaluate the proposed ideas for the 22 nm technology node with a nominal $V_{DD} = 1.0$ V. The variability parameters are set as follows. We divide the chip surface into a 100×100 grid and model variations in effective gate length on this grid. The standard deviation of process variations is set to be 10 % of the nominal process parameters ($\frac{\sigma_p}{\mu_p} = 0.1$). The spatial correlation parameter α is set such that spatial correlations become negligible (< 1 %) at a distance of half die length [15]. The technology specific parameters that model the relationship between the process parameter and delay/leakage (K' and K'') are set based on curve fitting SPICE simulations of ring oscillators in a 22 nm predictive technology model (PTM) [16].

Application Evaluated and Performance Modeling

We experiment with five multi-threaded applications from the SPLASH-2 [14] and PARSEC [2] benchmarks suites as shown in Table 4.1. The application parameters W_{seq} and W_{par} are extracted by demarcating the beginning and end of the sequential and parallel sections and measuring their respective execution times on the Sniper full-system multi-core simulator.

Algorithm 1: Optimal core selection, frequency scheduling, and frequency assignment

$E^* \leftarrow \infty$
for $i \in [1, N_{cores}]$ **do**
  for $j \in [1, N_{cores}] j \neq i$ **do**
    if $|Q^{ij}| \geq M - 1$ **then**
      $\Delta \leftarrow P_{max} - MP_D f_j^{max} - L_j - \sum_{k=1}^{M-1} L_{Q_k^{ij}}$
      if $\Delta \geq 0$ **then**
        $\theta \leftarrow 1$
        $E_{new} \leftarrow \frac{w_{seq}}{f_i} + \frac{w_{par}}{M f_j^{MAX}}$
      else
        $\theta \leftarrow \frac{\Delta}{MP_D}$
        $E_{new} \leftarrow \frac{w_{seq}}{f_i} + \frac{w_{par}}{M\theta f_j^{MAX}}$
      end if
      if $E_{new} < E^*$ **then**
        $E^* \leftarrow E_{new}$
        $x_{seq}^* \leftarrow i$
        $x_{par,1}^* \leftarrow j$
        $x_{par,k}^* \leftarrow Q_{k-1}^{ij} \forall k \in [2, M]$
        $f_k^* \leftarrow 0 \ \forall k \in [1, N_{cores}]$
        $f_i^* \leftarrow f_i^{MAX}$
        $f_j^* \leftarrow \theta f_j^{MAX}$
        if $\theta < 1$ **then**
          $f_{Q_k^{ij}}^* \leftarrow \theta f_j^{MAX} \ \forall k \in [1, M-1]$
        else
          $f_{Q_k^{ij}}^* \leftarrow f_j^{MAX} \ \forall k \in [1, M-1]$
        end if
      end if
    end if
  end for
end for
return $\{x^*, f^*\}$

Table 4.1 Details of the applications simulated

Application	Description	$W_{par} : W_{seq}$
blackscholes	Financial option pricing	5.59 : 1
fft	Fast Fourier transform	2.90 : 1
fluid-animate	Fluid dynamics	1.18 : 1
radix	Image processing	37 : 1
swaptions	Portfolio pricing	13.36 : 1

Micro-Architectural Parameters

We model homogeneous CMPs of varying size with $N_{core} = \{16, 24, 32\}$ and in all cases, we set the dark silicon TDP constraint to be 110 W.

Table 4.2 Details of the micro-architectural configurations simulated

	Value
Core parameters	
Nominal frequency	3.0 GHz
Area	10.3 mm^2
Peak dynamic power	4.08 W
Peak leakage power	2.1 W
L2 cache parameters (per core)	
Size	2 MB
Area	11.2 mm^2
Peak dynamic power	0.76 W
Peak leakage power	1.56 W

Table 4.2 shows the architectural configurations of a single tile (core+L2) in the homogeneous CMP architectures that we model. Experiments are run on the **Sniper** multi-core simulator [4] which accurately models the core, network, and memory sub-system components of a CMP using a trace-driven timing model. The simulator is extended to allow for online binding of threads to cores, in order to validate the proposed scheduling algorithms. Finally, McPAT [10] is used to estimate the area and power consumption of the on-chip components.

Note that the 16-core CMP has a nominal TDP dissipation of 108 W in the absence of process variations, which is just within the power budget of 110. Thus, the 16-core CMP serves as the baseline design in our experiments since it does not have any dark silicon. On the other hand, the 24-core and 32-core CMPs have 33 and 50 % dark silicon, respectively.

Experimental Results

Performance Model Validation

We begin by validating the proposed performance model against full-system Sniper simulations. To do so, we ran 30 simulations with randomly selected frequency values and thread to core mappings for the 16-core, 24-core, and 32-core CMPs. The number of parallel threads in the application was set to $M = 16$. Figure 4.3 shows the scatter plot of predicted versus measured performance for each CMP architecture. The average error over all Monte Carlo runs and across all five applications is only 7.2 % while the root mean square (RMS) error is 11.9 %.

Performance Improvements

We performed experiments using the proposed core selection, thread scheduling, and frequency scaling algorithms for a power budget of 110 W for three different

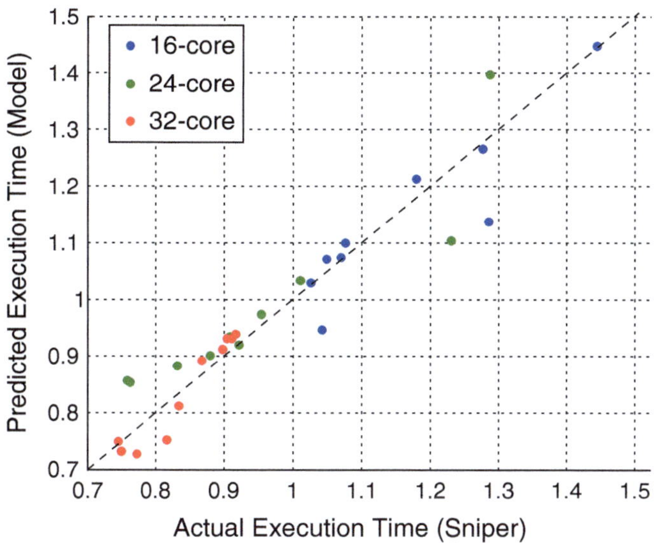

Fig. 4.3 Accuracy of proposed performance model. Ideally, all points should lie along the *dotted line*

CMP architectures: $N_{core} = 16$ (baseline design, no dark silicon), $N_{core} = 24$ (33 % dark silicon), and $N_{core} = 32$ (50 % dark silicon). The process variation parameters used are the ones described in the previous section. Note that since in all cases the power budget remains the same, all benefits in performance come from the ability of dark silicon chips to exploit process variations. For each experiment, we performed 5 runs of Monte Carlo simulation.

As shown in Fig. 4.4, the average performance improvement of the 32-core CMP (50 % dark silicon) is 24 % and that of the 24-core CMP (33 % dark silicon) is 17 %. The maximum performance improvement with respect to the 16-core baseline chip is 28 % for the 32-core chip (on the swaptions benchmark) and 24 % for the 24-core chip (on the radix benchmark). Note that in all these cases, the performance improvement is a result of the greater ability to exploit process variations in the 24- and 32-core chips and not a result of increasing parallelism. The results also, as expected, indicate that cherry-picking provides more benefits for applications with higher percentage parallelism.

To understand the reason for the performance benefits that arise from cherry-picking cores in dark silicon homogeneous CMPs, we show in Fig. 4.5 a scatter plot of frequency and power values for a 16-core and a 32-core homogeneous CMP. Also shown in the figure are the cores that are picked for sequential and parallel execution and the cores left dark (only for the 32-core CMP) when scheduling blackscholes on these platforms. It can be seen that because there are more cores to pick from and more core-to-core variability in the 32-core CMP case (1) the sequential core picked on the 32-core CMP is 6.4 % faster than the sequential core picked on the 16-core CMP; and (2) the slowest parallel core on the 32-core CMP is 33 % faster than the slowest parallel core on the 16-core CMP.

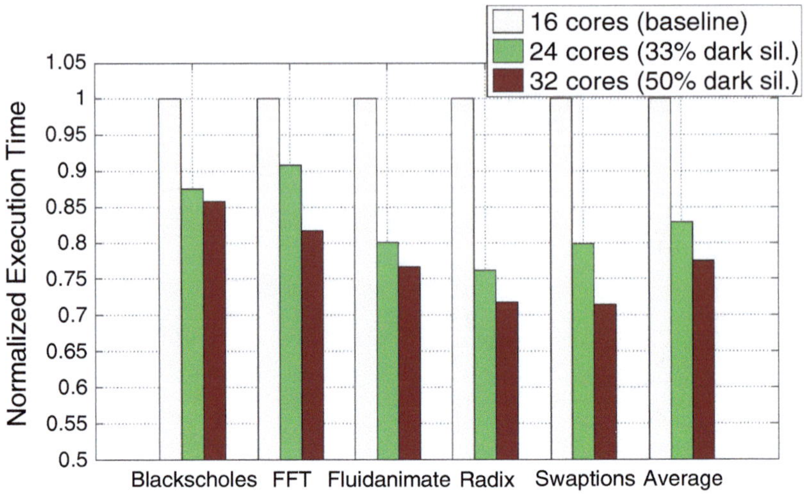

Fig. 4.4 Average execution time for the 16-core baseline, 24-core and 32-core chips over all runs of Monte Carlo simulations

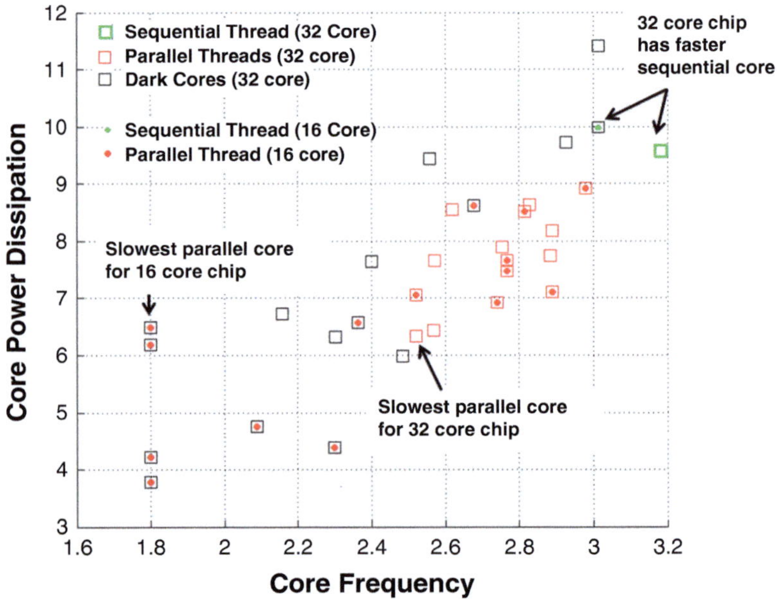

Fig. 4.5 Power and frequency scatter plots for a 16-core and a 32-core CMP. Also shown are the cores on which the sequential and parallel threads are scheduled and the dark cores for the 32-core CMP case. Note: best viewed in color

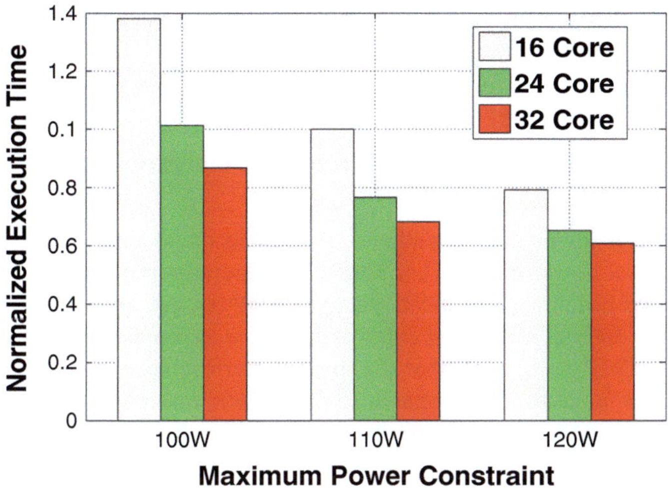

Fig. 4.6 Average execution time for the 16-core baseline, 24-core and 32-core chips with different power budgets

Table 4.3 Overhead of core selection, thread scheduling, and frequency assignment algorithm

$N_{core} = 16$	$N_{core} = 24$	$N_{core} = 32$
3.3 ms	11.1 ms	21.7 ms

Figure 4.6 shows the impact of varying the power budget on the execution time of the radix benchmark on the 16-core, 24-core, and 32-core CMPs. As expected, the execution latency decreases in all cases with increasing power budgets, since higher leakage cores can be selected for execution. It can be observed that cherry-picking consistently provides performance benefits over the 16-core baseline over a range of power budget values.

Overhead of Scheduling Algorithm

Although the proposed core selection, thread scheduling and frequency assignment algorithm is polynomial time, it is still important to characterize the overhead associated with executing this algorithm before application execution. The average execution time (averaged over all experiments) for the 16-, 24-, and 32-core CMPs is tabulated in Table 4.3 and is in the order of milliseconds for all benchmarks. Note that the algorithm is executed only *once* before the application is executed and therefore the overhead is negligible if the application execution time is large compared to the time the core selection, thread scheduling, and frequency assignment algorithm.

Hades: Exploiting Micro-Architectural Heterogeneity

The problem of architectural synthesis of heterogeneous dark silicon CMPs can be defined as follows: given (1) a library of general-purpose cores, (2) a set of multi-threaded benchmark applications, (3) a TDP budget, and (4) an area budget; determine the optimal number of cores of each type to provision the heterogeneous dark silicon CMP with, such that the average performance over all benchmarks applications is maximized. Compared to prior work on architectural synthesis for application-specific multi-processor systems [1, 9], architectural synthesis for dark silicon CMPs introduces a number of new challenges and metrics of interest. *First*, accurate analytical models for the execution time of multi-threaded benchmark applications running on heterogeneous cores are not readily available, although these are essential for optimization.

This is in contrast to the application specific-domain where the performance models are well defined, often using formal models of computation such as data-flow graphs [9]. *Second*, typical multi-threaded applications, for example, the applications from the SPLASH-2 and PARSEC benchmark suites, can be executed with a variable degree of parallelism (DOP), i.e., a varying number of parallel threads, at run-time. The architectural synthesis algorithm must be aware of the optimal DOP and optimal run-time scheduling of threads to cores for each application. In fact, we show that assuming a fixed DOP for the benchmark applications can result in sub-optimal architectures. *Finally*, from an empirical perspective, an important metric of interest is the performance benefit of heterogeneous versus homogeneous CMP architectures with increasing dark silicon area. With increasing dark silicon area, a greater number of cores specialized for each application can be included. Therefore, the performance benefits of heterogeneous CMPs over homogeneous CMPs should increase with increasing dark silicon—empirical validation of this trend is of immense interest to system designers.

In our prior work [13], we have propose Hades—a framework for architectural synthesis of heterogeneous dark-silicon CMPs. The Hades framework consists of the following novel features:

- A new analytical performance model for multi-threaded applications from the SPLASH-2 and PARSEC-2.1 benchmark suites. The proposed performance model is shown to be accurate for applications scheduled on both homogeneous and heterogeneous CMPs, and across a wide range of DOP values for each application. The proposed model has less then 5.2 % average absolute error with respect to detailed simulations.
- An efficient, iterative algorithm to determine the optimal number of cores of each type to provision the heterogeneous dark silicon CMP with. The algorithm

takes into account run-time optimization of the DOP and mapping of application threads to cores. We observe empirically that the iterative algorithm is 430× faster than the optimal ILP-based solution with only a 2.5 % loss in optimality.
- Comprehensive validation and evaluation of the proposed models and optimization technique on Sniper [4], a detailed software simulator for multi-core systems. Our results show that heterogeneous dark silicon CMPs provide up to 60 and 19 % performance gains over homogeneous dark silicon CMPs with fixed and optimized DOPs, respectively. The performance gains are greater for more dark silicon.

To the best of our knowledge, Hades was the first architectural synthesis framework targeted at optimal synthesis of heterogeneous, dark silicon CMP architectures. Figure 4.7 provides an overview of the Hades framework.

Preliminaries and Assumptions

We begin by discussing the assumptions and mathematical notation relevant to our work.

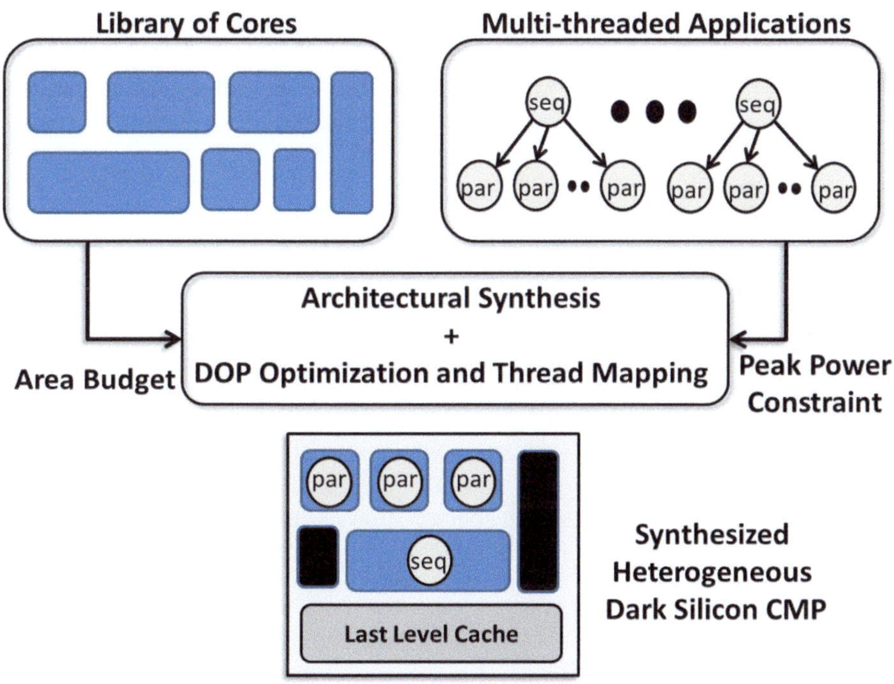

Fig. 4.7 Overview of the Hades framework. In the synthesized heterogeneous dark silicon CMP, a subset of cores are dark

Applications

We assume that we are given a set of N representative benchmark applications. Each application is multi-threaded and consists of both sequential and parallel phases. A single sequential thread executes in the sequential phase, while multiple parallel threads execute in the parallel phase.

The DOP of an application, i.e., the number of threads in the parallel phase can be determined at compile time and is represented as D_i. Without loss of generality, we assume that $D_i \in \mathbb{N}$ and $1 \leq D_i \leq D_{max}$. Note that, in practice, the DOP for some applications can be restricted to a certain subset of values, values that are powers of two, for example.

We assume that each multi-threaded application runs independently on the CMP. The run-time scheduler determines the optimal DOP and the optimal mapping of threads to cores to minimize the execution time within a TDP budget, P_{budget}. Each core executes only one application thread.

Core Library

We assume a library of M different, general-purpose cores—in other words, each core is capable of executing each application. Each core consists of the micro-architectural pipeline and private instruction and data caches. In the experimental results section, we discuss the micro-architectural parameters, e.g., issue width, cache size, etc., that we vary to generate a library of cores.

The TDP dissipation of core j ($j \in [1, M]$) executing one of the parallel threads of application i ($i \in [1, N]$) is denoted by P_{ij}, and includes both the peak dynamic and leakage power components. When a core is idle, i.e., no thread is mapped to it, its leakage power dissipation is denoted by P_j^{idle}. The area of core of type j is A_j, including the pipeline and private caches. Since our goal is to maximize performance, we assume that each core runs at its highest voltage and frequency level.

Uncore Components

We focus only on optimizing the number of cores (including their private caches) of each type. We assume that all cores share a single, global last-level cache (LLC) with a uniform access penalty. For fairness, the LLC size is kept constant over all experiments. By the same token, the configuration of other uncore components, for example, the number and bandwidth of the off-chip DRAM memory controllers, is fixed.

Since the uncore parameters are kept constant, we assume, without loss of generality, that the area budget, A_{budget} and TDP constraint, P_{budget} are in terms of the total core area and total core power, respectively. The uncore components add a fixed cost to the area and power budgets.

Hades Framework

Hades enables the efficient exploration of the vast design space of heterogeneous dark silicon CMP architectures to pick the optimal design that maximizes application performance within area and power budgets. In addition, the optimization is inherently aware of and accounts for run-time decisions including the optimal DOP for each application and mapping of threads to cores—i.e., Hades performs joint design-time and run-time optimization.

We now discuss the two components of the Hades framework: we first discuss the proposed application performance model, and then the iterative optimization used to determine the optimal heterogeneous dark silicon CMP architecture.

Application Performance Model

We focus on multi-threaded applications from the scientific computing domain, as in our prior cherry-picking approach.

Homogeneous CMPs

As the DOP of the application is increased, the execution latency of the parallel phase decreases and is ideally inversely dependent on the DOP. However, due to increased contention for shared hardware and software resources, for example, LLC capacity (hardware) or shared data-structures (software), the benefits of increasing the DOP begin to saturate. In fact, for some applications, the execution time of the parallel phase might actually *increase* beyond a certain DOP. This can be observed in Fig. 4.8a, which shows the execution time of the *radix* benchmark on a homogeneous CMP with a varying DOP. Observe that execution time increases while going from DOP=32 to DOP=64.

Based on this observation, we propose the following performance model for multi-threaded applications executing on homogeneous cores, and then generalize the model for heterogeneous CMPs. The execution time E_{ij} of an application i ($i \in [1,N]$) running on a homogeneous CMP with cores of type j ($j \in [1,M]$) is expressed as:

$$E_{ij} = t^{s}_{ij} + \frac{t^{p}_{ij}}{D_i} + D_i K_{ij} \quad i \in [1,N],\ j \in [1,M] \tag{4.13}$$

In this equation, t^{s}_{ij} is the execution time of the sequential phase, t^{p}_{ij} is the execution time of the parallelizable part of the parallel phase, and K_{ij} represents the increase in execution latency in the parallel phase because of resource contention.

The model parameters of Eq. (4.13) are learned by executing each application with different DOP values on homogeneous CMPs and using standard regression

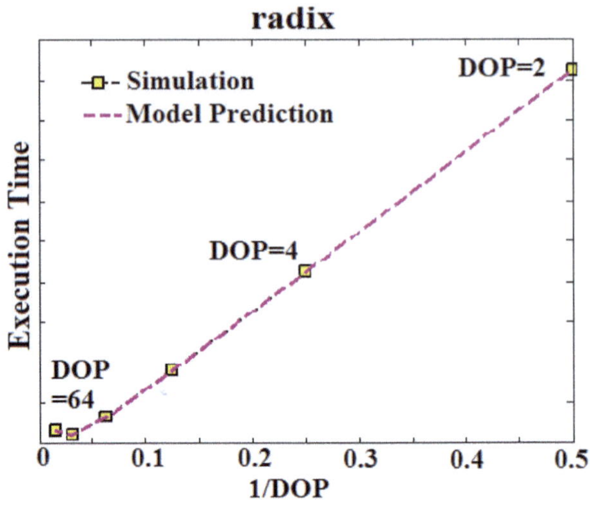

Fig. 4.8 Execution time for the *radix* benchmark on a homogeneous 64-core CMP with varying DOP

techniques to minimize the error between the execution times obtained from simulation and from analytical prediction. Therefore, at most $N \times M \times D_{max}$ simulations are required to learn the model parameters. Figure 4.8 shows that the proposed model is, in fact, able to provide very accurate estimates of the actual execution time for varying DOPs.

Heterogeneous CMPs

In a heterogeneous CMP, the execution time of each parallel thread is different, since threads execute on heterogeneous cores. As has been empirically validated in Section "Performance Modeling for Multi-Threaded Applications", the slowest core determines the application execution time.

Based on this observation and Eq. (4.13), we propose the following model for the execution time of an application executing on a homogeneous CMP:

$$E_{ij} = t^{s}_{im^{s}} + \max_{v \in [1,D_i]} \left(\frac{t^{p}_{im^{p}(v)}}{D_i} + D_i K_{im^{p}(v)} \right) \tag{4.14}$$

In this equation, m^s represents the type of core that the sequential thread executes on, while $m^p(v)$ represents the type of core that the vth parallel thread executes on.

In the experimental results section, we provide detailed results that validate the accuracy of the proposed performance model for a number of different synthesized homogeneous and heterogeneous CMP architectures.

Architectural Synthesis

We now formulate the architectural synthesis problem as an integer program. As we will show, the direct formulation that utilizes Eq. (4.14) results in a non-linear, integer program that can be converted, in polynomial time, to an integer linear program (ILP) without loss of optimality. Although powerful commercial-off-the-shelf ILP solvers exist, we observe empirically that for even reasonable problem instances, the computational cost of using an ILP solver is significant. To address this concern, we then propose an iterative optimization procedure that provides orders-of-magnitude speed-up without significant loss in optimality.

Non-Linear Integer Programming Formulation

We begin by discussing the performance maximization objective function and then incorporate the constraints.

Objective Function The goal is to minimize the weighted sum of execution time for each benchmark application,

$$\min\left(\sum_{i=1}^{N}\rho_i(l_i^s + l_i^p)\right) \tag{4.15}$$

where l_i^s and l_i^p are the sequential and parallel execution times for each benchmark. The weight ρ_i is a designer specified constant.

Sequential Execution Time Let $s_{ij} \in \{0, 1\}$ be an indicator variable that is 1 if the sequential thread of application i executes on a core of type j. The sequential execution time of application i can be written as:

$$l_i^s = \sum_{j=1}^{M} s_{ij} t_{ij}^s \quad \forall i \in [1, N] \tag{4.16}$$

In addition, each application is allowed to use exactly one sequential core. Therefore:

$$\sum_{j=1}^{M} s_{ij} = 1 \quad \forall i \in [1, N] \tag{4.17}$$

Parallel Execution Time Let $b_{ij} \in \{0, 1\}$ be an indicator variable that is 1 if at least one parallel thread of application i executes on a core of type j. The execution time of the parallel phase is determined by the slowest parallel thread. Therefore:

$$l_i^p \geq b_{ij}\left(\frac{t_{ij}^p}{D_i} + K_{ij}D_i\right) \quad \forall i \in [1,N], j \in [1,M] \tag{4.18}$$

Note that since the DOP values, D_i, are also variables in the formulation, Eq. (4.18) is a *non-linear* constraint.

Let $r_{ij} \in \mathbb{N}$ be the number of cores of type j used by parallel threads of application i. The DOP of the application must be equal to the total number of parallel cores utilized:

$$\sum_{j=1}^{M} r_{ij} = D_i \quad \forall i \in [1,N] \tag{4.19}$$

and must be less than the maximum value:

$$1 \leq D_i \leq D_{max} \quad \forall i \in [1,N] \tag{4.20}$$

Finally, r_{ij} should be zero when b_{ij} is zero and at most D_{max} when $b_{ij} = 1$:

$$r_{ij} - D_{max}b_{ij} \leq 0 \quad \forall i \in [1,N], j \in [1,M] \tag{4.21}$$

Number of Cores of Each Type The design vector $Q \in \mathbb{Z}^M$ represents the number of cores of each type: $Q = \{Q_1, Q_2, \ldots, Q_M\}$, and is the ultimate objective of the architectural synthesis problem.

The number of cores of type j should be at least larger than the number of cores of that type used by the sequential and parallel threads of any application. Therefore:

$$Q_j \geq s_{ij} + r_{ij} \quad \forall i \in [1,N] \tag{4.22}$$

TDP Constraint The TDP dissipation of the dark silicon heterogeneous CMP must be below the TDP budget for each application:

$$\sum_{j=1}^{M} r_{ij}P_{ij} + \left(Q_j - r_{ij}\right) P_j^{idle} \leq P_{budget} \quad \forall i \in [1,N] \tag{4.23}$$

Dark Silicon Area Constraint All cores must fit in the prescribed area budget.

$$\sum_{j=1}^{M} Q_jA_j \leq A_{budget} \quad \forall i \in [1,N] \tag{4.24}$$

The objective function in Eq. (4.15) and the constraints in Eqs. (4.16)–(4.24) represent a non-linear integer optimization problem which we refer to as **NILP-OPT**. Solving the problem yields the optimal number of cores of each type, i.e., the vector Q, *and* the optimal DOP for each application, D_i, $\forall i \in [1, N]$.

ILP Formulation

We now show that the non-linear constraint in NILP-OPT, i.e., Eq. (4.18), can be readily linearized to result in a standard ILP problem.

We introduce an indicator variable, $y_{iw} \in \{0, 1\}$ ($i \in [1, N], w \in [1, D_{max}]$) that is 1 if and only if application i has an optimal DOP of w. Equation (4.18) can be re-written as:

$$l_i^p \geq \sum_{w=1}^{D_{max}} b_{ij} y_{iw} \left(\frac{t_{ij}^p}{w} + K_{ij} w \right) = \sum_{w=1}^{D_{max}} m_{ijw} \left(\frac{t_{ij}^p}{w} + K_{ij} w \right) \tag{4.25}$$

where $m_{ijw} = b_{ij} y_{iw} = \min(b_{ij}, y_{iw})$ and $m_{ijw} \in \{0, 1\}$. The following three linear equations express the relationship between m_{ijw}, b_{ij}, and y_{iw}:

$$m_{ijw} \leq b_{ij} \quad \forall i \in [1, N], j \in [1, M], w \in [1, D_{max}] \tag{4.26}$$

$$m_{ijw} \leq y_{iw} \quad \forall i \in [1, N], j \in [1, M], w \in [1, D_{max}] \tag{4.27}$$

$$m_{ijw} \geq b_{ij} + y_{iw} - 1 \quad \forall i \in [1, N], j \in [1, M], w \in [1, D_{max}] \tag{4.28}$$

Equations (4.25)–(4.28) are the linearized versions of Eq. (4.18). This completes the ILP formulation. Note that the ILP formulation has $\mathcal{O}(MND_{max})$ variables and $\mathcal{O}(MND_{max})$ constraints. We refer to this formulation as **ILP-OPT**.

Iterative Optimization

Although ILP-OPT guarantees optimality, we find that empirically the computational time of running the ILP optimization to completion can be significant. To address this issue, we propose an iterative optimization approach, **ITER-OPT**, that separates the architectural optimization from DOP optimization.

ITER-OPT operates as follows: it starts with an initial guess for the optimal architectural design vector Q^*, determines the optimal DOP for each application for this heterogeneous design, re-synthesizes the optimal heterogeneous architecture for this *fixed* DOP, and iterates till convergence, i.e., till no further improvement in performance is observed. The solution to which ITER-OPT converges represents a *local* optima in the design space. Figure 4.9 as an illustration of how the ITER-OPT algorithm works. We now discuss the two primary components of the ITER-OPT algorithm: (1) the architecture optimizer and (2) the DOP optimizer.

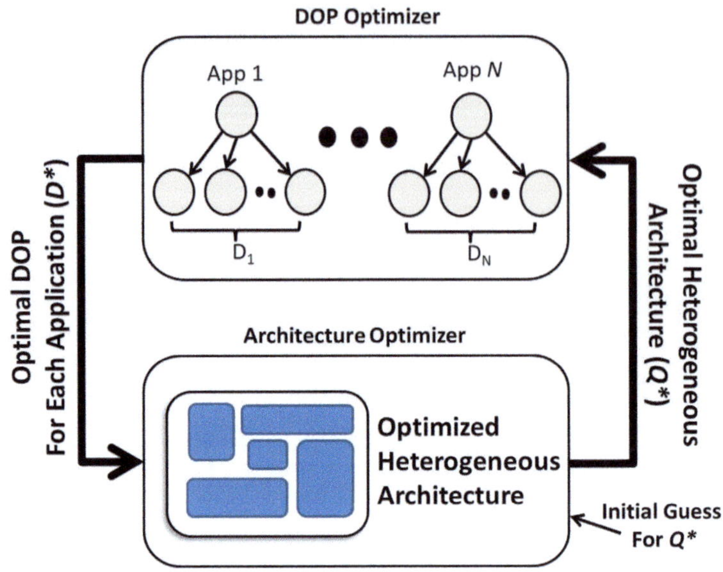

Fig. 4.9 Illustration of the **ITER-OPT** algorithm

Architecture Optimizer Note that if the DOP of each application is fixed, then Eq. (4.18) is, in fact, linear. Thus NILP-OPT is converted to an ILP problem with only $\mathcal{O}(MN)$ variables, and can therefore be solved significantly faster than ILP-OPT. The architecture optimizer takes a fixed DOP, D^*, vector as input and outputs the optimal heterogeneous architecture Q^* for these input DOPs.

DOP Optimizer The DOP optimizer determines the optimal DOP for each application, given a heterogeneous architecture, Q^*, i.e., the number of cores of each type. As we show in Algorithm 2, the time complexity of determining the optimal DOP for each application is $\mathcal{O}(M^3 D_{max})$, i.e., *polynomial* time.

Intuitively, Algorithm 2 is based on the fact that the execution time of any scheduling of threads to cores depends only on the performance of two cores—the sequential core and the slowest parallel core. Thus, it is sufficient to search over all possible pairs of cores types in the heterogeneous architecture, which is polynomial in the number of core types.

We note that ITER-OPT is guaranteed to converge since: (a) the objective function, i.e., the application execution time, always decreases in each iteration; and (b) the globally optimal execution time is bounded. As an initial guess for Q^*, the available area budget is divided equally between all core types.

Algorithm 2: Optimal DOP for an application $i \in [1, N]$ given heterogeneous architecture Q^*

1: $L \leftarrow$ List of cores in ascending order of power dissipation
2: $t^* \leftarrow \infty$ $D_i^* \leftarrow 1$
3: **for** $d \in [1, D_{max}]$ **do**
4:   **for** $j \in [1, M]$ **do**
5:     $Q^{used} \leftarrow Q^*$
6:     **if** $Q_j^{used} \geq 1$ **then**
7:       $t^{seq} \leftarrow t_{ij}^s$
8:       $Q_j^{used} \leftarrow Q_j^{used} - 1$
9:     **else**
10:       $t^{seq} \leftarrow \infty$
11:     **end if**
12:     **for** $k \in [1, M]$ **do**
13:       **if** $Q_k^{used} \geq 1$ **then**
14:         $t^{par} \leftarrow \frac{t_{ik}^p}{d} + K_{ik}d$
15:         $Q_k^{used} \leftarrow Q_k^{used} - 1$
16:         $P_{used} \leftarrow P_{used} + P_{ik}$
17:       **else**
18:         $t^{par} \leftarrow \infty$
19:       **end if**
20:       $req \leftarrow d - 1$
21:       **for** $l \in [1, M]$ **do**
22:         $c \leftarrow L_l$ /* l^{th} element of list L */
23:         **if** $\frac{t_{il}^p}{d} + K_{il}d \leq \frac{t_{ik}^p}{d} + K_{ik}d$ **then**
24:           $add \leftarrow \min(Q_c^{used}, req)$
25:           $req \leftarrow req - add$
26:           $P_{used} \leftarrow P_{used} + add \times P_c^{ik}$
27:           **if** $P_{used} > P_{budget} | req = 0$ **then**
28:             Go to line 37
29:           **end if**
30:         **end if**
31:       **end for**
32:       **if** $req = 0$ **then**
33:         $t_{jk} \leftarrow t^{seq} + t^{par}$
34:       **end if**
35:     **end for**
36:   **end for**
37:   **if** $\min_{jk}(t_{jk}) \leq t^*$ **then**
38:     $t^* \leftarrow \min_{jk}(t_{jk})$ $D_i^* \leftarrow d$
39:   **end if**
40: **end for**
41: **return** D_i^*

Experimental Methodology

All our experiments are run on the Sniper [4] multi-core simulator that provides the ability to model heterogeneous core configurations and detailed models for the memory hierarchy.

A library of 108 cores was generated by varying a number of key micro-architectural parameters as shown in Table 4.1, yielding a rich design space of area, power, and performance values. These are obtained from the McPAT tool [11] for a 22 nm technology node and a 1.0 V nominal supply voltage.

The last-level L3 cache size is set to 32 MB and has a uniform access latency. We model a 1 GB DRAM main memory that is accessed through a DRAM memory controller with aggregate bandwidth of 7.6 GBps.

We experiment with five multi-threaded applications from the SPLASH-2 [14] and PARSEC [2] benchmarks suites—*blackscholes*, *fft*, *fluid animate*, *radix*, and *swaptions*. The maximum DOP is set to 32 for each application. Simulating these applications on the entire core library reveals that only 15 cores are Pareto optimal in terms of area, power, or performance for at least one application. These 15 cores are retained for further experimentation. We have used the Gurobi ILP tool-box [7] to implement both ILP-OPT and ITER-OPT (Table 4.4).

Experimental Results

We conduct our first set of experimental results with an area budget $A_{budget} = 180\,\text{mm}^2$ and for different TDP budgets: $P_{budget} = \{40\,\text{W}, 60\,\text{W}, 80\,\text{W}\}$. Note that these represent the area and power budgets for the cores only. The globally shared L3 cache, memory controller, and other peripherals consume additional, but fixed, area and power.

ITER-OPT vs. ILP-OPT

We begin by noting that the proposed iterative optimization scheme (ITER-OPT) compares favorably with the optimal ILP (ILP-OPT) solution. For $A_{budget} = 180\,\text{mm}^2$ and $P_{budget} = 60\,\text{W}$, ITER-OPT took only 17 s to provide a solution

Table 4.4 Micro-architectural parameters of core library

Core parameter	Values
Dispatch width	{1,2,4}
ROB window size	{8,16,32,64}
L1-I/D cache size	{64, 128, 256} KB
L2 cache (private)	256 KB
Frequency	{2.5, 3.5, 4.5} GHz

within 2.5 % of the solution provided by ILP-OPT in 2 h. This represents a 430× reduction in run-time with only marginal loss in optimality.

Performance Model Validation

Figure 4.10 shows the scatter plot of predicted performance using the proposed model and the simulated performance for a variety of homogeneous and heterogeneous dark silicon CMP architectures that we experimented with. As it can be seen, the performance model agrees very well with simulated values and provides 5.2 % error on average over 180 experiments.

Heterogeneous vs. Homogeneous

We synthesized heterogeneous dark silicon CMPs using the Hades framework for the three power budgets mentioned above. These are compared with: (1) a homogeneous dark silicon CMP where every application has a fixed DOP=16 and (2) a homogeneous dark silicon CMP where the DOP for every application was optimized for the homogeneous architecture. The heterogeneous CMP always uses DOPs optimized for its architecture. Since we have shown that the DOP optimization has polynomial time complexity, we believe that this optimization can be performed by scheduler at run-time (Fig. 4.11).

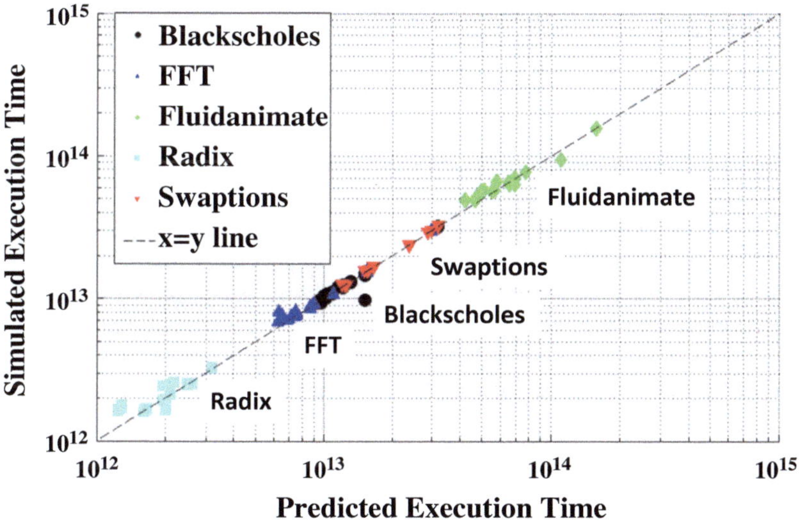

Fig. 4.10 Validation of proposed analytical performance model against sniper simulations for both homogeneous and heterogeneous dark silicon CMPs

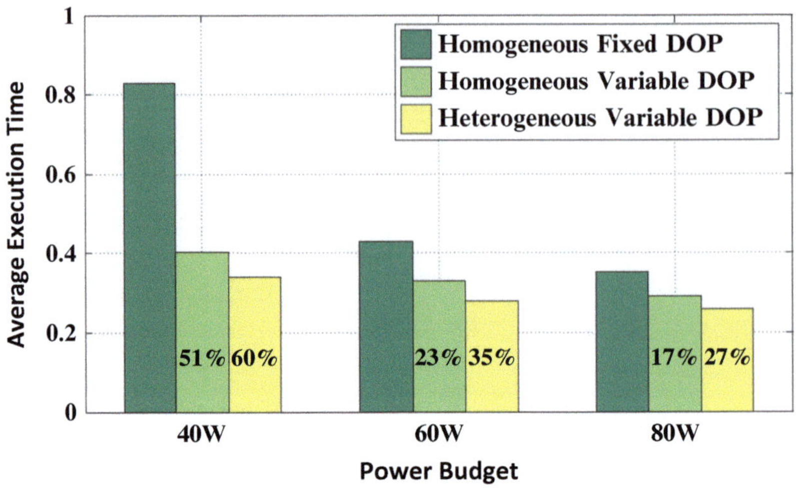

Fig. 4.11 Average execution time over all benchmarks for homogeneous architecture with a fixed DOP=16; homogeneous architecture with optimal DOP for each application; and optimal heterogeneous architecture synthesized using the Hades framework

For power budgets of 40, 60, and 80 W, the heterogeneous CMP has 60, 35, and 27 % higher performance than a homogeneous CMP with fixed DOPs, respectively. Assuming a homogeneous CMP with variable DOPs, the heterogeneous CMP still has 19, 16, and 12 % higher performance. Low power budgets correspond to more dark silicon, and we observe that the benefits of heterogeneity are, as expected, greater with increasing dark silicon. For the 40 W budget, we observe that the heterogeneous CMP has > 50 % dark silicon.

Performance Benefits with Increasing Dark Silicon

To study the benefits of heterogeneity with increasing dark silicon, we plot in Fig. 4.12 the performance improvement of heterogeneous versus homogeneous for increasing area budgets, from 90 mm^2 to 300 mm^2. Also shown is the % dark silicon for each area budget. It is evident that the benefits of heterogeneity saturate after a certain point. This would suggest that for very large dark silicon areas, techniques beyond core level heterogeneity would be required to reap the benefits of the available dark silicon.

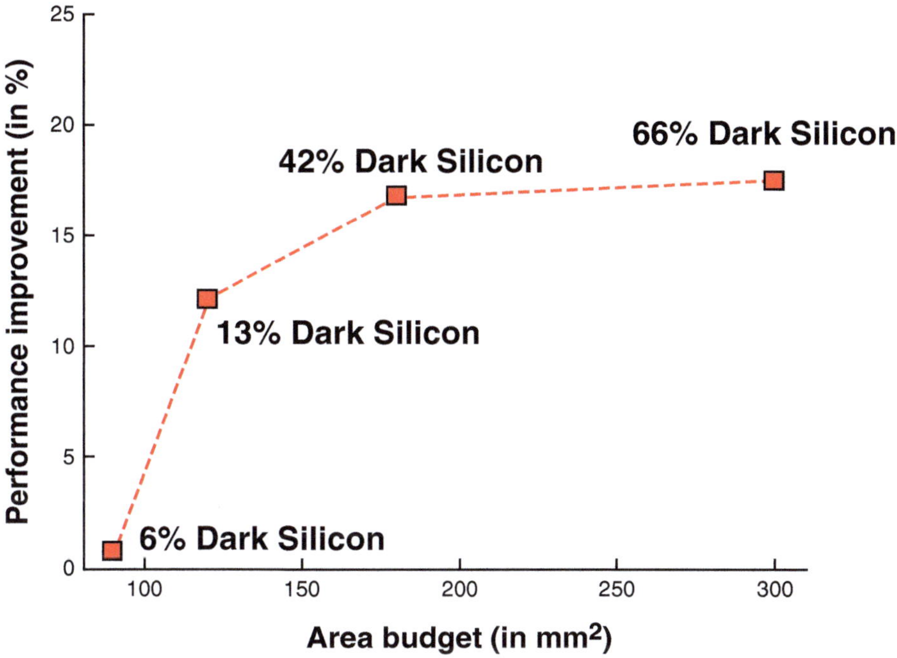

Fig. 4.12 Performance improvement of a heterogeneous dark silicon CMP with increasing area budgets and increasing proportion of dark silicon transistors

Conclusion

In conclusion we believe that the dark silicon problem necessitates a fundamental re-design of conventional assumptions and approaches in processor and micro-architecture design. We present two such examples in this chapter. First, we show how redundant cores in a homogeneous setting can provide increase variability tolerance and opportunities to improve performance within a given power budget. Second, we show how micro-architectural heterogeneity can be leveraged in the dark silicon era to utilize the abundant silicon resources. Yet, several questions remain to be explored. For instance, how does micro-architectural heterogeneity interact with process variation induced heterogeneity. And, how can we co-synthesize heterogeneous cores with a sea of accelerators?

References

1. F. Angiolini, J. Ceng, R. Leupers, F. Ferrari, C. Ferri, L. Benini, An integrated open framework for heterogeneous MPSoC design space exploration, in *Proceedings of DATE'06* (2006)
2. C. Bienia, S. Kumar, J.P. Singh, K. Li, The parsec benchmark suite: characterization and architectural implications, in *Proceedings of the 17th International Conference on Parallel Architectures and Compilation Techniques* (ACM, 2008), pp. 72–81
3. K.A. Bowman, S.G. Duvall, J.D. Meindl, Impact of die-to-die and within-die parameter fluctuations on the maximum clock frequency distribution for gigascale integration. IEEE J. Solid-State Circuits **37**(2), 183–190 (2002)
4. T.E. Carlson, W. Heirmant, L. Eeckhout, Sniper: exploring the level of abstraction for scalable and accurate parallel multi-core simulation, in *2011 International Conference for High Performance Computing, Networking, Storage and Analysis (SC)* (IEEE, 2011), pp. 1–12
5. H. Esmaeilzadeh, E. Blem, R.S. Amant, K. Sankaralingam, D. Burger, Dark silicon and the end of multicore scaling, in *Proceeding of the 38th annual international symposium on Computer architecture* (ACM, 2011), pp. 365–376
6. N. Goulding, J. Sampson, G. Venkatesh, S. Garcia, J. Auricchio, J. Babb, M.B. Taylor, S. Swanson, Greendroid: a mobile application processor for a future of dark silicon. Hot Chips (2010), pp. 1–39
7. Gurobi Optimizer (2016) (www.gurobi.com)
8. S. Herbert, D. Marculescu, Characterizing chip-multiprocessor variability-tolerance, in *45th ACM/IEEE Design Automation Conference, 2008 (DAC, 2008)* (IEEE, 2008), pp. 313–318
9. A. Kumar, B. Mesman, B. Theelen, H. Corporaal, Y. Ha, Analyzing composability of applications on MPSoC platforms. J. Syst. Archit. **54**(3–4), 369–383 (2008)
10. S. Li, J.H. Ahn, R.D. Strong, J.B. Brockman, D.M. Tullsen, N.P. Jouppi, Mcpat: an integrated power, area, and timing modeling framework for multicore and manycore architectures, in *42nd Annual IEEE/ACM International Symposium on Microarchitecture, 2009 (MICRO-42)* (IEEE, 2009), pp. 469–480
11. S. Li, J.H. Ahn, R.D. Strong, J.B. Brockman, D.M. Tullsen, N.P. Jouppi, Mcpat: an integrated power, area, and timing modeling framework for multicore and manycore architectures, in *Microarchitecture, 2009* (IEEE, 2009)
12. R. Teodorescu, J. Torrellas, Variation-aware application scheduling and power management for chip multiprocessors, in *ACM SIGARCH Computer Architecture News*, vol. 36 (IEEE Computer Society, 2008), pp. 363–374
13. Y. Turakhia, B. Raghunathan, S. Garg, D. Marculescu, Hades: Architectural synthesis for heterogeneous dark silicon chip multi-processors, in *Proceedings of the 50th Annual Design Automation Conference* (ACM, 2013), p. 173
14. S.C. Woo, M. Ohara, E. Torrie, J.P. Singh, A. Gupta, The splash-2 programs: characterization and methodological considerations, in *ACM SIGARCH Computer Architecture News*, vol. 23 (ACM, 1995), pp. 24–36
15. J. Xiong, V. Zolotov, L. He, Robust extraction of spatial correlation. IEEE Trans. Comput. Aided Des. Integr. Circuits Syst. **26**(4), 619–631 (2007)
16. W. Zhao, Y. Cao, New generation of predictive technology model for sub-45 nm early design exploration. *IEEE Trans. Electron Devices* **53**(11), 2816–2823 (2006)

Part II
Run-Time Resource Management: Computational Perspective

Chapter 5
Thermal Safe Power: Efficient Thermal-Aware Power Budgeting for Manycore Systems in Dark Silicon

Santiago Pagani, Heba Khdr, Jian-Jia Chen, Muhammad Shafique, Minming Li, and Jörg Henkel

Introduction

For a specific chip, the common industry practice is to provide the system designers with the Thermal Design Power (TDP). According to [11], TDP "is the highest expected sustainable power while running known power intensive real applications" and it should be a safe power level in which to run the system. Therefore, manufacturers recommend that the cooling solution should be designed to dissipate TDP, such that the heat sink is not over-dimensioned and running at this target power level does not cause thermal problems. However, TDP is not the maximum achievable power. To avoid the system from possible overheating, chips are provided with Dynamic Thermal Management (DTM) techniques. DTM can power-down cores, gate their clocks, reduce their supply voltage and frequency, boost-up the fan speed, etc. By directly measuring the temperature, if the system heats up above a certain threshold, DTM is triggered such that the temperature is reduced [11]. An abstract example of a DTM technique that uses voltage and frequency scaling can be seen in Fig. 5.1.

A power constraint (or power budget) is an abstraction that allows system designers to indirectly deal with thermal issues and at the same time avoid excessive

S. Pagani (✉) • H. Khdr • M. Shafique • J. Henkel
Chair for Embedded Systems (CES), Karlsruhe Institute of Technology (KIT), Karlsruhe, Germany
e-mail: pagani@kit.edu; heba.khdr@kit.edu; shafique@kit.edu; henkel@kit.edu

J.-J. Chen
Department of Informatics, TU Dortmund, Dortmund, Germany
e-mail: jian-jia.chen@cs.uni-dortmund.de

M. Li
Department of Computer Science, City University of Hong Kong (CityU), Hong Kong, China
e-mail: minming.li@cityu.edu.hk

A.M. Rahmani et al. (eds.), *The Dark Side of Silicon*,
DOI 10.1007/978-3-319-31596-6_5

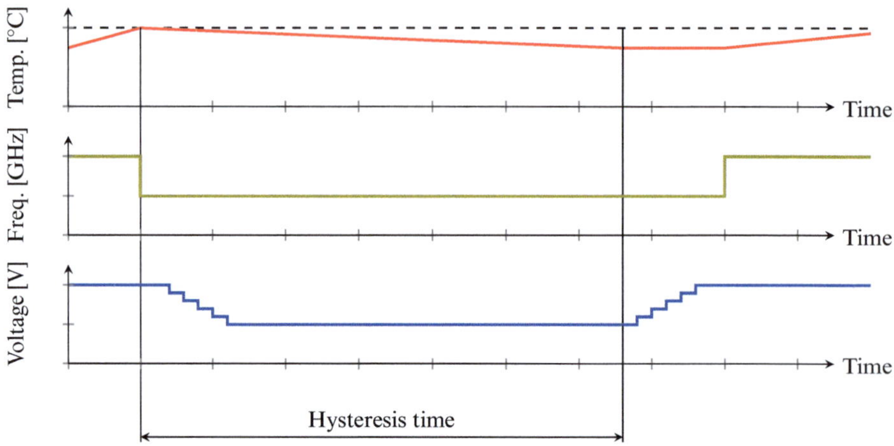

Fig. 5.1 Example of a DTM technique based on voltage and frequency scaling

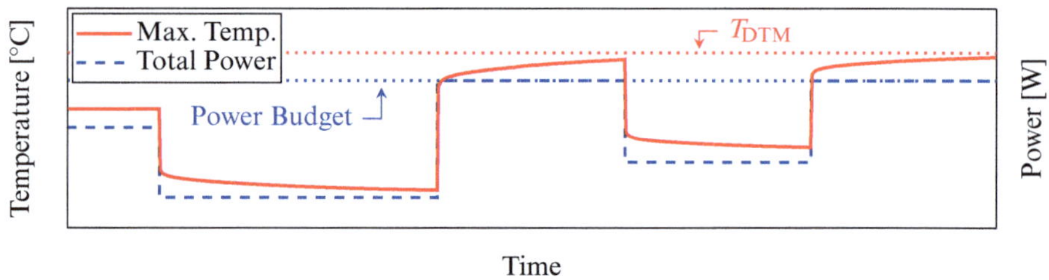

Fig. 5.2 Abstract example of an *efficient* power budget

triggers of DTM. Namely, keeping the system under the specified power budget should avoid that the temperature anywhere on the chip exceeds the threshold temperature that triggers DTM, denoted as T_{DTM}. A power budget that fails to achieve this goal cannot be considered *safe*, and using it would result in frequent triggers of DTM. Contrarily, a power budget is said to be *pessimistic* if it results in maximum temperatures that are far away from T_{DTM}. Figure 5.2 shows an abstract example of an *efficient* power budget, such that when the systems executes just under the budget the maximum temperatures are just below T_{DTM}. Usually, system designers use TDP as a power constraint. Nevertheless, on homogeneous and heterogeneous manycore systems, using a *single* and *constant* value as a power constraint for each core or for the entire chip (e.g., TDP), can result in significant performance losses, or in frequent triggers of DTM and thus a total performance much smaller than expected by the task partitioning and core mapping algorithms.

Furthermore, given that the continuous increasing performance demands and power consumption issues have led to the emergence of heterogeneous architectures [27], power budgeting techniques should be developed to natively handle core heterogeneity. Moreover, different applications consume different amounts of power depending on the number of executing threads, selected frequency levels, and types of cores. This makes task partitioning and core mapping a very challenging process, specially when involving core heterogeneity or timing guarantees.

In this chapter we present a power budget concept called Thermal Safe Power (TSP) [20, 22]. TSP is an abstraction that provides safe and efficient power constraint values as a function of the number of simultaneously active cores. Executing cores at power consumptions below TSP results in maximum temperatures just below the threshold level that triggers DTM. TSP can also serve as a fundamental tool for guiding task partitioning and core mapping decisions.

Based on the RC thermal network [10] of a specific chip, this chapter presents polynomial-time algorithms to compute TSP for both *homogeneous* and *heterogeneous* systems. For simplicity of presentation and introduction of the TSP concept, we first present the special case for handling homogeneous cores. We then present the algorithms for the general case, which are suitable for heterogeneous systems. Sections "TSP for a Given Core Mapping (Homogeneous Systems)" and "TSP for a Given Core Mapping (Heterogeneous Systems)" present the algorithms that compute TSP for a particular mapping of active cores, which can be used online, thus accounting both for changes in the mapping decisions and ambient temperature. Sections "TSP for Worst-Case Mappings (Homogeneous Systems)" and "TSP for Worst-Case Mappings (Heterogeneous Systems)" present the computation of TSP for the worst-case mappings of active cores. That is, a mapping of active cores, for every possible number of simultaneously active cores, that results in the lowest TSP values. Such worst-case core mappings are the most pessimistic cases, which result in TSP values that are safe for any other mapping scenario. Therefore, this allows the system designers to abstract from core mapping decisions.

Using gem5 [2], McPAT [17], and HotSpot [10], we run simulations to compare the total system performance for using six different power constraints: TSP for given mappings, worst-case TSP, two constant power budgets per-chip, a constant power budget per-core, and a boosting technique [4, 6, 12, 25]. We also show how TSP can be used to estimate the amount of dark silicon [3, 8, 26], and how such estimations are less pessimistic than those for constant power budgets.

Open-Source Tools: The algorithms to compute TSP are implemented as an open-source tool available for download at http://ces.itec.kit.edu/download.

Motivational Example

This example provides some insight on the drawbacks of using *single* and *constant* power constraints. For simplicity of presentation, consider a hypothetical manycore system with 16 homogeneous cores of size 2.31 × 2.31 mm (*simple in-order* Alpha 21264 cores in 45 nm, simulated with McPAT [17]), arranged in four rows and four columns. Assume a threshold temperature that triggers DTM of 80 °C, and a cooling solution taken from the default configuration of HotSpot [10] (detailed in section "Setup"). Now, consider a *constant* power budget of 90.2 W *per-chip*. Running simulations with HotSpot, for different number of active cores, Fig. 5.3 shows the resulting steady-state temperatures on the chip when equally distributing the 90.2 W per-chip budget among the active cores. From the figure, we can clearly observe that this per-chip power budget is only efficient when equally distributing the budget among 8 cores. However, this power budget is pessimistic or not thermally safe when equally distributing the budget among more than 8 cores or among less than 8 cores, respectively.

In order to show that similar effects occur for other power budget values, Fig. 5.4 presents simulations conducted in HotSpot that show the maximum temperature among all cores (in the steady-state) as a function of the number of simultaneously

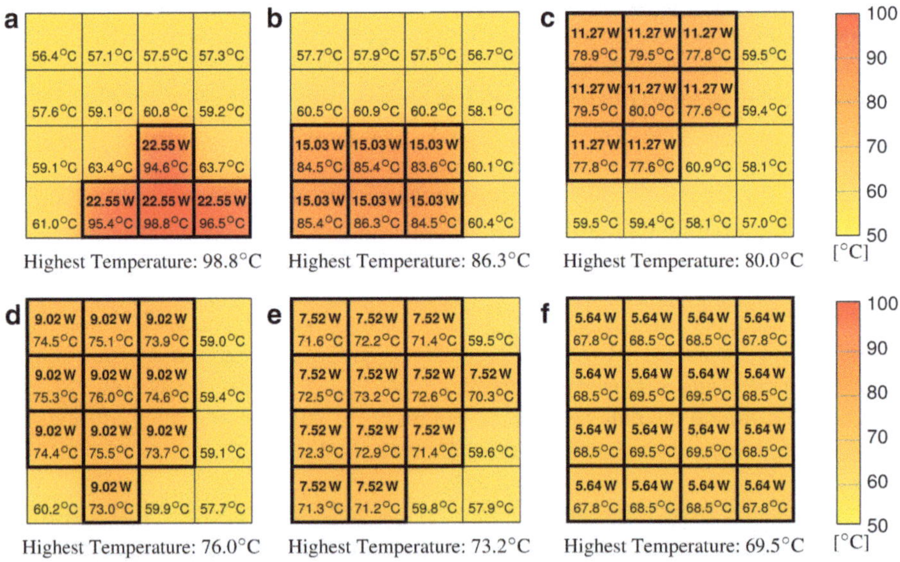

Fig. 5.3 Temperature distribution example for using a 90.2 W per-chip power budget. *Top numbers* are the power consumptions of each active core (*boxed in black*). *Bottom numbers* are the temperatures in the center of each core. Detailed temperatures are shown according to the color bar. (**a**) 4 active cores. (**b**) 6 active cores. (**c**) 8 active cores. (**d**) 10 active cores. (**e**) 12 active cores. (**f**) 16 active cores

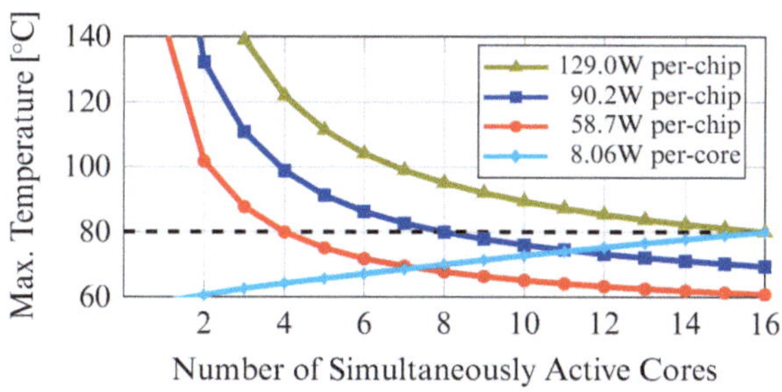

Fig. 5.4 Maximum steady-state temperature (with DTM deactivated) among all cores as a function of the number of simultaneously active cores, when using different *single* and *constant* per-chip and per-core power budgets

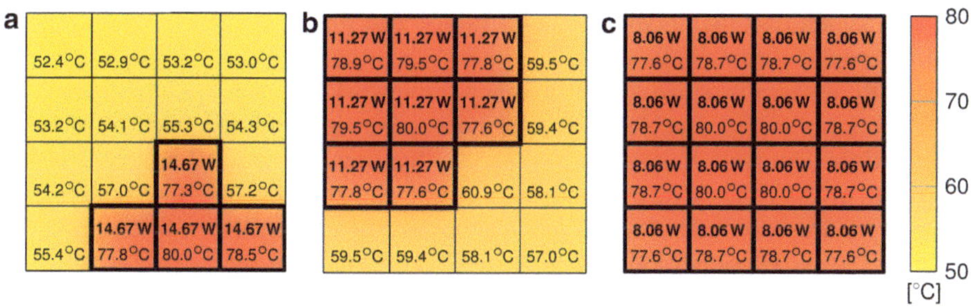

Fig. 5.5 Example for a maximum temperature of 80 °C. *Top numbers* are the power consumptions of each active core (*boxed in black*). *Bottom numbers* are the temperatures in the center of each core. Detailed temperatures are shown according to the color bar. (**a**) 4 active cores. (**b**) 8 active cores. (**c**) 16 active cores

active cores, for three *per-chip* power budgets and one *per-core* power budget. Specifically, 58.7 W per-chip, 90.2 W per-chip, 129.0 W per-chip, and 8.06 W per-core. From Fig. 5.4, it becomes clear that using a *single* and *constant* power budget can either be a pessimistic approach (for small power budgets), or it can result in frequent triggers of DTM (for large power budgets). Using *constant per-core* power budgets, e.g., 8.06 W for the example in Fig. 5.4, can be a good compromise, because the temperature never exceeds 80 °C, but without being too far from it. However, there is still room for improvement.

A more efficient power budgeting technique would use different per-core power budgets depending on the number of active cores, such that the maximum steady-state temperature among all cores is 80 °C for all cases, and *that is precisely the power budget concept called Thermal Safe Power (TSP)*. For example, running simulations with HotSpot, Fig. 5.5a shows the resulting steady-state temperatures on the chip when 4 cores consume 14.67 W each (58.7 W in total). Similarly, Fig. 5.5b and c show the resulting steady-state temperatures on the chip when 8 cores consume 11.27 W each (90.2 W in total) and when 16 cores consume 8.06 W each (129.0 W in total), respectively. For all cases, although different power budgets are considered, at least one core reaches 80 °C in the steady-state.

State-of-the-Art in Power Budgeting Techniques

There are many works in the literature that focus on improving performance under a given (constant) power budget [15, 23, 24], including several that specifically use TDP as the power constraint [16, 18]. There is also work on energy efficiency [5] and reliability [14] under TDP constraints, and on power budgeting based on reinforcement learning [7].

In [18], authors propose a control-based framework to obtain the optimal trade-off between power and performance for homogeneous multicore systems under a TDP budget. Dynamic Voltage and Frequency Scaling (DVFS) and task migrations are used at different levels: per task, per cluster and on chip controllers. The controllers are coordinated to throttle down the power when the system exceeds TDP, and to assign tasks onto cores to optimize the performance.

The work in [24] exploits process variations between cores in a homogeneous multicore system to pick the more suitable cores for an application to improve its performance. Due to the proportional increment of process variations, their results show that the performance efficiency can be increased along with the increase in the dark silicon area.

The work in [16] presents throughput analysis for the impact of applying per-core power gating and DVFS to thermal and power constrained multicore processors, using TDP as the power constraint. The authors also exploit power and thermal headroom resulting from power-gated idle cores, allowing active cores to increase their frequency, optimizing the throughput for applications with limited parallelism.

In [15], authors build a 32-core TFET-CMOS heterogeneous multicore and propose a runtime scheme to improve the performance of applications running on such cores, operating under a fixed power budget. The scheme combines heterogeneous thread-to-core mapping, dynamic work partitioning, and dynamic power partitioning.

In [5], the authors evaluate the energy efficiency of different processors from Intel's i5 and i7 family, using selected benchmarks with variable core counts and vectorization techniques, while using TDP as the power constraint.

All of the above mentioned work uses a *single* and *constant* value as the power constraint to abstract from thermal problems. However, the motivational example in section "Introduction" shows that, depending on the amount of simultaneously active cores, different power consumptions result in the same maximum temperatures. This means that using a single value as the power constraint, e.g., TDP, can result in performance losses for multicore and manycore systems.

A few other technologies, like Intel's Turbo Boost [4, 6, 12, 25] and AMD's Turbo CORE [19], leverage this temperature headroom allowing power consumptions above the TDP constraint during *short time intervals* by increasing the voltage and frequency of the cores in an online manner. Due to the increases in power consumption, boosting normally results in an increment of the temperature through time. Once the highest temperature among all cores reaches a predefined threshold, the system must either return to nominal operation (needing some cool-down time before another boosting interval), or use some closed-loop control to oscillate around the threshold (prolonging the boosting time). Unlike these techniques, in TSP the increases in power consumption are constrained (according to the number of cores) such that DTM is not activated, allowing the system to remain *indefinitely* in such power states.

> The introduction of TSP may motivate researchers to revisit the related work, possibly resulting in extensions of the existing literature, that achieve a better system performance.

System Model

Hardware Model

The computation of TSP is based on an RC thermal network modeled from the floorplan of any given architecture. The detail of the blocks that conform the floorplan can be freely chosen. For example, consider the 64 cores system presented in Fig. 5.6, which is based on simulations conducted with gem5 [2] and McPAT [17] for *out-of-order* (OOO) Alpha 21264 cores in 22 nm technology. Each core has an area of $9.6\,\mathrm{mm}^2$, and there is a shared L2 cache and a memory controller every 4 cores. The area of the L2 blocks together with the memory controllers is $4.7\,\mathrm{mm}^2$, which is comparable to the area of the cores. Therefore, it is reasonable to have independent blocks for the L2 cache and the memory controllers. For the rest of the chip, a practical approach is to consider blocks at a core level, and compute the TSP values for such a granularity, specifically, because the power consumptions of the internal blocks of the cores are tightly related with the execution frequency on each core.

For simplicity of presentation, throughout this chapter we focus on a manycore system with M cores. However, we consider that there are Z blocks in the floorplan, such that $Z - M$ is the amount of blocks that correspond to other types of components, e.g., L2 caches and memory controllers. When computing TSP, we consider that each one of these $Z - M$ blocks always consumes its highest power.

We refer to a core being *active*, whenever the core is in its execution mode. To maintain a core active at its lowest speed a minimum power consumption is needed, which we define as $P^{\mathrm{core}}_{\mathrm{min}}$. Contrarily, a core is *inactive* when it is in some low-power mode, e.g., sleep or turned off (power-gated). We define the power consumption

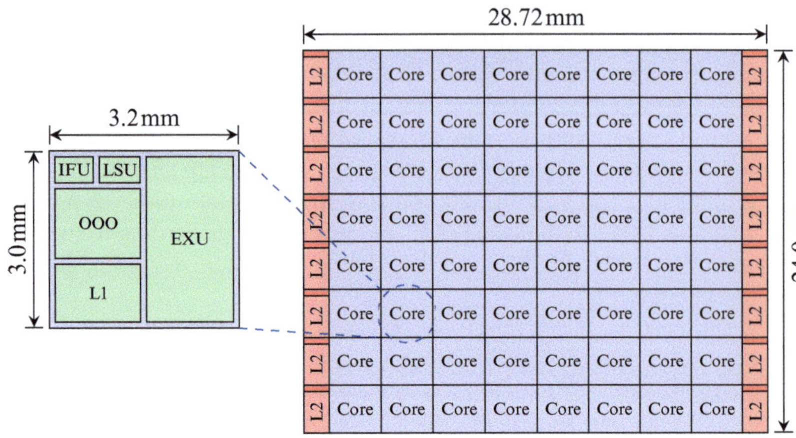

Fig. 5.6 Floorplan for a 64 core system based on simulations in McPAT [17]. In McPAT, cores are composed by several units: an instruction fetch unit (IFU), an execution unit (EXU), a load and store unit (LSU), an out-of-order (OOO) issue/dispatch, and a private L1 cache

of an inactive core as $P^{\text{core}}_{\text{inact}}$, and it holds that $P^{\text{core}}_{\min} \geq P^{\text{core}}_{\text{inact}}$. Moreover, when simultaneously activating m cores, there are $\binom{M}{m}$ combinations of *which* specific cores in the floorplan to activate. Throughout this chapter, we refer to a specific decision of *which* cores to activate as *core mapping* or *mapping of cores*.

The power consumption in a CMOS core is the summation of its dynamic power consumption (mainly generated by switching activities) and its static power consumption (mainly generated by leakage currents). Part of the static power consumption depends on the temperature of the core, which means that higher temperatures cause higher power consumptions. In order to consider *safe margins for offline power profiles* of tasks and their scheduling decisions for safe operation, we assume that the power consumptions of cores are modeled at temperatures near the threshold level that triggers DTM, which we define as T_{DTM}. That is, when experimentally measuring the power consumption of a core executing a task at a specific voltage and frequency, this is done at temperatures near T_{DTM}. If not, we may have underestimated power models of tasks. When making online decisions, if cores are equipped with power meters, then no over- or underestimation occurs, because the system can measure the actual power consumption of each task (including leakage effects), and act accordingly.

Thermal Model

For our thermal model, we consider the well-known duality between thermal and electrical circuits, i.e., we use an RC thermal network [10]. Such a network is composed by N thermal nodes, where $N \geq Z$. In an RC thermal network, thermal

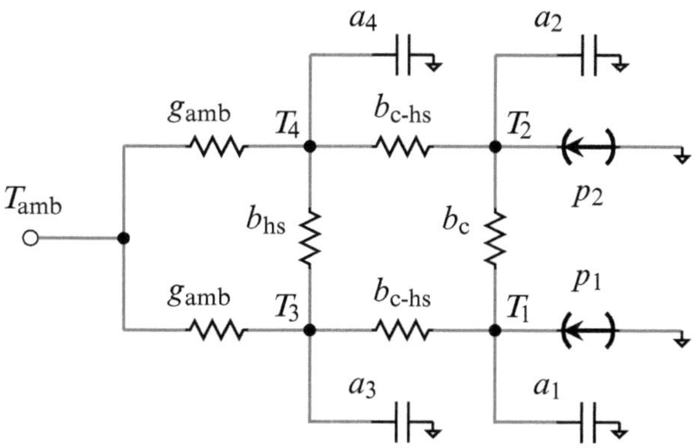

Fig. 5.7 Simple RC thermal network example for 2 cores. This is a simplified example because we consider that cores are in direct contact with the heat sink, and that there is no other connection between cores and the ambient temperature. In a more detailed example, between a core and the heat sink there are many layers, e.g., the ceramic packaging substrate, the thermal paste, and the heat spreader; and there are more paths that lead to the ambient temperature, e.g., through the circuit board

nodes are interconnected between each other through thermal conductances. Each thermal node also has a thermal capacitance associated to it, which accounts for the transient temperatures. The ambient temperature, denoted as T_{amb}, is considered to be constant, and thus there is no capacitance associated with it. The power consumptions of cores and other blocks correspond to heat sources. With these considerations, the temperature of every thermal node is a function of its power consumption, the temperatures of the neighbouring nodes, and the ambient temperature.

Figure 5.7 presents a simplified example of an RC thermal network for a chip with two cores. In Fig. 5.7, T_1 and T_2 are the temperatures on core 1 and core 2, respectively; T_3 and T_4 are the temperatures on the heat sink nodes above core 1 and core 2, respectively; p_1 and p_2 are the power consumptions on core 1 and core 2, respectively; b_c is the thermal conductance between core 1 and core 2; $b_{c-\text{hs}}$ is the thermal conductance between a core and the heat sink; b_{hs} is the thermal conductance between heat sink nodes; g_{amb} is the thermal conductance between a heat sink node and the ambient temperature; and a_i is the thermal capacitance of node i. Through Kirchoff's first law, we derive a system of differential equations for the example in Fig. 5.7. That is,

$$\begin{cases} p_1 - (T_1 - T_3)\, b_{\text{c}-\text{hs}} + (T_2 - T_1)\, b_{\text{c}} - a_1 \dfrac{\mathrm{d}T_1}{\mathrm{d}t} = 0 \\ p_2 - (T_2 - T_4)\, b_{\text{c}-\text{hs}} - (T_2 - T_1)\, b_{\text{c}} - a_2 \dfrac{\mathrm{d}T_2}{\mathrm{d}t} = 0 \\ (T_1 - T_3)\, b_{\text{c}-\text{hs}} + (T_4 - T_3)\, b_{\text{hs}} - a_3 \dfrac{\mathrm{d}T_3}{\mathrm{d}t} - (T_3 - T_{\text{amb}})\, g_{\text{amb}} = 0 \\ (T_2 - T_4)\, b_{\text{c}-\text{hs}} - (T_4 - T_3)\, b_{\text{hs}} - a_4 \dfrac{\mathrm{d}T_4}{\mathrm{d}t} - (T_4 - T_{\text{amb}})\, g_{\text{amb}} = 0. \end{cases}$$

This system of differential equations can be rewritten as

$$\begin{bmatrix} (b_{\text{c}-\text{hs}} + b_{\text{c}}) & -b_{\text{c}} & -b_{\text{c}-\text{hs}} & 0 \\ -b_{\text{c}} & (b_{\text{c}-\text{hs}} + b_{\text{c}}) & 0 & -b_{\text{c}-\text{hs}} \\ -b_{\text{c}-\text{hs}} & 0 & (b_{\text{c}-\text{hs}} + b_{\text{hs}} + g_{\text{amb}}) & -b_{\text{hs}} \\ 0 & -b_{\text{c}-\text{hs}} & -b_{\text{hs}} & (b_{\text{c}-\text{hs}} + b_{\text{hs}} + g_{\text{amb}}) \end{bmatrix} \begin{bmatrix} T_1 \\ T_2 \\ T_3 \\ T_4 \end{bmatrix}$$

$$+ \begin{bmatrix} a_1 & 0 & 0 & 0 \\ 0 & a_2 & 0 & 0 \\ 0 & 0 & a_3 & 0 \\ 0 & 0 & 0 & a_4 \end{bmatrix} \begin{bmatrix} T'_1 \\ T'_2 \\ T'_3 \\ T'_4 \end{bmatrix} = \begin{bmatrix} p_1 \\ p_2 \\ 0 \\ 0 \end{bmatrix} + T_{\text{amb}} \begin{bmatrix} 0 \\ 0 \\ g_{\text{amb}} \\ g_{\text{amb}} \end{bmatrix}.$$

Hence, in matrix form, for any RC thermal network with N thermal nodes, we can build a system of N differential equations associated with it, that is,

$$\mathbf{A}\mathbf{T}' + \mathbf{B}\mathbf{T} = \mathbf{P} + T_{\text{amb}}\mathbf{G},$$

where for a system with N thermal nodes, matrix $\mathbf{A} = \left[a_{i,j}\right]_{N\times N}$ contains the thermal capacitance values, matrix $\mathbf{B} = \left[b_{i,j}\right]_{N\times N}$ contains the thermal conductance values in $\left[\frac{\text{Watt}}{\text{Kelvin}}\right]$, column vector $\mathbf{T} = [T_i]_{N\times 1}$ represents the temperature on each node, column vector $\mathbf{T}' = \left[T'_i\right]_{N\times 1}$ accounts for the first order derivative of the temperature on each node with respect to time, column vector $\mathbf{P} = [p_i]_{N\times 1}$ contains the power consumption on each node, and column vector $\mathbf{G} = [g_i]_{N\times 1}$ contains the thermal conductance between each node and the ambient. If node i is not in contact with the ambient temperature, e.g., the temperature of a core or an internal node, the value of g_i is set to zero. The thermal conductance values in matrix $\mathbf{B}$ include the thermal conductances between vertical and lateral neighboring nodes. Matrix $\mathbf{A}$ is generally a diagonal matrix, since thermal capacitances are modeled to ground. For specific floorplans, the values for matrices and vector $\mathbf{A}$, $\mathbf{B}$, and $\mathbf{G}$ can be computed through HotSpot [10], deriving a detailed RC thermal network which is the same one used internally by HotSpot to conduct thermal simulations.

When only considering the *steady-state*, we have that

$$\mathbf{B}\mathbf{T} = \mathbf{P} + T_{\mathrm{amb}}\mathbf{G} \qquad \text{or} \qquad \mathbf{T} = \mathbf{B}^{-1}\mathbf{P} + T_{\mathrm{amb}}\mathbf{B}^{-1}\mathbf{G},$$

where from $\mathbf{B}^{-1}$ we have that ${b^{-1}}_{i,j} \cdot p_j$ represents the amount of heat contributed by node j into the steady-state temperature of node i, i.e., T_i.

Regarding vector $\mathbf{P}$, we can divide it into three sub-vectors: $\mathbf{P}^{\mathrm{cores}}$ for the power consumption on the cores; $\mathbf{P}^{\mathrm{blocks}}$ for the power consumption on blocks of a different type, which we consider to be always active at their highest power consumption values, e.g., the L2 caches for the manycore system in Fig. 5.6 as explained in section "Hardware Model"; and $\mathbf{P}^{\mathrm{int}}$ for internal nodes in which $p_i^{\mathrm{int}} = 0$ for all i. It holds that $\mathbf{P} = \mathbf{P}^{\mathrm{cores}} + \mathbf{P}^{\mathrm{blocks}} + \mathbf{P}^{\mathrm{int}}$. Decomposing $\mathbf{P}$, with $p_i^{\mathrm{int}} = 0$ for all i, we have that

$$\mathbf{T} = \mathbf{B}^{-1}\mathbf{P}^{\mathrm{cores}} + \mathbf{B}^{-1}\mathbf{P}^{\mathrm{blocks}} + T_{\mathrm{amb}}\mathbf{B}^{-1}\mathbf{G}, \tag{5.1}$$

where by only focusing on one equation from the system of equations, the steady-state temperature on node i is

$$T_i = \sum_{j=1}^{N} {b^{-1}}_{i,j} \cdot p_j^{\mathrm{cores}} + \sum_{j=1}^{N} {b^{-1}}_{i,j} \left(p_j^{\mathrm{blocks}} + T_{\mathrm{amb}} \cdot g_j\right). \tag{5.2}$$

For notational brevity, we define set $\mathbf{L} = \{\ell_1, \ell_2, \ldots, \ell_Z\}$, such that the elements in $\mathbf{L}$ include all indexes of the thermal nodes that correspond to blocks of the floorplan; as opposed to thermal nodes that represent the heat sink, internal nodes of the heat spreader, the thermal interface material, etc. Similarly, we define set $\mathbf{K} = \{k_1, k_2, \ldots, k_M\}$, such that $\mathbf{K}$ contains all indexes of nodes that correspond to cores.

Furthermore, we also define column vector $\mathbf{Q} = [q_i]_{M \times 1}$ for a particular mapping of *active* cores. Vector $\mathbf{Q}$ is a binary vector: $q_i = 1$ means that core i corresponds to an *active* core; $q_i = 0$ means that core i corresponds to an *inactive* core.

Problem Definition

This chapter presents a power budget concept called Thermal Safe Power (TSP). TSP is an abstraction that provides power constraint values as a function of the number of simultaneously active cores, according to the definition of *active*/*inactive* from section "Hardware Model". The values of TSP vary according to the floorplan and which cores are simultaneously active. Some specific core mappings result in the lowest TSP values, and we define such core mappings as the *worst-case mappings*. Executing cores at power consumptions below TSP, for the corresponding mapping and number of active cores, results in maximum temperatures below T_{DTM}.

For a specific chip and its corresponding RC thermal network, the *first objective* of this chapter is to provide a numerical method to compute TSP for a given mapping of cores. The mapping of cores is typically determined by an operating/run-time system or by an offline system software. This method should have polynomial-time complexity, such that TSP can be computed online for a particular mapping of cores and ambient temperatures. Formally, for a given core mapping $\mathbf{Q}$, this means obtaining a uniform power constraint for the active cores, defined as $P_{\text{TSP}}(\mathbf{Q})$, such that $T_i \leq T_{\text{DTM}}$ (in the steady-state) for all $i \in \mathbf{L}$.

The *second objective* is to derive an algorithm to compute the most pessimistic TSP values for a given number of simultaneously active cores, i.e., for the worst-case mappings. Such TSP values can be used as safe power constraints for any possible mapping of cores, thus allowing the system designers to abstract from core mapping decisions. Formally, this means obtaining the most pessimistic uniform power constraint for any m active cores, defined as $P_{\text{TSP}}^{\text{worst}}(m)$, such that $T_i \leq T_{\text{DTM}}$ (in the steady-state) for all $i \in \mathbf{L}$.

> The algorithms presented in sections "Thermal Safe Power (TSP) for Homogeneous Systems" and "Thermal Safe Power (TSP) for Heterogeneous Systems" are derived considering the steady-state. Section "Transient State Considerations" explains a method to further consider the transient temperatures.

Thermal Safe Power (TSP) for Homogeneous Systems

TSP for a Given Core Mapping (Homogeneous Systems)

This subsection presents a polynomial-time algorithm to compute TSP in an online manner for a particular core mapping and ambient temperature, which results in a uniform value of TSP per-core, for all active cores in mapping $\mathbf{Q}$. That is, one power constraint value for each active core in the specified mapping, that results in a maximum temperature (in the steady-state) among all cores which does not exceed T_{DTM}. We define such a power constraint as $P_{\text{TSP}}(\mathbf{Q})$.

> Note that this does not mean that all active cores consume the same power, as this would be an unrealistic assumption. In fact, this means that each active core can consume any amount of power, which can be different for each core, as long as the power consumption of each core is no more than $P_{\text{TSP}}(\mathbf{Q})$.

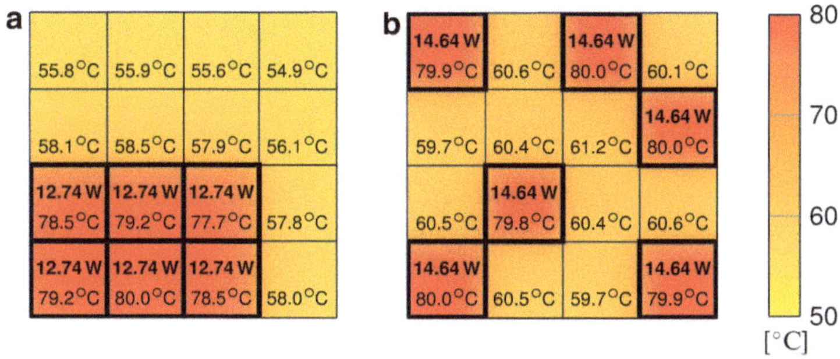

Fig. 5.8 Example of TSP for two different mappings for a maximum temperature of 80 °C. *Top numbers* are the power consumptions of each active core (*boxed in black*). *Bottom numbers* are the temperatures in the center of each core. Detailed temperatures are shown according to the color bar. (**a**) Concentrated mapping example with 6 active cores. (**b**) Distributed mapping example with 6 active cores

Due to the heat transfer among cores, the mapping of cores, i.e., *which* cores are active and *which* cores are inactive, plays a major role in the computation of the maximum temperatures. This can be seen in the following example. Consider a manycore system of 16 cores with the same settings as in the motivational example from section "Introduction". Figure 5.8 shows two possible mappings when simultaneously activating 6 cores. For Fig. 5.8a, the maximum temperature among all cores reaches 80 °C when each core consumes 12.74 W. However for Fig. 5.8b, this happens when each core consumes 14.64 W. We have a total of 11.4 W difference between these two core mappings, but with the same maximum temperature.

For a given mapping executing under power budget $P_{\text{TSP}}(\mathbf{Q})$, the maximum temperatures will clearly occur when all cores consume the entire power budget. Therefore, by considering vector $\mathbf{Q}$, set $\mathbf{K}$, and assuming that all inactive cores consume $P_{\text{inact}}^{\text{core}}$ and all active cores consume equal power P_{equal} at a given time, we start by rewriting Eq. (5.2) as shown in Eq. (5.3).

$$T_i = P_{\text{equal}} \cdot \sum_{j=1}^{M} {b^{-1}}_{i,k_j} \cdot q_j + P_{\text{inact}}^{\text{core}} \cdot \sum_{j=1}^{M} {b^{-1}}_{i,k_j} \left(1 - q_j\right) + \sum_{j=1}^{N} {b^{-1}}_{i,j} \left(p_j^{\text{blocks}} + T_{\text{amb}} \cdot g_j\right) \tag{5.3}$$

In Eq. (5.3) P_{equal} is not yet $P_{\text{TSP}}(\mathbf{Q})$, but the expression represents the resulting temperature on node i when all active cores in the given mapping consume equal power. Furthermore, for a given $\mathbf{Q}$, $\mathbf{P}^{\text{blocks}}$, T_{amb}, and floorplan, the only variables in Eq. (5.3) are T_i and P_{equal}. As seen in Eq. (5.3), that is,

$$T_i = P_{\text{equal}} \cdot \sum_{j=1}^{M} b^{-1}{}_{i,k_j} \cdot q_j + \underbrace{P^{\text{core}}_{\text{inact}} \cdot \sum_{j=1}^{M} b^{-1}{}_{i,k_j} \left(1 - q_j\right) + \sum_{j=1}^{N} b^{-1}{}_{i,j} \left(p_j^{\text{blocks}} + T_{\text{amb}} \cdot g_j\right)}_{\textbf{Constant for node i}}.$$

Given that the temperature which we do not want to exceed is T_{DTM}, the goal is now to find what value of P_{equal} would make T_i reach such a temperature. This can be easily done from Eq. (5.3) by setting T_i to T_{DTM} and deriving P_{equal} as shown in Eq. (5.4).

$$P_{\text{equal}} = \frac{T_{\text{DTM}} - P^{\text{core}}_{\text{inact}} \cdot \sum_{j=1}^{M} b^{-1}{}_{i,k_j} \left(1 - q_j\right) - \sum_{j=1}^{N} b^{-1}{}_{i,j} \left(p_j^{\text{blocks}} + T_{\text{amb}} \cdot g_j\right)}{\sum_{j=1}^{M} b^{-1}{}_{i,k_j} \cdot q_j} \tag{5.4}$$

The most pessimistic value of P_{equal} is a safe power constraint for all cores in mapping $\mathbf{Q}$. Therefore, the resulting $P_{\text{TSP}}(\mathbf{Q})$ for the given $\mathbf{Q}$, $\mathbf{P}^{\text{blocks}}$, T_{amb}, T_{DTM}, $P^{\text{core}}_{\text{inact}}$, and floorplan, can be computed by finding the minimum P_{equal} for all blocks in the floorplan, that is, $\forall i \in \mathbf{L}$. The computation of $P_{\text{TSP}}(\mathbf{Q})$ is presented in Eq. (5.5), with a total time complexity of $O(ZN)$. The value of i that results in $P_{\text{TSP}}(\mathbf{Q})$ corresponds to the block with the highest temperature for such a case.

$$P_{\text{TSP}}(\mathbf{Q}) = \min_{\forall i \in \mathbf{L}} \left\{ \frac{T_{\text{DTM}} - P^{\text{core}}_{\text{inact}} \cdot \sum_{j=1}^{M} b^{-1}{}_{i,k_j} \left(1 - q_j\right) - \sum_{j=1}^{N} b^{-1}{}_{i,j} \left(p_j^{\text{blocks}} + T_{\text{amb}} \cdot g_j\right)}{\sum_{j=1}^{M} b^{-1}{}_{i,k_j} \cdot q_j} \right\} \tag{5.5}$$

A brief example of how TSP was computed through Eq. (5.5) for Fig. 5.8a can be seen in Fig. 5.9, considering a manycore system of 16 cores with the same settings as in the motivational example from section "Introduction". Namely, Fig. 5.9a shows what power consumption value should be consumed by the 6

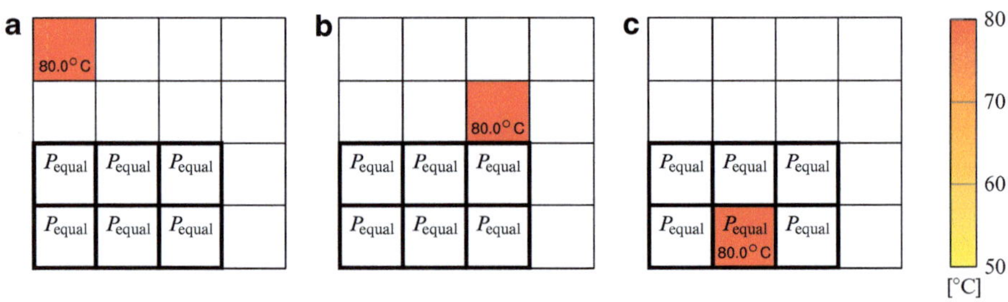

Fig. 5.9 TSP computation example for the given mapping from Fig. 5.8a. (**a**) For T_1, P_{equal}=41.30 W. (**b**) For T_7, $P_{\text{equal}} = 33.11$ W. (**c**) For T_{14}, $P_{\text{equal}} = 12.74$ W

active cores, that is, P_{equal}, such that T_1 reaches 80 °C, momentarily ignoring the temperature on other blocks. The same thing is shown for Fig. 5.9b and c when focusing on T_7 and T_{14}, respectively (the computation of P_{equal} for the other blocks is not shown, but the same principle applies). After computing P_{equal} for all T_i, particularly $T_1, T_2, \ldots, T_{16}$ for this example, $P_{\text{TSP}}(\mathbf{Q})$ is set to the smallest value of P_{equal}, which was the case when computing T_{15}. Moreover, since the final value of $P_{\text{TSP}}(\mathbf{Q})$ is found when computing T_{14}, it holds that this block will have the highest temperature if all cores consume equal power, shown in Fig. 5.8a.

TSP for Worst-Case Mappings (Homogeneous Systems)

This subsection presents a polynomial-time algorithm to compute TSP for the worst-case core mappings, for m active cores. The algorithm results in a uniform value of TSP per-core for all active cores. That is, one power constraint value for each active core in *any* possible core mapping with m simultaneously active cores, that results in a maximum temperature (in the steady-state) among all cores which does not exceed T_{DTM}. We define such a power constraint as $P_{\text{TSP}}^{\text{worst}}(m)$.

As in section "TSP for a Given Core Mapping (Homogeneous Systems)", note that this does not mean that all active cores consume the same power, as this would be an unrealistic assumption. It means that each active core, *for any given mapping of m active cores*, can consume any amount of power, which can be different for each core, as long as the power consumption of each core is no more than $P_{\text{TSP}}^{\text{worst}}(m)$. The purpose behind this is to allow system designers to abstract themselves from mapping decisions, as opposed to the TSP computations from section "TSP for a Given Core Mapping (Homogeneous Systems)".

As mentioned in section "TSP for a Given Core Mapping (Homogeneous Systems)" and shown in Fig. 5.8, the mapping of cores plays a major role in the resulting maximum temperatures due to the heat transfer among cores. According to Eq. (5.3), if all the active cores in the system run at the same power consumption, one or more of the active cores will heat up the most with respect to the rest of the cores. For the same value of P_{equal}, different $\mathbf{Q}$ mappings result in different maximum temperatures. Dually, for the same maximum temperature among all cores, different mappings will result in different power consumption values to produce such a temperature. This duality is what allows us to compute $P_{\text{TSP}}(\mathbf{Q})$ through Eq. (5.5) for a given $\mathbf{Q}$.

Generally, as shown in Fig. 5.8, for the same maximum temperature values with $p_i^{\text{blocks}} = 0$ for all i, activating cores together in a corner of the chip results in lower P_{equal} values, compared to dispersing them throughout the chip. *This happens because* active *cores have higher chances of transferring heat to* inactive *cores when they are dispersed.* The worst-case mappings for TSP are those that produce the lowest power constraints, while no block in the floorplan exceeds (in the steady-state) the threshold temperature that triggers DTM, i.e., $T_i \leq T_{\text{DTM}}$ for all $i \in \mathbf{L}$.

By computing TSP for such cases, system designers can abstract themselves from core mapping decisions. This happens because such worst-case core mappings are the most pessimistic cases. When having m active cores, executing cores at power consumptions below $P_{\text{TSP}}^{\text{worst}}(m)$ will result in maximum temperatures (in the steady-state), among all blocks in the floorplan, below the threshold level that triggers DTM, for any possible mapping of m active cores.

For example, considering a manycore system of 16 cores with the same settings as in the motivational example from section "Introduction", Fig. 5.10 shows TSP for the worst-case core mappings when having and ambient temperature of 45 °C and $m = 1, 2, \ldots, 16$ active cores. The figure shows how TSP is a per-core power budget that results in a decreasing function with respect to the number of active cores. For presentation purposes, Fig. 5.10 also shows the resulting power consumptions at a chip level when all active cores consume the entire power budget, by multiplying the TSP values with the number of active cores for each case, resulting in a non-decreasing function with respect to the number of active cores.

In order to reinforce the concept that computing TSP for the worst-case mappings with m active cores results in a power constraint that is safe for *any* possible core mapping with m simultaneously active cores, considering a manycore system of 16

Active Cores	1	2	3	4	5	6	7	8	9	10	11	12	13	14	15	16
TSP worst cases (per-core) [W]	20.79	18.08	16.00	14.67	13.64	12.74	11.97	11.27	10.67	10.17	9.72	9.33	8.96	8.63	8.33	8.06

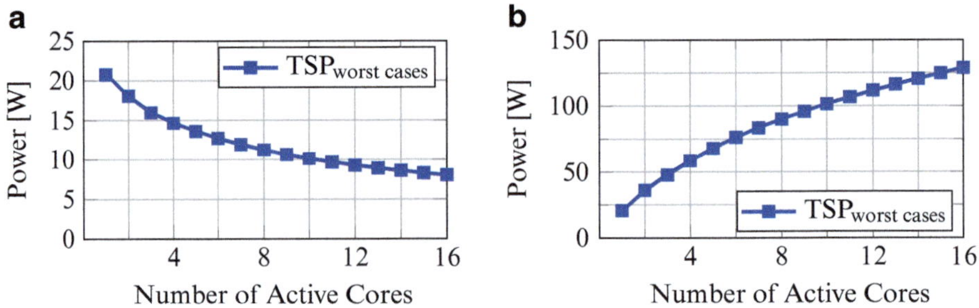

Fig. 5.10 Example of TSP results for the worst-case core mappings. (**a**) Per-core budget. (**b**) Estimated per-chip budget

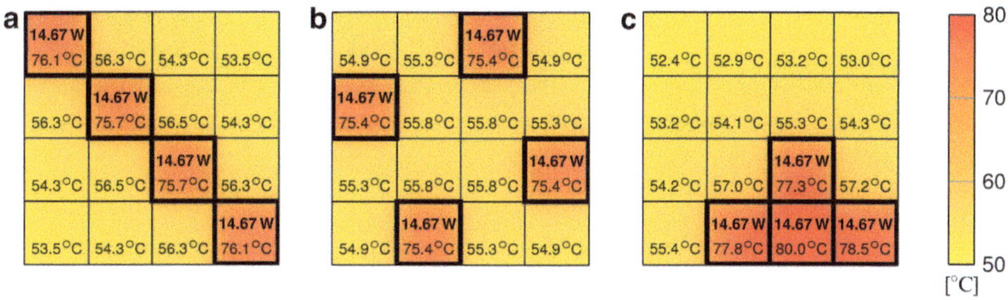

Fig. 5.11 Example of different core mappings executing under TSP for the worst-case mappings with 4 active cores. *Top numbers* are the power consumptions of each active core (*boxed in black*). *Bottom numbers* are the temperatures in the center of each core. Detailed temperatures are shown according to the color bar. (**a**) Highest temperature: 76.1 °C. (**b**) Highest temperature: 75.4 °C. (**c**) Highest temperature: 80.0 °C

cores with the same settings as in the motivational example from section "Introduction", Fig. 5.11 shows an example for different core mappings with 4 active cores executing under the computed worst-case TSP of 14.67 W.

One approach for computing TSP for the worst-case mappings would consist on first finding one of such worst-case mappings and then computing the TSP value for the mapping through Eq. (5.5). Nevertheless, it would be more efficient if we can find a method to directly compute $P_{\text{TSP}}^{\text{worst}}(m)$ without troubling on finding a worst-case mapping first.

For a given node i, we know from Eq. (5.3) that there is one or more mappings $\mathbf{Q}$ that result in the maximum T_i for a given P_{equal}. Specifically, we can distinguish which part of Eq. (5.3) is constant for a given node i and which part depends on the mapping as

$$T_i = \underbrace{P_{\text{equal}} \cdot \sum_{j=1}^{M} {b^{-1}}_{i,k_j} \cdot q_j + P_{\text{inact}}^{\text{core}} \cdot \sum_{j=1}^{M} {b^{-1}}_{i,k_j} \left(1 - q_j\right)}_{\textbf{Depends on the mapping}} + \underbrace{\sum_{j=1}^{N} {b^{-1}}_{i,j} \left(p_j^{\text{blocks}} + T_{\text{amb}} \cdot g_j\right)}_{\textbf{Constant for node i}} .$$

Namely, from Eq. (5.3), given that $\mathbf{B}^{-1}$, $\mathbf{G}$, $\mathbf{P}^{\text{blocks}}$, and T_{amb} are constant, for a given i and P_{equal}, the value of T_i is maximized when $P_{\text{equal}} \cdot \sum_{j=1}^{M} {b^{-1}}_{i,k_j} \cdot q_j + P_{\text{inact}}^{\text{core}} \cdot \sum_{j=1}^{M} {b^{-1}}_{i,k_j} \left(1 - q_j\right)$ is maximized. Furthermore, assuming that any active core consumes more power than an inactive core, even if the active cores execute at minimum frequency, it holds that $P_{\text{equal}} \geq P_{\text{min}}^{\text{core}} \geq P_{\text{inact}}^{\text{core}}$. Since q_j and $\left(1 - q_j\right)$ are mutually exclusive, for a given i and P_{equal}, the value of T_i is maximized when $\sum_{j=1}^{M} {b^{-1}}_{i,k_j} \cdot q_j$ is maximized, and this happens when mapping $\mathbf{Q}$ activates the m cores with the highest ${b^{-1}}_{i,k_j}$, for row i.

Therefore, if all active cores consume equal power P_{equal}, the m highest values for $P_{\text{equal}} \cdot b^{-1}{}_{i,j}$ such that $j \in \mathbf{K}$, correspond to the amount of heat contributed to temperature T_i by the m cores that contribute more heat into T_i.

At this point, we define auxiliary matrix $\mathbf{H} = \left[h_{i,j}\right]_{Z\times M}$ which will be later used to compute the maximum amount of heat that *any* m cores can contribute to the temperature on node i. Matrix $\mathbf{H}$ is built by making a partial copy of matrix $\mathbf{B}^{-1}$, such that $h_{i,j} = b^{-1}{}_{\ell_i,k_j}$ for all $i = 1,2,\ldots,Z$ and for all $j = 1,2,\ldots,M$, and then reordering each row of $\mathbf{H}$ in a decreasing manner. For a given chip and RC thermal network, matrix $\mathbf{H}$ needs to be build and ordered only one time, which has a time complexity $O\left(ZM \log M\right)$. The pseudo-code to build matrix $\mathbf{H}$ is presented in Algorithm 1.

Algorithm 1: Computation of matrix $\mathbf{H}$

Input: Matrix $\mathbf{B}^{-1} = \left[b_{i,j}\right]_{N\times N}$, set $\mathbf{L}$, and set $\mathbf{K}$;
Output: Auxiliary matrix $\mathbf{H} = \left[h_{i,j}\right]_{Z\times M}$;

for $i = 1$ **to** Z **do**
  for $j = 1$ **to** M **do**
    $h_{i,j} \leftarrow b^{-1}{}_{\ell_i,k_j}$;
  end for
  Re-order row i of matrix $\mathbf{H}$ in a decreasing manner;
end for
return Matrix $\mathbf{H}$;

Based on the definition of auxiliary matrix $\mathbf{H}$, the first m elements in row i of matrix $\mathbf{H}$ correspond to the m highest $b^{-1}{}_{\ell_i,j}$ values, for node ℓ_i, such that $j \in \mathbf{K}$. Therefore, if all active cores consume equal power P_{equal}, multiplying the power consumption on each core with the summation of the first m elements in row i of auxiliary matrix $\mathbf{H}$, that is, computing $P_{\text{equal}} \cdot \sum_{j=1}^{m} h_{i,j}$, results in the maximum amount of heat that any m cores can contribute to the steady-state temperature on node ℓ_i, that is, T_{ℓ_i}.

In order to derive $P_{\text{TSP}}^{\text{worst}}(m)$, we first start by rewriting Eq. (5.3) considering matrix $\mathbf{H}$ as

$$T_{\ell_i} \leq P_{\text{equal}} \cdot \sum_{j=1}^{m} h_{i,j} + P_{\text{inact}}^{\text{core}} \cdot \sum_{j=m+1}^{M} h_{i,j} + \sum_{j=1}^{N} b^{-1}{}_{\ell_i,j} \left(p_j^{\text{blocks}} + T_{\text{amb}} \cdot g_j\right).$$

Similar to section "TSP for a Given Core Mapping (Homogeneous Systems)", by setting T_{ℓ_i} to T_{DTM} for a given i, we can compute what value of P_{equal} would make T_{ℓ_i} reach T_{DTM} as

$$P_{\text{equal}} = \frac{T_{\text{DTM}} - P^{\text{core}}_{\text{inact}} \cdot \sum_{j=m+1}^{M} h_{i,j} - \sum_{j=1}^{N} b^{-1}{}_{\ell i,j} \left(p^{\text{blocks}}_j + T_{\text{amb}} \cdot g_j\right)}{\sum_{j=1}^{m} h_{i,j}}.$$

The most pessimistic value of P_{equal} is a safe power constraint for any m active cores. Therefore, the resulting $P^{\text{worst}}_{\text{TSP}}(m)$ for the given $\mathbf{P}^{\text{blocks}}$, T_{amb}, T_{DTM}, $P^{\text{core}}_{\text{inact}}$, and floorplan, can be computed by finding the minimum P_{equal} for all blocks in the floorplan, that is, for every $i = 1, 2, \ldots, Z$. The computation of $P^{\text{worst}}_{\text{TSP}}(m)$ is presented in Eq. (5.6), and the value of i that results in $P^{\text{worst}}_{\text{TSP}}(m)$ corresponds to the block with the highest temperature in the worst-case mapping.

$$P^{\text{worst}}_{\text{TSP}}(m) = \min_{1 \le i \le Z} \left\{ \frac{T_{\text{DTM}} - P^{\text{core}}_{\text{inact}} \cdot \sum_{j=m+1}^{M} h_{i,j} - \sum_{j=1}^{N} b^{-1}{}_{\ell i,j} \left(p^{\text{blocks}}_j + T_{\text{amb}} \cdot g_j\right)}{\sum_{j=1}^{m} h_{i,j}} \right\} \tag{5.6}$$

Given that matrix $\mathbf{H}$ only needs to be build once for a given chip by using Algorithm 1, the total time complexity for computing $P^{\text{worst}}_{\text{TSP}}(m)$ for a given m is $O(ZN)$, and for computing $P^{\text{worst}}_{\text{TSP}}(m)$ for all $m = 1, 2, \ldots, M$ is $O(MZN)$.

Similar to section "TSP for a Given Core Mapping (Homogeneous Systems)", a brief example of how TSP was computed through Eq. (5.6) for the worst-case mappings with 4 active cores can be seen in Fig. 5.12, considering a manycore system of 16 cores with the same settings as in the motivational example from section "Introduction". Namely, Fig. 5.12a shows what power consumption value should be consumed by the 4 cores that contribute the most heat to T_1, such that T_1 reaches 80 °C, momentarily ignoring the temperature on other blocks. The same thing is shown for Fig. 5.12b and c when focusing on T_7 and T_{15}, respectively (the computation of P_{equal} for the other blocks is not shown, but the same principle applies). Is important to note that for each T_i, there is a different set of m active cores that contributes more heat into temperature T_i. After computing P_{equal} for

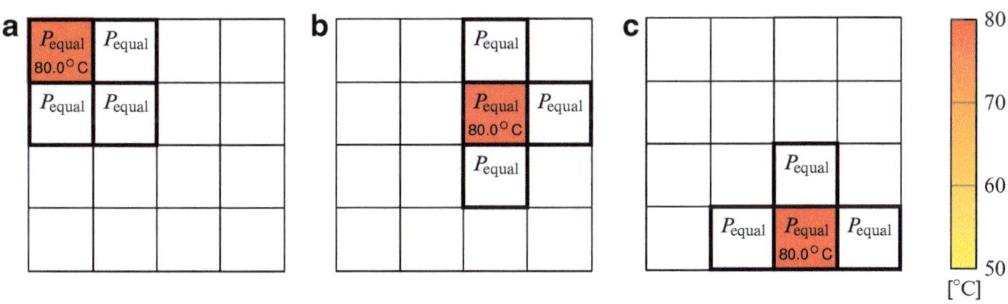

Fig. 5.12 TSP computation example for the worst-case mappings with 4 active cores. (**a**) For T_1, $P_{\text{equal}} = 14.77$ W. (**b**) For T_7, $P_{\text{equal}} = 15.07$ W. (**c**) For T_{15}, $P_{\text{equal}} = 14.67$ W

all T_i, particularly $T_1, T_2, \ldots, T_{16}$ for this example, $P_{\text{TSP}}^{\text{worst}}(m)$ is set to the smallest value of P_{equal}, which was the case when computing T_{15}. Moreover, since the final value of $P_{\text{TSP}}^{\text{worst}}(m)$ is found when computing T_{15}, it holds that this block will have the highest temperature if all cores consume equal power and under the worst-case mapping, shown in Fig. 5.5a. For symmetrical chips, there is clearly more than one worst-case mapping, but computing TSP for one such mapping is sufficient to derive safe power constraints for *any* possible mapping.

Thermal Safe Power (TSP) for Heterogeneous Systems

Given that in heterogeneous systems cores have different areas and consume different amounts of power, an efficient power budgeting technique would not use the same power constraint for different types of cores. The two algorithms in section "Thermal Safe Power (TSP) for Homogeneous Systems" provide the foundations of TSP for homogeneous cores. In this section, we briefly explain how to extend these concepts for the general case of heterogeneous systems.

TSP for a Given Core Mapping (Heterogeneous Systems)

In this subsection we revise Eq. (5.5) from section "TSP for a Given Core Mapping (Homogeneous Systems)" in order to consider heterogeneous systems. It is important to note that the temperature on a chip is directly related to the *power density*, and only indirectly related to the power consumption. That is, for two cores with different areas, if both cores consume the same power, the core with smaller area will have a higher temperature than the other core. Hence, we should handle core heterogeneity by focusing on power density instead of power consumption, thus deriving a *power density constraint*. This could be potentially done by dividing the cores of a heterogeneous floorplan into smaller homogeneous sub-blocks of a unique size, and then computing TSP through Eq. (5.5) for these smaller sub-blocks. However, such an approach would not be very efficient, specially considering the required size of the sub-blocks (and corresponding RC thermal network) to perfectly fit in all types of cores. A more efficient method is therefore to directly compute the power density constraints.

Therefore, we derive a uniform *power density* constraint for each active core in the specified mapping (independent of the type of core) that results in a maximum temperature (in the steady-state) among all cores that does not exceed T_{DTM}. We define such a power density constraint as $P_{\text{TSP}}^{\rho}(\mathbf{Q})$. Furthermore, the area of core j is defined as $\text{area}_j^{\text{core}}$, for $j = 1, 2, \ldots, M$. In order to compute the corresponding TSP value for core j, we simply multiply the power density constraint $P_{\text{TSP}}^{\rho}(\mathbf{Q})$ with the area of core j. Note that for heterogeneous systems, the power consumption of inactive cores varies among different types of cores. Therefore, we define the power consumption of core j when the core is inactive as $p^{\text{core}}_{\text{inact}j}$, for $j = 1, 2, \ldots, M$.

Considering the above, Eq. (5.2) can be rewritten as shown in Eq. (5.7).

$$T_i = P^{\rho}_{\text{equal}} \sum_{j=1}^{M} b^{-1}{}_{i,k_j} \cdot \text{area}^{\text{core}}_j \cdot q_j + \sum_{j=1}^{M} b^{-1}{}_{i,k_j} \cdot p^{\text{core}}_{\text{inact}j} (1 - q_j) + \sum_{j=1}^{N} b^{-1}{}_{i,j} \left(p^{\text{blocks}}_j + T_{\text{amb}} \cdot g_j \right) \tag{5.7}$$

From this expression, by following a similar procedure as in section "TSP for a Given Core Mapping (Homogeneous Systems)", we can compute $P^{\rho}_{\text{TSP}}(\mathbf{Q})$ as presented in Eq. (5.8), with a total time complexity of $O(ZN)$. The value of i that results in $P^{\rho}_{\text{TSP}}(\mathbf{Q})$ corresponds to the block with the highest temperature for such a case.

$$P^{\rho}_{\text{TSP}}(\mathbf{Q}) = \min_{\forall i \in \mathbf{L}} \left\{ \frac{T_{\text{DTM}} - \sum_{j=1}^{M} b^{-1}{}_{i,k_j} \cdot p^{\text{core}}_{\text{inact}j} (1 - q_j) - \sum_{j=1}^{N} b^{-1}{}_{i,j} \left(p^{\text{blocks}}_j + T_{\text{amb}} \cdot g_j \right)}{\sum_{j=1}^{M} b^{-1}{}_{i,k_j} \cdot \text{area}^{\text{core}}_j \cdot q_j} \right\} \tag{5.8}$$

TSP for Worst-Case Mappings (Heterogeneous Systems)

This subsection revises Eq. (5.6) presented in section "TSP for Worst-Case Mappings (Homogeneous Systems)" in order to consider heterogeneous systems. Similar to section "TSP for a Given Core Mapping (Heterogeneous Systems)", we focus on deriving a safe *power density* constraint, rather than focus on power consumption. Moreover, in section "TSP for Worst-Case Mappings (Homogeneous Systems)" the value of the power constraint for the worst-case core mappings depends on the number of active cores. Nevertheless, for the heterogeneous case, since different core types have different areas, the value of the power density constraint will depend on the *number of active cores for each type of core*. That is, assume that there are W core types (arbitrarily ordered) and that for every type of core w there are a total of M_w cores, such that $M = \sum_{w=1}^{W} M_w$. We assume that all cores of the same type have equal area and inactive power, and we denote the area and inactive power of core type w as $\text{area}^{\text{type}}_w$ and $p^{\text{type}}_{\text{inact}w}$, respectively. We define sets $\mathbf{K}^w = \left\{k^w_1, k^w_2, \ldots, k^w_{M_w}\right\}$ for all core types $w = 1, 2, \ldots, W$, such that $\mathbf{K}^w$ contains the indexes of the thermal nodes that correspond to cores of type w. Furthermore, set $\mathbf{m} = \{m_1, m_2, \ldots, m_W\}$ represents the number of active cores for core types $\{1, 2, \ldots, W\}$. Hence, in this section we derive a safe *power density* constraint for the worst-case core mappings with $\mathbf{m} = \{m_1, m_2, \ldots, m_W\}$ active cores, and we define such a power density constraint as $P^{\rho\text{worst}}_{\text{TSP}}(\mathbf{m})$. For example, if we have a system with three types of cores and we would like to activate 4 cores of type 1 and 7 cores of type 3, the safe power density constraint for such a case is defined as $P^{\rho\text{worst}}_{\text{TSP}}(\{4, 0, 7\})$. We then simply multiply the power density constraint by the area of each core, thus obtaining a safe power constraint value for each type of core for *any* possible core mapping with $\mathbf{m} = \{m_1, m_2, \ldots, m_W\}$ simultaneously active cores, that results in a maximum temperature (in the steady-state) among all cores which does not exceed T_{DTM}.

Similar to section "TSP for Worst-Case Mappings (Homogeneous Systems)", we define auxiliary matrix $\mathbf{H}^\rho = \left[h^\rho_{w,i,j}\right]_{W \times Z \times M_w}$ which is used to compute the maximum amount of heat that any m_w cores of type w can contribute to the temperature on node i, for all core types $w = 1, 2, \ldots, W$. Matrix $\mathbf{H}^\rho$ is built by making a partial copy of matrix $\mathbf{B}^{-1}$, considering the type and the area of the cores, such that $h^\rho_{w,i,j} = b^{-1}{}_{\ell_i,k^w_j} \cdot \mathrm{area}^{\mathrm{type}}_w$ for all $w = 1, 2, \ldots, W$, for all $i = 1, 2, \ldots, Z$, and for all $j = 1, 2, \ldots, M_w$, and then, for every w and every i, reordering each row of $\mathbf{H}^\rho$ in a decreasing manner. For a given chip and RC thermal network, matrix $\mathbf{H}^\rho$ needs to be built and ordered only one time, which requires a time complexity of $O(ZM_w \log M_w)$ for every core type w. The pseudo-code to build matrix $\mathbf{H}^\rho$ is presented in Algorithm 2.

Algorithm 2: Computation of matrix $\mathbf{H}^\rho$

Input: Matrix $\mathbf{B}^{-1} = \left[b_{i,j}\right]_{N \times N}$, set $\mathbf{L}$, number of cores M_w and sets $\mathbf{K}^w$ for $w = 1, 2, \ldots, W$;
Output: Auxiliary matrix $\mathbf{H}^\rho = \left[h^\rho_{w,i,j}\right]_{W \times Z \times M_w}$;

```
for w = 1 to W do
    for i = 1 to Z do
        for j = 1 to M_w do
            h^ρ_{w,i,j} ← b^-1_{ℓ_i,k^w_j} · area^type_w;
        end for
        For these w and i, re-order H^ρ (with respect to j) decreasingly;
    end for
end for
return Matrix H^ρ;
```

In order to derive $P^{\rho\mathrm{worst}}_{\mathrm{TSP}}(\mathbf{m})$, we first start by rewriting Eq. (5.2) considering matrix $\mathbf{H}^\rho$ as

$$T_{\ell_i} = P^\rho_{\mathrm{equal}} \sum_{w=1}^{W} \sum_{j=1}^{m_w} h^\rho_{w,i,j} + \sum_{w=1}^{W} \frac{p^{\mathrm{type}}_{\mathrm{inact}w}}{\mathrm{area}^{\mathrm{type}}_w} \cdot \sum_{j=m_w+1}^{M_w} h^\rho_{w,i,j} + \sum_{j=1}^{N} b^{-1}{}_{\ell_i,j} \left(p^{\mathrm{blocks}}_j + T_{\mathrm{amb}} \cdot g_j\right).$$

From this expression, by following a similar procedure as in section "TSP for Worst-Case Mappings (Homogeneous Systems)", we can compute $P^{\rho\mathrm{worst}}_{\mathrm{TSP}}(\mathbf{m})$ as presented in Eq. (5.9).

$$P_{\text{TSP}}^{\rho\text{worst}}(\mathbf{m}) = \min_{1\leq i\leq Z}\left\{\frac{T_{\text{DTM}} - \sum_{w=1}^{W}\frac{p_{\text{inact}w}^{\text{type}}}{\text{area}_w^{\text{type}}}\sum_{j=m_w+1}^{M_w} h_{w,i,j}^{\rho}}{\sum_{w=1}^{W}\sum_{j=1}^{m_w} h_{w,i,j}^{\rho}} + \frac{-\sum_{j=1}^{N} b^{-1}{}_{\ell_i,j}\left(p_j^{\text{blocks}} + T_{\text{amb}}\cdot g_j\right)}{\sum_{w=1}^{W}\sum_{j=1}^{m_w} h_{w,i,j}^{\rho}}\right\} \tag{5.9}$$

Given that matrix $\mathbf{H}^{\rho}$ only needs to be built once for a given chip using Algorithm 2, the total time complexity for computing $P_{\text{TSP}}^{\rho\text{worst}}(\mathbf{m})$ for a given $\mathbf{m} = \{m_1, m_2, \ldots, m_W\}$ is $O(ZN)$, and for computing $P_{\text{TSP}}^{\rho\text{worst}}(\mathbf{m})$ for all possible combinations of active cores of different types is $O\left(ZN\prod_{w=1}^{W} M_w\right)$.

Note that the solutions for homogeneous systems presented in section "Thermal Safe Power (TSP) for Homogeneous Systems" are a special case of the solutions presented in this section. Specifically, when the area of all cores is constant, that is, $\text{area}_j^{\text{core}} = \text{area}_{j+1}^{\text{core}}$ for $j = 1, 2, \ldots, M-1$, and when the power consumption of all inactive cores is also constant, that is, $p_{\text{inact}j}^{\text{core}} = p_{\text{inact}j+1}^{\text{core}}$ for $j = 1, 2, \ldots, M-1$, then multiplying the derived power density constraint with the area of the cores results in the same TSP values computed through the algorithms from section "Thermal Safe Power (TSP) for Homogeneous Systems".

Transient State Considerations

Due to the effect of transient temperatures, if TSP is computed for T_{DTM}, the temperatures of some cores may exceed the value of T_{DTM} during short time intervals when there are changes or increases in power that reach TSP. When this happens, DTM is triggered, resulting in some performance losses. *When using a constant power constraint, e.g., TDP, these effects due to transient temperatures are also present.*

The following example uses HotSpot [10] with its default configuration (detailed in section "Setup") to show this effect for systems constrained both by (1) TDP and (2) TSP. The same example applies for both cases. Consider a manycore system of 16 cores with the same settings as in the motivational example from section "Introduction", with $P_{\text{inact}}^{\text{core}}$ of 0 W, and (1) TDP of 90.2 W per-chip, or (2) $P_{\text{TSP}}^{\text{worst}}(4) = 14.67\text{W}$ and $P_{\text{TSP}}^{\text{worst}}(8) = 11.27\,\text{W}$, i.e., TSP of 14.67 and

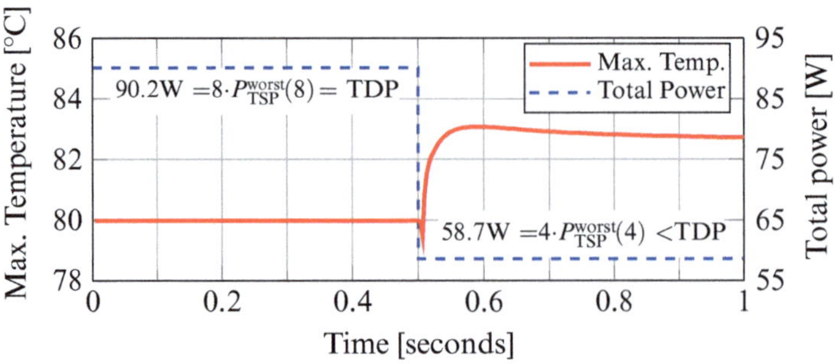

Fig. 5.13 Transient example for both TSP and TDP (the *red line* shows the maximum temperature among all cores). During $t = [0\,\text{s}, 0.5\,\text{s}]$ there are 8 active cores according to Fig. 5.5b, consuming 11.27 W each. During $t = [0.5\,\text{s}, 1\,\text{s}]$, these cores are shut-down and we activate 4 cores according to Fig. 5.5a, consuming 14.67 W each

11.27 W per-core when activating 4 and 8 cores, respectively. Figure 5.13 presents simulations in which during $t = [0\,\text{s}, 0.5\,\text{s}]$ there are 8 active cores according to Fig. 5.5b, consuming 11.27 W each (90.2 W in total). During $t = [0.5\,\text{s}, 1\,\text{s}]$, these cores are shut-down and we activate 4 cores according to Fig. 5.5a, consuming 14.67 W each (58.7 W in total). That is, if constrained by (1) TDP, the system consumes TDP during $t = [0\,\text{s}, 0.5\,\text{s}]$, and less than TDP during $t = [0.5\,\text{s}, 1\,\text{s}]$. Moreover, if constrained by (2) TSP, the system consumes the corresponding TSP according to the number of active cores all the time. Nevertheless, although for both cases the power budgets are not violated, Fig. 5.13 shows that during $t = [0.5\,\text{s}, 1\,\text{s}]$ the temperature of at least one core exceeds the 80 °C threshold for triggering DTM.

If the frequency of the changes in power that produce this effect is too high, then DTM could be triggered frequently, and such performance losses would be noticeable. For such cases, one strategy is to quantify the maximum values that these temperatures can actually reach during the transient state. The difference between such maximum transient temperatures and T_{DTM} is denoted as $\Delta T^{\max}_{\text{transient}}$, with value 3.08 °C for the example in Fig. 5.13. Therefore, by computing TSP for $T_{\text{DTM}} - \Delta T^{\max}_{\text{transient}}$, we can make sure that the transient temperatures never reach T_{DTM}. Nevertheless, depending on the floorplan and the resulting thermal capacitances, it can happen that the transient temperatures for this new TSP are too pessimistic compared to T_{DTM}. For such cases, with just a few iterations, a near optimal value for which to compute TSP can be achieved. This method should be applied offline, due to the overheads for obtaining $\Delta T^{\max}_{\text{transient}}$ for each case. *A similar method should be adopted for systems that use constant power budgets, e.g., TDP.*

Procedure Example Consider a manycore system of 16 cores with the same settings as in the motivational example from section "Introduction", and an ambient temperature of 45 °C. For the power consumptions, consider a scenario in which the power of the cores changes every 0.1 s. The changes in power are for a random number of active cores. The mapping and power values adopted in all cases are those of TSP for the worst-case, computed through Eq. (5.6), according to the

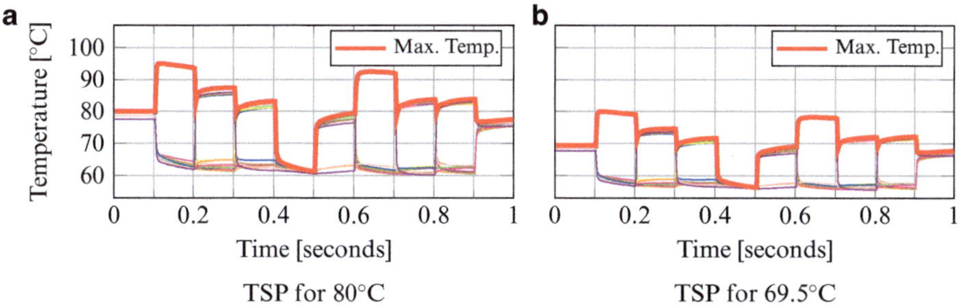

Fig. 5.14 Transient example for 16 cores. The number of active cores and their power consumption changes every 0.1 s. The mapping and power values adopted are those of worst-case TSP computed for (**a**) 80 °C and (**b**) 69.5 °C

number of active cores at each time instant. We run simulations with HotSpot [10] or MatEx [21] for such a case, and present the results in Fig. 5.14a. Figure 5.14a presents the temperature on each core with a different color, and the maximum temperature among all cores at any given time is highlighted by the bold-red curve. We can observe that for this floorplan the thermal capacitances are not negligible, which results in long transient state periods. Therefore, for such a case, TSP should be recomputed for some temperature below T_{DTM}.

Looking at Fig. 5.14a, we can quantify $\Delta T^{\max}_{\text{transient}}$, which we set to 15 °C, and then recompute TSP for 65 °C. Given that this results in a maximum transient temperature of 74 °C, which is too pessimistic, we need a higher value. Thus, we iterate computing TSP and running transient temperature simulations. After just 5 iterations, specifically, computing TSP for 80, 65, 71, 69, and 69.5 °C, we reach a near optimal value for which to compute TSP, which for this floorplan is 69.5 °C. Clearly, the new TSP values are smaller than those for a TSP computed for 80 °C.

Finally, we run the transient simulations with the same settings, but with power states according to the new TSP values. The results are presented in Fig. 5.14b, which shows that when computing TSP for 69.5 °C, the temperature is always below T_{DTM}. When computing TSP online for particular mapping scenarios, the target temperature should also be 69.5 °C, and not the original 80 °C.

If, *unlike* Fig. 5.14, the changes in power are *not* too frequent, such that DTM is triggered with low frequency during short time intervals, then computing TSP for T_{DTM} is still a better approach that results in higher total performance.

Experimental Evaluations

This section presents evaluations conducted with gem5 [2], McPAT [17], and HotSpot [10], for homogeneous systems. We compare the total system performance of six different power budget techniques: TSP for given mappings, worst-case

TSP, two constant power budgets per-chip, a constant power budget per-core, and boosting over a constant power budget per-chip, specifically, Intel's Turbo Boost [11, 25].

Setup

For our hardware platform, we consider a system very similar to Intel's Single Chip Cloud computer (SCC) [13], particularly, the homogeneous 64 cores system presented in Fig. 5.6, based on simulations conducted on gem5 [2] and McPAT [17] for *out-of-order* Alpha 21264 cores in 22 nm technology. The cores are composed by several units: an instruction fetch unit (IFU), an execution unit (EXU), a load and store unit (LSU), an out-of-order (OOO) issue/dispatch, and a private L1 cache. According to our simulations, as shown in Fig. 5.6, each core has an area of $9.6\,\mathrm{mm}^2$. There is a shared L2 cache of 2 MB and a memory controller every 4 cores, with a combined area of $4.7\,\mathrm{mm}^2$. We assume available frequencies $\{0.2, 0.4, \ldots, 4.0\}$ GHz, and the voltage settings for each frequency are taken from [9].

For such a floorplan, we consider one thermal block for each core and independent thermal blocks for the L2 caches and other hardware. We then obtain the values for $\mathbf{B}$, $\mathbf{B}^{-1}$, and $\mathbf{G}$, by using HotSpot [10] v5.02 with its default configuration in the block model mode. That is, chip thickness of 0.15 mm, silicon thermal conductivity of $100\,\frac{\mathrm{W}}{\mathrm{m \cdot K}}$, silicon specific heat of $1.75 \cdot 10^6\,\frac{\mathrm{J}}{\mathrm{m^3 \cdot K}}$, a heat sink of 6×6 cm and 6.9 mm thick, heat sink convection capacitance of $140.4\,\frac{\mathrm{J}}{\mathrm{K}}$, heat sink convection resistance of $0.1\,\frac{\mathrm{K}}{\mathrm{W}}$, heat sink and heat spreader thermal conductivity of $400\,\frac{\mathrm{W}}{\mathrm{m \cdot K}}$, heat sink and heat spreader specific heat of $3.55 \cdot 10^6\,\frac{\mathrm{J}}{\mathrm{m^3 \cdot K}}$, a heat spreader of 3×3 cm and 1 mm thick, interface material thickness of 20 um, interface material thermal conductivity of $4\,\frac{\mathrm{W}}{\mathrm{m \cdot K}}$, and interface material specific heat of $4 \cdot 10^6\,\frac{\mathrm{J}}{\mathrm{m^3 \cdot K}}$.

We consider that the ambient temperature is 45 °C, and the threshold temperature that triggers DTM, T_{DTM}, is 80 °C. For all the evaluated power constraints, except when a boosting technique is being applied, we consider a standard reactive control-based closed-loop DTM technique [11] that, when the threshold temperature is exceeded anywhere in the chip, it reduces the voltage/frequency levels of all active cores, one step at a time, by using a control period of 1 ms. When the maximum temperature in the chip is below T_{DTM}, the voltage/frequency on the cores is increased one step at a time, by using the same 1 ms control period, until all cores reach their nominal operation according to the established power constraint.

For benchmarks, we use applications from the PARSEC benchmark suite [1]. Specifically, an x264 application (H.264 video encoder), a body track application, an option pricing application with Black-Scholes Partial Differential Equation, and a pricing application of a portfolio of swaptions. All applications can run $1, 2, \ldots, 8$ parallel dependent threads. For each experiment, we conduct closed-loop evaluations involving simulations with gem5 [2], McPAT [17], and HotSpot [10].

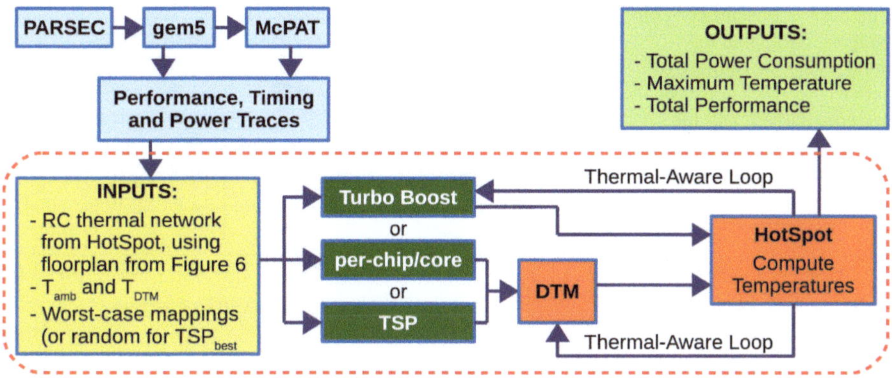

Fig. 5.15 Overview of our simulation framework

Fig. 5.16 Worst- and best-case TSP for the floorplan in Fig. 5.6, compared to a constant per-core budget, and estimations of constant per-chip budgets equally distributed among active cores

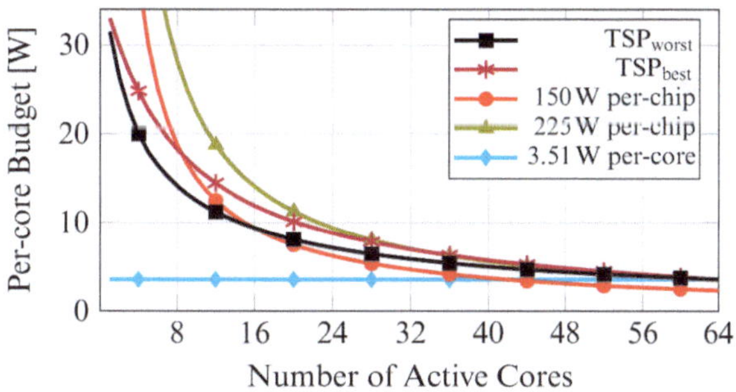

We consider that every time an application instance finishes, another instance is immediately executed. Figure 5.15 shows an overview of our simulation framework.

Power Constraints

In this subsection we compute the power constraints used in our simulations. Using Eq. (5.6), we compute TSP for the worst-case mappings, i.e., $P_{\mathrm{TSP}}^{\mathrm{worst}}(m)$ for all $m = 1, 2, \ldots, M$. Figure 5.16 presents the computed TSP values per-core as $\mathrm{TSP}_{\mathrm{worst}}$, resulting in a decreasing function with respect to the number of active cores. For presentation purposes, Fig. 5.17 shows TSP estimations at a chip level, by multiplying the $\mathrm{TSP}_{\mathrm{worst}}$ values from Fig. 5.16 with the number of active cores for each case. In Fig. 5.17, TSP results in a non-decreasing function with respect to the number of active cores. Moreover, to compare TSP for different core mappings, we consider $64 \cdot 10^5$ random mappings (10^5 for each number of active cores), compute TSP using Eq. (5.5) for every given mapping, and present the highest resulting TSP values as $\mathrm{TSP}_{\mathrm{best}}$ in Figs. 5.16 and 5.17.

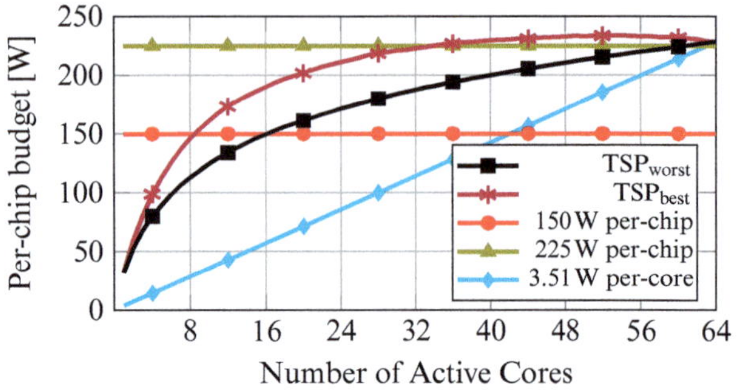

Fig. 5.17 Constant per-chip budgets, compared to multiplying the number of active cores by a constant per-core budget, and the worst- and best-case TSP for the floorplan from Fig. 5.6

For the constant per-chip power constraints, since this is a simulated platform, we cannot simply consider TDP because we have no datasheet with that information. Therefore, we consider two different per-chip power budgets, which coincide with $m \cdot P_{\text{TSP}}^{\text{worst}}(m)$ for $m = 16$, and $m = 64$. Specifically, these power budgets are 150.0 W per-chip, and 225.0 W per-chip. These constant per-chip budgets are also representative TDP values for current technologies [11], and are shown as horizontal lines in Fig. 5.17. In Fig. 5.16, we estimate the maximum power per core when these constraints are equally distributed among the active cores. For the constant per-core power constraint, we consider it to be equal to TSP when simultaneously activating all cores, i.e., $P_{\text{TSP}}^{\text{worst}}(m)$ for $m = 64$. This results in a constant power budget of 3.51 W per-core, which is represented by a horizontal line in Fig. 5.16 and by an increasing linear function in Fig. 5.17. *Note that representing TSP and the constant per-core power constraint in Fig. 5.17 does not mean that either constraint should be considered as a per-chip constraint. Both budgets should be strictly considered at a per-core level. We include them in Fig. 5.17 only to compare their resulting total system power to the per-chip power budgets. Similarly, the opposite applies when representing the constant per-chip budgets in Fig. 5.16.*

With regards to temperature, Fig. 5.18 presents simulations conducted in HotSpot that show the maximum temperature (in the steady-state) among all blocks as a function of the number of simultaneously active cores, for the discussed power budgets, in case that DTM is deactivated. As expected, when consuming TSP in all active cores, the maximum temperature among all blocks is 80 °C. Moreover, from Fig. 5.18 we can conclude that whenever TSP is greater than any constant power budget, if such a power budget is used as the power constraint, the system could actually consume more power without reaching T_{DTM}, which accounts for performance losses. Contrarily, when a constant power budget exceeds TSP, this means that if the cores consume more power than TSP, most likely DTM will be triggered almost the entire time. Note that, when activating just a few cores, the maximum temperatures do not reach the theoretical values because the power consumption on the cores is limited by their maximum frequency, and thus there is no individual core that can consume more than 20 W under the given setup.

It should be noted in reference to Fig. 5.18, that although we only plot the maximum temperature among all cores, it is not just one core that reaches such high

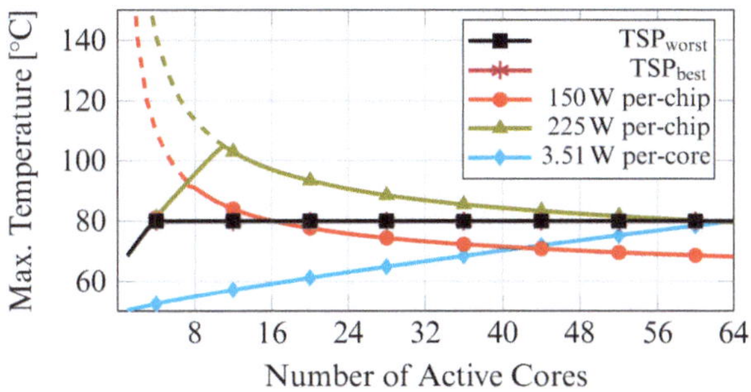

Fig. 5.18 Maximum steady-state temperatures among all blocks (DTM deactivated), when using TSP, a constant per-core budget, and equally distributed constant per-chip budgets

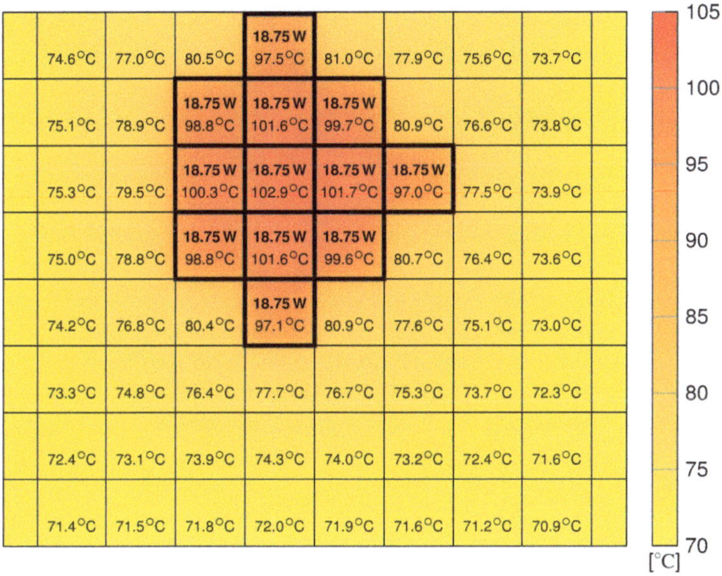

Fig. 5.19 Temperature snapshot for 12 active cores when equally distributing the 225 W per-chip power budget, with a resulting highest temperature of 102.9 °C

temperatures. For example, Fig. 5.19 shows a temperature snapshot for the worst-case mapping when activating 12 cores and equally distributing the 225 W per-chip power budget among the active cores. For such a case, the highest temperature among all cores is 102.9 °C. Nevertheless, the lowest temperature among the active cores is 97.0 °C, which is still much higher than 80.0 °C. This point is very relevant from the perspective of DTM, as otherwise it could be interpreted that DTM might only be needed in a few cores.

Execution Time of Online TSP Computation

In order to verify the applicability of TSP computations at runtime, on a desktop computer with a 64-bit quad-core Intel Sandybridge i5-2400 CPU running at 3.10 GHz, we measure the execution time required by an algorithm implementing Eq. (5.5). The algorithm is implemented in C++ as a single-threaded application. We consider several floorplans with different number of cores in each floorplan. For every floorplan, we consider 25,000 random mappings and present the maximum measured execution time for each case in Fig. 5.20, from which we can conclude that TSP is *suitable for online usage*. Furthermore, TSP can be easily implemented as a multi-threaded application, further reducing the execution time.

Dark Silicon Estimations

Through Fig. 5.21 (a summarized version of Fig. 5.16), we can easily estimate the amount of dark silicon. For example, we assume that the minimum power consumption for activating a core at its lowest speed is 4 W (*this is not a real practical setting, but we choose such a value for presentation purposes*). Then, considering the values of TSP, no more than 54 cores could be active simultaneously, which results in 15.63 % of the chip being dark at all times. If a constant per-chip power budget is used for the same example estimations, e.g., 150.0 W per-chip, then only 37 cores could be activated simultaneously, resulting in 42.19 % of the chip being dark, which is much higher than the estimations for TSP.

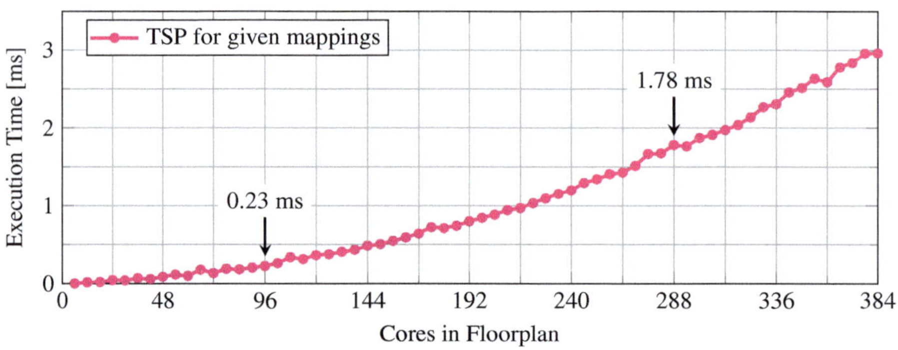

Fig. 5.20 Execution time for computing TSP for a given mapping, considering several floorplans with different number of cores in each floorplan

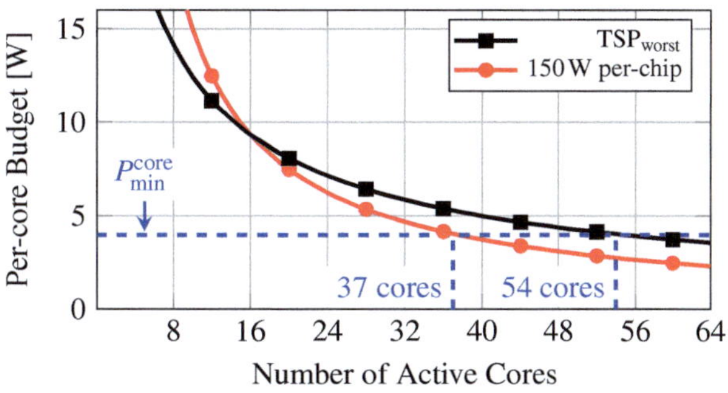

Fig. 5.21 Example of dark silicon estimations for using TSP for the worst-case mappings and a constant per-chip power budget

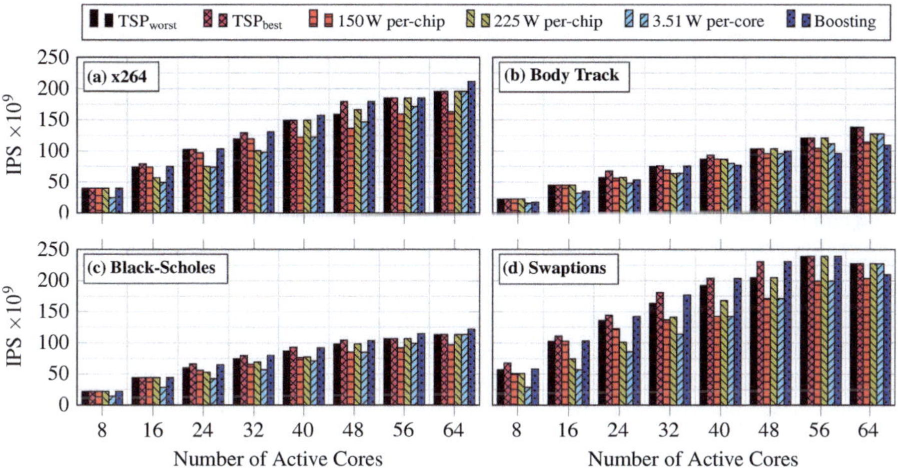

Fig. 5.22 Evaluation results: average total system performance for *homogeneous* systems when using different power budgets

Performance Simulations

Figure 5.22 presents the average total IPS count, for considering different numbers of active cores and the different power budgets described in section "Power Constraints". We assume that at nominal operation every active core runs at the highest possible frequency and performance such that the power budget under consideration is not exceeded.

Furthermore, in order to also compare with a boosting technique, in Fig. 5.22 we also consider Intel's Turbo Boost [11, 25] over one of the constant per-chip power budgets described in section "Power Constraints", specifically, 150 W. When the temperature on all blocks is below T_{DTM} the cores boost their voltage and frequency levels in single steps (within a control period of 1 ms). Similarly, if some temperature exceeds T_{DTM} the cores reduce their voltage and frequency levels in single steps until the thermal constraints are satisfied.

In Fig. 5.22, we can see that when we activate all 64 cores, the per-core budget and the 225 W per-chip budget achieve the same performance as TSP_{worst}. This was an expected result, as in both cases these power budgets coincide with TSP. However, there are many cases in which the fixed power budgets are pessimistic and thus the chip remains underutilized, as explained in section "Power Constraints" (Figs. 5.16, 5.17, and 5.18). Specifically, this happens with all other cases for the per-core power budget, and with the 150 W per-chip budget when activating more than 16 cores. On the other hand, there are many other cases in which the fixed power budgets constantly trigger DTM, specifically, when activating less than 16 cores with the 150 W per-chip budget, and when activating less than 64 cores with the 225 W per-chip budget. Here, since DTM is constantly triggered, thermal violations are avoided at the cost of reducing the frequency of the cores. Therefore, although the resulting performance is in just a few cases higher than that of TSP, the achieved performance is much worse than expected by the task partitioning and mapping algorithms. Furthermore, for such cases there is no simple way of predicting such performance losses, making it almost impossible for the system to provide performance and timing guarantees. Contrarily, TSP never triggers DTM and can thus achieve the expected performance of the task partitioning and mapping decisions. The *average percentage increase in performance* (among all number of active cores and all applications) for using TSP_{worst} results in a 12 % higher average total IPS compared to all constant power budgets.

Intel's Turbo Boost is a simple but very efficient online technique, achieving a higher performance than TSP_{worst} for several cases, such that TSP_{worst} results in the same average total IPS than Turbo Boost (among all number of active cores and all applications). However, similar to having frequent triggers of DTM, there is no guarantee and no simple offline performance predictions, and thus Turbo Boost cannot guide the task partitioning and mapping algorithms to make intelligent decisions. Contrarily, this can be done with TSP, helping to simplify task partitioning and mapping algorithms to achieve a high *predictable* performance without thermal violations. Intel's Turbo Boost could be easily combined with TSP to further exploit the thermal headroom *after* the applications are mapped to the system with the help of TSP.

Conclusions

Using a *single* and *constant* power constraint, for example, TDP, is a pessimistic approach for homogeneous and heterogeneous manycore systems. Thus, this chapter presents a power budget concept called Thermal Safe Power (TSP), which results in a high total system performance, while the maximum temperature among all cores remains below the threshold level that triggers DTM. TSP is a fundamental new step and advancement towards dealing with the dark silicon problem as it alleviates the pessimistic bounds of TDP and enables system designers and architects to explore new avenues for performance improvements in the dark silicon era.

For a specific floorplan and ambient temperature, TSP can be computed offline to obtain safe power and power density constraints for the worst cases, allowing the system designers to abstract from mapping decisions. Moreover, TSP can also be computed online, for a particular mapping of active cores and ambient temperature. The simulations show the validity of our arguments, comparing the total performance of using TSP, several constant power constraints, and a boosting technique. TSP can also be used to estimate the amount of dark silicon, which results in less pessimistic estimations than those using constant power budgets.

Acknowledgements This work was partly supported by the German Research Foundation (DFG) as part of the Transregional Collaborative Research Centre *Invasive Computing* [SFB/TR 89], by Baden Württemberg MWK Juniorprofessoren-Programms, and by a grant from the Research Grants Council of the Hong Kong Special Administrative Region, China [Project CityU 117913].

References

1. C. Bienia, S. Kumar, J.P. Singh, K. Li, The PARSEC benchmark suite: characterization and architectural implications, in *PACT* (2008), pp. 72–81
2. N. Binkert, B. Beckmann, G. Black, S.K. Reinhardt, A. Saidi, A. Basu, J. Hestness, D.R. Hower, T. Krishna, S. Sardashti, R. Sen, K. Sewell, M. Shoaib, N. Vaish, M.D. Hill, D.A. Wood, The gem5 simulator. ACM SIGARCH Comput. Archit. News **39**(2), 1–7 (2011)
3. H. Bokhari, H. Javaid, M. Shafique, J. Henkel, S. Parameswaran, darkNoC: designing energy-efficient network-on-chip with multi-Vt cells for dark silicon, in *DAC* (2014), pp. 161:1–161:6
4. J. Casazza, First the tick, now the tock: Intel microarchitecture (Nehalem). White paper, Intel Corporation, 2009
5. J. Cebrian, L. Natvig, J. Meyer, Improving energy efficiency through parallelization and vectorization on intel core i5 and i7 processors, in *SC Companion* (2012), pp. 675–684
6. J. Charles, P. Jassi, N.S. Ananth, A. Sadat, A. Fedorova, Evaluation of the intel core i7 turbo boost feature, in *IISWC* (2009), pp. 188–197
7. T. Ebi, D. Kramer, W. Karl, J. Henkel, Economic learning for thermal-aware power budgeting in many-core architectures, in *CODES+ISSS* (2011), pp. 189–196
8. H. Esmaeilzadeh, E. Blem, R. St. Amant, K. Sankaralingam, D. Burger, Dark silicon and the end of multicore scaling, in *ISCA* (2011), pp. 365–376
9. A. Grenat, S. Pant, R. Rachala, S. Naffziger, 5.6 adaptive clocking system for improved power efficiency in a 28 nm x86-64 microprocessor, in *ISSCC* (2014), pp. 106–107
10. W. Huang, S. Ghosh, S. Velusamy, K. Sankaranarayanan, K. Skadron, M.R. Stan, HotSpot: a compact thermal modeling methodology for early-stage VLSI design. IEEE Trans. VLSI Syst. **14**(5), 501–513 (2006)
11. Intel Corporation, Dual-core intel xeon processor 5100 series datasheet, revision 003, August 2007.
12. Intel Corporation, Intel turbo boost technology in intel CoreTM microarchitecture (nehalem) based processors. White paper, November 2008
13. Intel Corporation, Single-chip cloud computer (SCC) (2009). www.intel.com/content/www/us/en/research/intel-labs-single-chip-cloud-overview-paper.html
14. F. Kriebel, S. Rehman, D. Sun, M. Shafique, J. Henkel, ASER: adaptive soft error resilience for reliability-heterogeneous processors in the dark silicon era, in *DAC* (2014), pp. 12:1–12:6
15. E. Kultursay, K. Swaminathan, V. Saripalli, V. Narayanan, M.T. Kandemir, S. Datta, Performance enhancement under power constraints using heterogeneous CMOS-TFET multicores, in *CODES+ISSS* (2012), pp. 245–254

16. J. Lee, N.S. Kim, Optimizing throughput of power- and thermal-constrained multicore processors using DVFS and per-core power-gating, in *DAC* (2009), pp. 47–50
17. S. Li, J.-H. Ahn, R. Strong, J. Brockman, D. Tullsen, N. Jouppi, McPAT: an integrated power, area, and timing modeling framework for multicore and manycore architectures, in *MICRO* (2009), pp. 469–480
18. T. Muthukaruppan, M. Pricopi, V. Venkataramani, T. Mitra, S. Vishin, Hierarchical power management for asymmetric multi-core in dark silicon era, in *DAC* (2013), pp. 174:1–174:9
19. S. Nussbaum, AMD trinity APU, in *Hot Chips* (2012)
20. S. Pagani, H. Khdr, W. Munawar, J.-J. Chen, M. Shafique, M. Li, J. Henkel, TSP: thermal safe power - efficient power budgeting for many-core systems in dark silicon, in *The International Conference on Hardware/Software Codesign and System Synthesis (CODES+ISSS)*, October 2014, pp. 10:1–10:10
21. S. Pagani, J.-J. Chen, M. Shafique, J. Henkel, MatEx: efficient transient and peak temperature computation for compact thermal models, in *The 18th Design, Automation and Test in Europe (DATE)*, March 2015, pp. 1515–1520
22. S. Pagani, J.-J. Chen, M. Shafique, J. Henkel, Thermal-aware power budgeting for dark silicon chips, in *the 2nd Workshop on Low-Power Dependable Computing (LPDC), Part of the 6th International Green and Sustainable Computing Conference (IGSC)*, December 2015
23. B. Raghunathan, S. Garg, Job arrival rate aware scheduling for asymmetric multi-core servers in the dark silicon era, in *CODES+ISSS* (2014)
24. B. Raghunathan, Y. Turakhia, S. Garg, D. Marculescu, Cherry-picking: Exploiting process variations in dark-silicon homogeneous chip multi-processors, in *DATE* (2013), pp. 39–44
25. E. Rotem, A. Naveh, D. Rajwan, A. Ananthakrishnan, E. Weissmann, Power-management architecture of the intel microarchitecture code-named sandy bridge. Micro, IEEE **32**(2), 20–27 (2012)
26. M. Shafique, S. Garg, J. Henkel, D. Marculescu, The EDA challenges in the dark silicon era: temperature, reliability, and variability perspectives, in *DAC* (2014), pp. 185:1–185:6
27. C. Tan, T. Muthukaruppan, T. Mitra, L. Ju, Approximation-aware scheduling on heterogeneous multi-core architectures, in *ASP-DAC*, January 2015, pp. 618–623

Chapter 6
Power Management of Asymmetric Multi-Cores in the Dark Silicon Era

Tulika Mitra, Thannirmalai Somu Muthukaruppan, Anuj Pathania, Mihai Pricopi, Vanchinathan Venkataramani, and Sanjay Vishin

Introduction

Computing in the dark silicon era [19] is fundamentally constrained by the available chip power budget known as the *Thermal Design Power (TDP)*. The TDP, expressed in watt, is the sustained maximum power that a chip is allowed to dissipate without overheating, while running bona-fide tasks. The cooling solution for a chip is designed to comfortably handle the TDP so as to prevent thermal emergencies. Thus, most tasks will be able to execute without pushing the thermal envelope. The TDP can be quite high for processors in desktop and laptop platforms due to the presence of active cooling mechanisms. For example, the latest *Intel* desktop processor *Core i5-6600* (*Skylake* micro-architecture, quad-core at 14 nm) has TDP of 65 W, while the low-power version of the same processor operates under 25 W TDP primarily due to the reduced clock frequency from 3.3 to 1.9 GHz. In contrast, the smart-phones and tablets can only afford passive cooling and hence are limited to deploying mobile processors with severely restrictive TDP budget of 4–5 W. This has resulted in *Intel* providing a different class of comparatively

T. Mitra (✉) • V. Venkataramani
School of Computing, National University of Singapore, Singapore, Singapore
e-mail: tulika@comp.nus.edu.sg

T.S. Muthukaruppan
Intel Microarchitecture Research Lab, Bangalore, India

A. Pathania
Chair of Embedded System, Karlsruhe Institute of Technology, Karlsruhe, Germany

M. Pricopi
ARM Cambridge, Cambridge, UK

S. Vishin
Cambridge Silicon Radio, San Jose, USA

A.M. Rahmani et al. (eds.), *The Dark Side of Silicon*,
DOI 10.1007/978-3-319-31596-6_6

under-performing *Core M* processors based on the same *Skylake* micro-architecture for the mobile market. A mobile processor is a System-on-a-Chip (SoC), which includes a Central Processing Unit (CPU), a Graphic Processing Unit (GPU), and a Memory Management Unit (MMU) among other components; the SoC needs to maintain the total power consumed by all the components below the hardware specified TDP.

The current processor landscape is dominated by symmetric multi-core architectures consisting of cores with identical Instruction-Set Architecture (ISA) and underlying micro-architecture. As the TDP remains constant while core counts increase, it will only be feasible to switch on a fraction of the available cores. It has been claimed that over 50 % of the chip will be dark [6] at 8 nm using International Technology Roadmap for Semiconductors (ITRS) device scaling projections. Sophisticated software resource management techniques [30] have allowed for less pessimistic projections [9]. Still, the ubiquitous symmetric multi-core architectures are not particularly well suited in the dark silicon context as they lead to severe wastage of the on-chip resources.

In the dark silicon era, the silicon area becomes an exponentially cheaper resource relative to power and the architects can potentially "spend" area to "buy" energy-efficiency [32]. This is the guiding principle behind heterogeneous multi-core architectures, where different kinds of cores co-exist in the same die. Each core type is useful for some tasks. Only the cores that are suited for a particular task need to be activated leading to faster and more energy-efficient computing, while still staying below the TDP budget. For a comprehensive overview of heterogeneous multi-core architectures, the reader may refer to [20].

As the power budget in smart-phones and tablets is much more restrictive compared to their desktop counterparts, the impact of dark silicon era has been manifested in mobile processors earlier. Mobile SoCs have embraced heterogeneous computing in two forms, which can be broadly categorized as *functional heterogeneity* and *performance asymmetry/heterogeneity* [14]. Functional heterogeneity is defined by the presence of cores with different functionality on the same chip. This is fairly commonplace in the embedded space where different software programmable cores (e.g., general-purpose cores, Digital Signal Processor (DSP), and GPU) coincide with various non-programmable hardware accelerators or Intellectual Property (IP) blocks (e.g., video encoder/decoder, imaging, modem, and communications such as WiFi and Bluetooth) on the same SoC. The presence of DSPs and GPUs is essential to accelerate graphics and signal processing. The hardware accelerators are introduced to perform certain demanding compute-intensive tasks such as video encoding [8] under strict power budgets; the performance per watt required for such tasks is simply not achievable on CPU cores.

Performance asymmetry is a relatively recent development prompted by the TDP constraint, where the cores share the same ISA but are distinguished by different micro-architectures. As the ISA is identical, the same software binary can execute on all the cores and no additional programming effort is required. The cores differ in terms of micro-architectural complexity (for example, in-order cores versus out-of-order cores, simple versus complex branch prediction, and small versus big cache among others) and/or process technology (low-power silicon process versus standard silicon process). These micro-architectural and device-level differences lead to differing power-performance characteristics of the cores. The simple cores are extremely power-efficient but offer lower performance whereas the complex cores exhibit higher performance alongside higher power consumption. SoCs based on these architectures are generally known as *performance-asymmetric multi-cores* or *single-ISA heterogeneous multi-cores* [15]. Examples of commercial asymmetric multi-cores include *ARM big.LITTLE* [7] integrating high-performance out-of-order cores with low-power in-order cores, Wearable Processing Unit (WPU) from *Ineda* consisting of cores with varying power-performance characteristics [12], and *nVidia Kal-El* (brand name *Tegra3*) [23] consisting of four high-performance cores in fast silicon process with one limited-performance core in low-power silicon process. The responsibility of allocating appropriate cores depending on the context and the performance demand of the task lies with the Operating System (OS).

Let us examine a state-of-the-art 14 nm mobile processor *Samsung Exynos 7 Octa (7420)* SoC introduced in 2015 for *Samsung Galaxy S6* devices that features both functional and performance heterogeneity. The SoC integrates performance-asymmetric *ARM big.LITTLE* architecture [7] as general-purpose processor and *ARM Mali-T760* GPU. In this SoC, the "LITTLE" cluster consists of four simple in-order *Cortex-A53* cores while the "big" cluster consists of four complex out-of-order *Cortex-A57* cores. All the cores have the same *ARMv8-A ISA*. The *A53* cluster frequency can be set between 400 and 1500 MHz, the *A57* cluster frequency ranges between 800 and 2100 MHz, while the GPU frequency can be scaled up to 772 MHz. An analysis of the SoC [2] reveals that even with 14 nm manufacturing process that reduces the power consumption dramatically, the little cluster can use up to 1 W when fully loaded at the highest frequency, the big cluster can consume up to 4.59 W power, while the maximum power consumption for the GPU is 4.86 W. The maximum total power for these cores (even excluding other on-chip components such as MMU) is 10.45 W, which far exceeds 4–5 W TDP budget for mobile devices. This SoC clearly demonstrates the reality and the challenges of the dark silicon era. In order for the chip to function effectively, at any point in time, a significant fraction of the cores have to be kept either idle (dark) or under-clocked, i.e., at a lower frequency (dim) through a runtime power management strategy.

This chapter focuses on power management for performance-asymmetric multi-cores. The asymmetry of the cores along with the Voltage-Frequency (V-F) setting of each cluster opens up a gamut of choices and trade-offs for efficient task execution. Let us first closely investigate the constraints that need to be enforced before proceeding to detail the available mechanisms or choices for power management. First, the dark silicon era implies that the chip has to operate under a strict TDP constraint; running all the clusters at the highest frequency level will violate the safe thermal thresholds. Thus, the power budget has to be judiciously allocated to the different clusters at runtime depending on the operating conditions and the current tasks. Secondly, the tasks, especially on mobile platforms such as smart-phones, demand a certain performance level or Quality of Service (QoS) that needs to be satisfied, while maintaining the power below the TDP. Finally, for battery-operated devices, energy is a first-class design consideration [29]. These constraints render the runtime management decidedly challenging.

We now introduce the knobs exposed for power management in asymmetric multi-cores. First of all, we may employ Dynamic Power Management (DPM) [1] and Dynamic Voltage-Frequency Scaling (DVFS) [28] per cluster. With DPM, a cluster, when idle, is switched to low-power state, leading to drastically reduced energy consumption. But switching to/from low-power state incurs non-negligible time and energy overhead and hence such state switching should be performed with care. DVFS allows the voltage and the clock frequency to be set to one of the available discrete V-F levels to trade time for energy. The power budget for each cluster essentially determines its power state and the frequency level. Thus, we need a coordinated power management strategy across the clusters so as to meet the performance demands of the currently executing tasks under the thermal constraint.

The core-level performance heterogeneity present additional mechanisms for power-performance trade-off. The runtime layer needs to orchestrate the execution by partitioning the workload among the cores so as to achieve the best energy-performance objective, while respecting the thermal constraints. The runtime management system is also responsible for mapping each task to the most appropriate core (small or big) at runtime. Finally, a task may have distinct phases that may benefit from different core complexity and the runtime system should perform migration to take advantage of the heterogeneity in improving energy-efficiency.

In summary, the runtime power management of an asymmetric multi-core under dark silicon constraints involves task partitioning, task mapping, and task migration in conjunction with DVFS and DPM per cluster with the goal of maximizing energy-efficiency of the entire system, while enforcing TDP and QoS constraints. In this chapter, we present a comprehensive power management framework for asymmetric multi-cores—in particular *ARM big.LITTLE* architecture in the context of mobile embedded platforms—that can provide satisfactory user experience, while minimizing energy consumption within the TDP constraint.

Experimental Infrastructure with Asymmetric Multi-Core

We first present in detail the experimental platform we employ to design, implement, and evaluate our power management strategies using *ARM big.LITTLE* performance-asymmetric multi-core. The understanding of the asymmetric architecture is crucial to appreciate the employed power management strategies. We use *Versatile Express* development platform [35] for our experiments. It is a flexible, configurable and modular development platform that allows quick prototyping of hardware and software projects. The system comprises of a motherboard on to which modular daughter boards can be plugged in. The *ARM big.LITTLE* processor is part of the daughter board. The motherboard handles the interconnection between the daughter board and the peripherals by using a Field-Programmable Gate Array (FPGA).

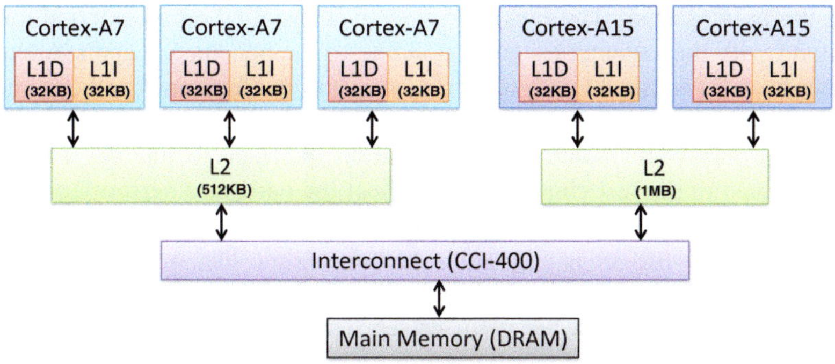

Fig. 6.1 *ARM big.LITTLE* asymmetric multi-core architecture in *Versatile Express* development board

Table 6.1 Micro-architectural parameters of *ARM Cortex-A7* and *ARM Cortex A15* cores

Parameter	*Cortex-A7*	*Cortex-A15*
Pipe-line	In-order	Out-of-order
Issue width	2	3
Fetch width	2	3
Pipeline stages	8–10	15–24
Branch predictor	512-entry BTB 2-way	2K-entry BTB 2-way
L1 I-cache	32 KB/2-way/64 B	32 KB/2-way/64 B
L1 D-cache	32 KB/4-way/64 B	32 KB/2-way/64 B
L2 unified-cache	512 KB/8-way/64 B	1 MB/16-way/64 B

The performance-asymmetric *ARM big.LITTLE* multi-core architecture consists of a high-performance *Cortex-A15* cluster and a power-efficient *Cortex-A7* cluster, as shown in Fig. 6.1. The evaluation platform we use contains a prototype chip with two *Cortex-A15* cores and three *Cortex-A7* cores at 45 nm technology. All the cores implement the same *ARM v7-A* ISA. The *Cortex-A15* is a complex out-of-order superscalar core that can execute high intensity workloads, while *Cortex-A7* is a power-efficient in-order core meant for low intensity workloads. Each core has private L1 instruction and data caches; but the L2 cache is shared across all the cores within a cluster. The L2 caches across the clusters are kept seamlessly coherent via the *CCI-400* cache coherent interconnect.

Table 6.1 summarizes the micro-architectural parameters of *Cortex-A15* and *Cortex-A7*, obtained from publicly released data. It should be evident that the cores are genuinely asymmetric in nature. The 2-way issue in-order pipeline of *A7* containing 8–10 stages is distinctly different from the 3-way issue out-of-order pipeline of *A15* containing 15–24 pipeline stages. Moreover, even the cache configurations and branch predictors are dramatically different between *A15* and *A7*.

The board boots an *Ubuntu 12.10* OS. The platform firmware runs on an *ARM* controller embedded on the motherboard. This firmware loads the *Linux* kernel during the boot process. The *Linux* file system is installed on an external memory card along with the executables corresponding to the benchmark applications. The daughter board is equipped with sensors to measure the frequency, voltage, current, power, and energy consumption per cluster. The *Linux* 3.8 kernel release we use provides the hardware monitor communication interface to interact with all the sensors located in the test chip. We also collect the hardware performance counter values from the individual cores using *ARM Streamline* gator kernel module and daemon. The gator driver is a dynamic kernel module that interrupts the core at periodic intervals to collect the performance counters. We compile and configure the *Linux* kernel to support the gator driver. The average CPU utilization of the gator daemon is less than 0.5 %, which indicates that the overhead of running gator daemon continuously in the background is minimal.

The *ARM big.LITTLE* architecture provides DVFS feature per cluster. The little *A7* cluster provides eight discrete frequency levels between 350 MHz and 1 GHz, while the big *A15* cluster also provides eight discrete frequency levels between 500 MHz and 1.2 GHz. Note that all the cores within a cluster should run at the same frequency level, i.e., the architecture does not provide per-core DVFS but rather per-cluster DVFS. An idle cluster can be powered down if necessary. The *Linux* kernel is booted in the little *A7* cluster. The powering down and modification of V-F levels are achieved using the drivers related to the oscillators. In particular, *cpufreq* utility is used to manipulate the frequency of a cluster. Given a frequency level set by the *cpufreq* utility, the voltage corresponding to the frequency level is automatically set by the hardware.

The task migration is handled by setting the task affinity through *sched_setaffinity* interface in the *Linux* scheduler. The migration cost among cores within *A15* cluster is 54–105 μs depending on the frequency level, while the cost within *A7* cluster is 71–167 μs. Thus intra-cluster migration overhead is relatively low. However, the migration costs between clusters are somewhat high: 1.88–2.16 ms for moving a task from *A7* to *A15* cluster at different frequency levels, and 3.54–3.83 ms for moving a task from *A15* to *A7* cluster. Due to high migration costs, the frequency of migrations across core types should be minimized.

The asymmetric cores exhibit different power-performance characteristics for the same workload due to the difference in micro-architecture. Figure 6.2 plots the execution time, energy consumption, and Energy-Delay Product (EDP) for 15 benchmarks (selected from *SD-VBS* [34], *SPEC CPU2000*, and *SPEC CPU2006* [10] benchmark suites) on *A15* normalized to *A7*. Clearly, the execution time on *A15* is always lower compared to *A7* leading to an average performance speedup of 1.86 for A15. But the speedup varies significantly across benchmarks from 1.45 to 2.30, i.e., the speedup is dependent on the workload and is not easily predictable. Even for the same application, the speedup can change from one program phase to another. In terms of power, as expected *A7* has lower average power compared to *A15* for all the benchmarks. The average power on *A7* is 1.44 W and shows little variation across benchmarks; the average power on *A15* varies from 4.20 to 5.15 W depending on the benchmark. Even though *A7* has worse performance, it can make up for it in terms of power to achieve 1.78 times lower energy consumption on average in comparison to *A15*.

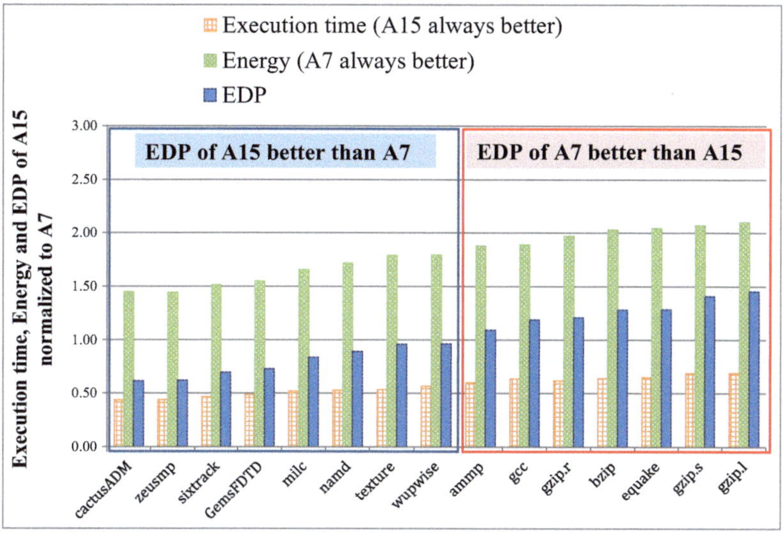

Fig. 6.2 Execution time, energy, and EDP of *A15* normalized to *A7*. *A15* is better suited for a task than *A7* when EDP value for the task is less than one

From the previous discussion, if the designer is interested in energy savings, *A7* is the better choice; while *A7* is the de facto choice for better performance. But in embedded systems, especially in interactive systems such as smart-phones, both battery life (energy consumption) and response time (performance) are equally important. So we need to use combination of energy and delay to decide on task-to-core mapping. This metric is captured in EDP (energy-delay product). Interestingly, in terms of EDP, there is no clear winner: *A15* is more efficient than *A7* for eight benchmarks due to faster execution time that overcomes the power inefficiency, while *A7* is superior for the remaining seven benchmarks due to lower power consumption. Thus, the scheduler needs to consider both power and performance behavior on a core type to decide on the appropriate mapping.

Power Management Overview

We design our power management framework for asymmetric multi-core architecture with the following objectives:

- The different power–performance behavior of the cores for the same application implies that we need to identify the right core for a task at runtime and migrate the task accordingly. Even for a single application, the appropriate core type can be different for different program phases. Thus a static task mapping is not suitable for our purpose.

- The power-hungry cores should be used only when absolutely necessary.
- In our architecture, DVFS as a control knob is available per cluster rather than per core necessitating appropriate strategies for load balancing. A cluster should run at the minimum frequency level required for maintaining QoS so as to conserve energy.
- The restricted TDP budget for the whole SoC precludes certain combination of frequencies for the different clusters. For example, it may be necessary to power down the small cluster when the big cluster is running at maximum frequency—a canonical example that illustrates the impact of the dark silicon era mentioned earlier. Thus, total power budget has to be allocated opportunistically among the clusters.
- If the power budget for the SoC is exceeded, the QoS for the tasks should degrade gracefully.
- The power management framework should be integrated in a commodity OS without altering any of its desirable properties for greater acceptability of the solution.

There exist solutions in the literature focusing primarily on mapping and scheduling on asymmetric multi-core architecture for improved performance [14, 17, 18, 27]. But these techniques do not consider power management. An energy-aware scheduling for a single task on Intel QuickIA heterogeneous platform with two cores has been developed in [4]. But the approach is quite restricted. In contrast, we need a general framework that can handle any number of tasks and cores, satisfy QoS and thermal constraints, minimize energy, and is implemented in a commodity operating system on a real platform.

The power management solutions for homogeneous multi-cores focus on a subset of the objectives mentioned earlier such as task mapping, DVFS for power management, thermal management, etc. But it should be clear that deploying these solutions together requires a carefully coordinated approach that is aware of the complex interplay among the individual solutions. For example, once the system exceeds the TDP of the entire chip, the power budgets for the clusters have to be reduced. The reduced power budget for a cluster leads to scaling down of the V-F levels of the cluster, and consequently degrading the QoS of the tasks that triggered the thermal emergency in the first place. However, once the system load is reduced (e.g., some tasks leave the system or move to a less demanding program phase), this process has to be reversed and the QoS of the tasks should be restored back to the original level. This requires synergistic interaction among the different solutions so as to ensure *safety* (operate under power budget) and *efficiency* (optimal trade-off between power and performance), while maintaining *stability*, i.e., avoiding oscillations.

We will first introduce a hierarchical control theory based reactive power management framework [21] that employs multiple Proportional-Integral-Derivative (PID) controllers in a synergistic fashion to achieve optimal power-performance efficiency, while respecting the TDP budget. This framework makes decisions regarding the initial mapping of a task to the appropriate core as well as migrating the task across cores in case of phase-change behavior within the task at a later point in time. The task mapping and migration are carefully orchestrated with the cluster-level DPM and DVFS to reach the full performance potential of the task without transgressing the thermal bounds. The drawbacks of this approach are slow response time and the poor scalability with increasing number of clusters as a centralized component allocates the power budgets for the entire system.

The first step towards moving from a reactive to a proactive approach for power management is to identify the appropriate core that fits the phase of a task, which in turn is achieved by estimating the power-performance behavior of the task on cores with different micro-architectural complexity and at different V-F levels. This estimation is challenging, as the cores can be dramatically different in terms of micro-architecture—not just in the pipeline organization but also in terms of memory hierarchy and the branch predictors. We will describe a solution proposed in [25] that overcomes these challenges through a combination of static (compile time) program analysis, mechanistic modeling that builds an analytical model from an understanding of the underlying architecture, and empirical modeling that employs statistical inference techniques like regression.

Given the power-performance estimation models, we will present a proactive, comprehensive, unified, distributed, and scalable power management strategy [22]. This strategy is based on price theory that strictly follows the supply and demand based market mechanisms to select the core for each task and the V-F level for each cluster. When the supply from the core (defined by the core type and its frequency) is equal to the demand of the task (defined in terms of QoS), the system reaches stability and is working at the most energy-efficient point. Otherwise, frequency scaling and/or task migration have to be invoked to achieve the price equilibrium [16].

Reactive Power Management

We first present a reactive power management approach for asymmetric multi-cores. It is a *Hierarchical Power Management (HPM)* framework that is based on the solid foundation of *control theory* and integrates multiple controllers to collectively achieve the goal of optimal energy-performance trade-off under restricted power budget in asymmetric multi-core architectures [21]. Moreover, the framework is built as an extension of *Linux' Completely Fair Scheduler* [24], while preserving all of its desirable properties such as fairness and non-starvation.

An overview of our *HPM* framework is presented in Fig. 6.3. We incorporate a number feedback-based controllers in our framework that interact with each other. A controller measures the output metric (measured value) and compares it with the reference metric (target value) as shown in Fig. 6.4. The error between the reference and the output is minimized by manipulating the actuators of the system. The actuation policy is determined by the model of the system being designed.

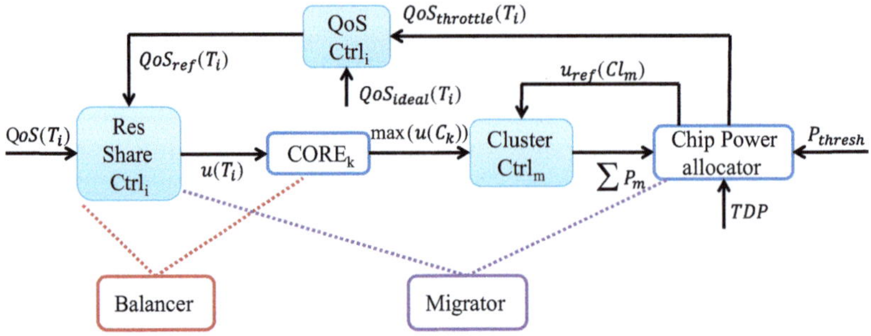

Fig. 6.3 Overview of the hierarchical power management (*HPM*) framework coordinating multiple controllers

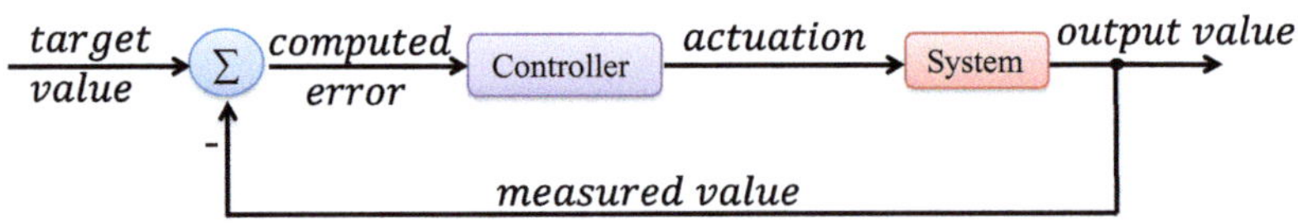

Fig. 6.4 Feedback-based controller

We employ PID controllers where the output is the weighted sum of the proportional (*P*), integral (*I*), and derivative (*D*) terms, respectively, as shown below:

$$z(t) = K_p e(t) + K_i \int e(t)dt + K_d \frac{de(t)}{dt}$$

where $z(t)$, $e(t)$, K_p, K_i, and K_d are the output of the controller, error, proportional gain, integral gain, and derivative gain, respectively.

Our framework can support two types of tasks: *QoS* and *non-QoS* tasks. A QoS task (e.g., video encoder and music player) is the one that demands certain user-defined throughput or QoS, while the non-QoS tasks do not specify any QoS requirement. In the rest of the chapter, we will only focus on the QoS tasks; the reader may refer to [21] for the treatment of the non-QoS tasks. We take advantage of the Heart Rate Monitor (HRM) [11] infrastructure in *Linux* to set the performance goal for a task and to monitor its execution progress as a QoS metric. HRM provides a simple and effective way to measure the performance of a task in terms of heartbeats per second, which is defined as the throughput of the critical kernel of the task. For example, the number of frames processed per second defines the heart rate of a video encoder. For each task, the user can define the QoS target in terms of reference heart rate range. The HPM framework attempts to maintain the heart rate within that range (thereby meeting the QoS requirement) while minimizing the energy. A task may impose variable computational demands to achieve the same QoS in different program phases and hence may need different V-F levels or even a cluster migration to keep the QoS (heart rate) within the specified range.

The HPM framework consists of three different types of controllers: per-task resource share controller, per-cluster DVFS controller, and per-task QoS controller as shown in Fig. 6.3. Each QoS task T_i in the system is assigned a resource share controller and a QoS controller. The resource share controller of a QoS task T_i manipulates the CPU share available to T_i so that it can meet the target QoS $QoS_{ref}(T_i)$. The distribution of compute resource among the tasks on a core is achieved by manipulating the *nice* values of each task. In *Linux* kernel, nice values are the indirect indications of task priorities for task management. For example, the lower nice value for a task manifests as higher priority and more resource consumption for that task.

The per-task QoS controller is inactive when the system is lightly loaded. In this case, the target QoS for a task is the same as its ideal value $QoS_{ideal}(T_i)$. However, when the total chip power exceeds the TDP, the QoS controller slowly throttles the target $QoS_{ref}(T_i)$ so that the system workload decreases to a sustainable level. The proportion by which the ideal QoS target is throttled ($QoS_{throttle}(T_i)$) per task is determined by the chip-level power allocator. The QoS controller returns the reference QoS to the original ideal level when the thermal emergency is over.

We have two cluster controllers corresponding to the *A7* and *A15* clusters. The objective of the cluster controller corresponding to the cluster Cl_m is to apply DVFS such that the cluster's utilization remains close to the target utilization $u_{ref}(Cl_m)$. A cluster's utilization is determined by the maximum utilization by any of its constituent cores. Thus, we periodically invoke a load balancer to ensure even utilization among the cores within a cluster so as to avoid overloading any core. We also invoke a migrator periodically (at a much longer interval compared to the load balancer) to migrate the tasks between the clusters if necessary. Finally, we have a chip-level power allocator that throttles the frequency of the clusters and forces QoS controller to degrade target QoS of the tasks when the total power exceeds the TDP.

The key challenge here is to coordinate the various controllers, load balancer, migrator, and chip-level power allocator. We achieve a synergistic coordination with two key mechanisms. First, the different components in our framework are invoked at different timescales. The per-task resource share controller and load balancer are invoked most frequently, followed by per-cluster DVFS controller and per-task QoS controller, then the migrator, and finally the chip-level power allocator. These invocation intervals ensure that a task attempts to reach its QoS target by first manipulating its share in a core or through migration within a cluster. If both attempts fail, then the task tries to change the cluster's frequency. As a last resort, the task is migrated to another cluster. The thermal emergency takes time to develop as temperature changes slowly; hence the power allocator is invoked least frequently.

The second mechanism we employ to coordinate the controllers is by ensuring that the controllers communicate with each other through designated channels. For example, the resource shares of the tasks within a core (both QoS and non-QoS) determines its utilization, which is provided as input to the cluster controller. More interestingly, when the power exceeds the TDP, the power allocator increases the target utilization levels of the clusters $u_{ref}(Cl_m)$. The cluster controller is forced to lower its frequency in order to meet the increased target utilization. The decreased

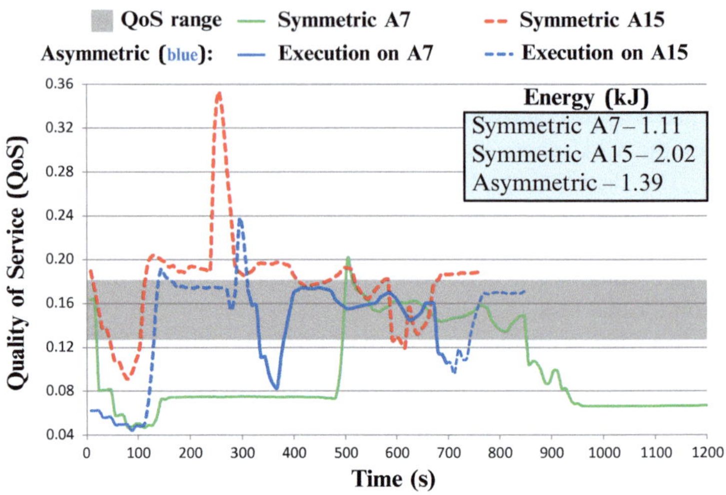

Fig. 6.5 x264: QoS on symmetric and asymmetric multi-core

frequency achieves the goal of decreasing power. In parallel, the power allocator also sends a QoS throttling factor $QoS_{throttle}(T_i)$ to each QoS controller, which makes them slowly degrade their target QoS. This reduced QoS is communicated to the resource share controller, which in turn, reduces the CPU share of the QoS tasks and hence the processor utilization to a more sustainable level. Overall, the system stabilizes to a level where the total power is just below the TDP.

The advantage of exploiting performance-asymmetric architecture is illustrated in Fig. 6.5. We use *x264* video encoder task from *PARSEC* benchmark suite [3] that exhibits phases with varying computational demands during execution. The symmetric architectures are emulated using only *A7* cluster or *A15* cluster. All the configurations use *HPM* framework; but inter-cluster migration is disabled for symmetric architectures. Figure 6.5 plots the QoS (expressed as heart rate) on the asymmetric and symmetric configurations. The gray shaded area shows the specified QoS range. The QoS line type specifies the cluster on which the task is running: continuous line corresponds to *A7* cluster and dashed line corresponds to *A15* cluster. For each symmetric configurations, only one line type (A7 or A15) is used. But in an asymmetric configuration, execution can move from A7 to A15 or vice versa depending on the program phases.

On symmetric configurations, the measured QoS (heart rate) is below the QoS range most of the time when executing on *A7* cluster, while the QoS mostly exceeds the QoS range when running on *A15* cluster. As expected, the energy consumption is very low (1.11 kJ) in *A7* cluster and quite high in *A15* cluster (2.02 kJ). On the asymmetric architecture, the task migrates to *A15* cluster for the demanding phases and moves back to *A7* cluster as the computational demand decreases. The *HPM* manages to maintain the QoS within the reference range with very low energy consumption (1.39 kJ), which is 68 % less than the energy consumption on *A15* cluster alone. The HPM can leverage the asymmetric architecture to provide the best of both the worlds where the QoS is similar to the execution on high-performance *A15* cluster, while the energy consumption is closer to the one on low-power *A7* cluster.

Power-Performance Modeling

As the reactive power management solution is based on control theory, it has somewhat slow response time when the system dynamics changes. The slow reaction time can be countered with a predictive approach. However, the predictive approach requires power-performance modeling. In particular, we need to estimate the power-performance behavior when a task migrates from one core type to another. In this section, we present a power-performance modeling approach for *ARM big.LITTLE architecture* [25]. However, the modeling is not restricted to this particular architecture and can be easily re-targeted to other asymmetric architectures. Note that this modeling assumes that both the cores are running at the same frequency (1 GHz in our experiments in this section). The performance, power, and energy prediction across different V-F levels on a particular core type, given the execution profile at a particular V-F level on the same core type, can be performed using various techniques such as the one proposed in [31].

Performance Modeling

Performance modeling estimates the performance of a task on a second core type (little/big) given its execution profile on the first core (big/little) type. Our model centers around Cycle per Instruction (CPI) stacks [13]. The idea behind the model is that the CPI follows a sustained performance CPI_{steady} punctuated by miss events that show up as temporary peaks. CPI_{steady} captures the cycles spent in the architectural events tightly coupled to the pipeline such as data dependency among instructions and structural hazards, while CPI_{misses} represents the cycles spent due to the external events such as cache miss and branch mis-predictions. The *CPI* is the summation of CPI_{steady} and CPI_{misses}.

$$CPI = CPI_{steady} + CPI_{miss}$$

The power-performance estimation framework is shown in Fig. 6.6. The performance prediction part comprises of three major steps. The first step is an off-line procedure where we build *intra-core CPI stack model* for each core type independently. CPI_{miss} can be easily expressed in terms of miss events and their latencies; but computing CPI_{steady} generally requires the presence of elaborate hardware mechanisms [33] that can collect inter-instruction dependencies and are

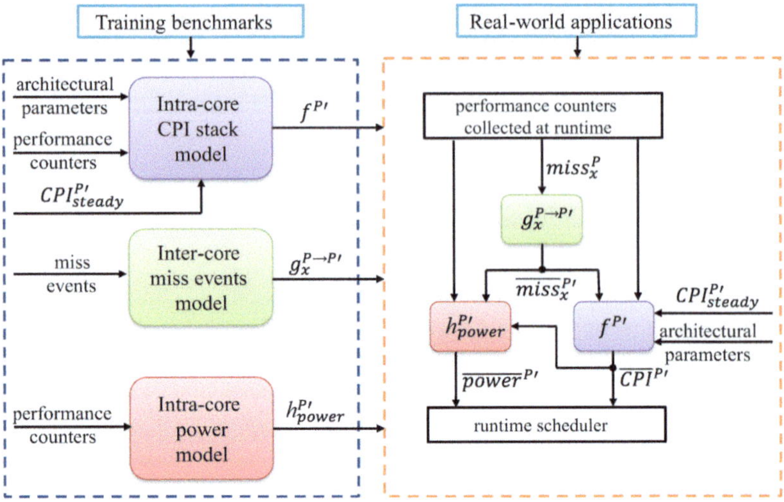

Fig. 6.6 Inter-core power, performance estimation from core P to core P'

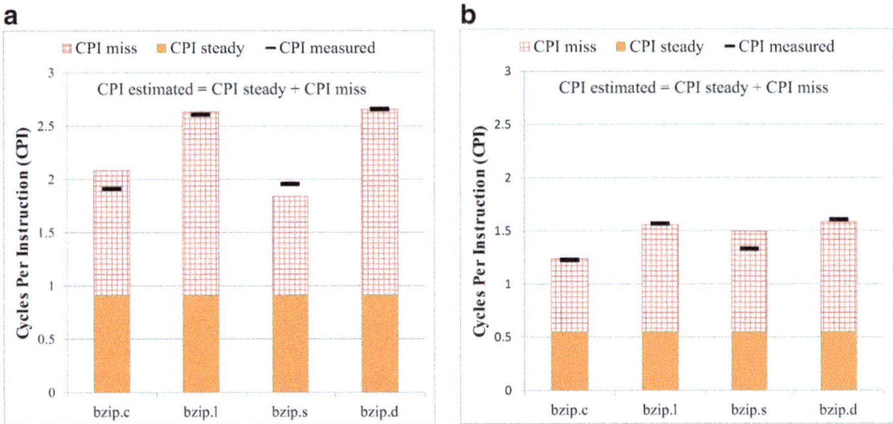

Fig. 6.7 Estimated CPI_{steady} and CPI_{miss} of different inputs for the same benchmark. (**a**) *Cortex-A7*; (**b**) *Cortex-A15*

not available in existing off-the-shelf processors. We avoid the requirement of additional hardware mechanism with the observation that CPI_{steady} is an intrinsic characteristics of a program on a core type and is stable across different program inputs, whereas CPI_{miss} is highly dependent on the program inputs. For example, Fig. 6.7a, b show the estimated CPI_{steady} and CPI_{miss} values of *bzip* benchmark for different program inputs on *A7* and *A15* cores. The estimated CPI is the summation of the estimated CPI_{steady} and CPI_{miss}. For reference, we have also plotted the measured CPI. The plots validate our conjecture that CPI_{steady} of a task on a core type is stable across different program inputs; but CPI_{miss} shows significant variation across inputs. Thus, the variation in CPI for different inputs can be captured accurately only through variation in CPI_{miss}. The CPI_{steady} value simply needs to be estimated once for each core type and task (irrespective of the input to the task). Note that CPI_{steady} is higher on *A7* than *A15* because *A15* can better exploit Instruction Level Parallelism (ILP) in the presence of data dependencies and structural hazards with its out-of-order execution engine.

We exploit this observation to estimate CPI_{steady} of a task on both core types at compile time and encode this information with the binary executable. In other words, we estimate both CPI^{big}_{steady} and CPI^{little}_{steady} for a program at compile time. For tasks with distinct program phases, i.e., multiple computation kernels with different behavior, we have to estimate separate CPI_{steady} value for each phase and embed this information with the binary executable.

We build the CPI stack on a core P by collecting the execution profiles of a set of training benchmarks through the hardware performance counters. We then combine analytical modeling with linear and non-linear regressions to derive the CPI stack model that accurately captures the contributions of the different events to performance. The CPI stack model can thus be expressed as the function f^P

$$CPI^P = f^P(CPI^P_{steady}, miss^P_X, latency^P_X)$$

where $miss^P_X$ and $latency^P_X$ are the number of occurrences and latency of each occurrence of the miss event X on processor P. The miss events include cache miss and branch mis-prediction.

The second step is another off-line procedure, where we develop regression models to estimate the number of occurrences of different miss events on processor P' given the frequency of the miss events on processor P. These *inter-core miss event estimation models* are built by collecting and correlating corresponding miss events on both cores using a set of training benchmarks. The inter-core estimation model from P to P' for an event X can be expressed by a function $g^{P \rightarrow P'}_X$

$$\overline{miss}^{P'}_X = g^{P \rightarrow P'}_X \left(miss^P_X\right)$$

where $\overline{miss}^{P'}_X$ is the predicted occurrence of miss event X on P'. We use $\overline{M}$ to indicate the estimated value of a metric across cores.

At runtime, when a new task is executing on core P, the OS collects the counter values at regular intervals to get information about the miss events on P. For each miss event X, we use inter-core miss event estimation model to predict $\overline{miss}^{P'}_X$ on core type P'. Finally, we plug in the estimated miss event counter values in the CPI stack model of P' to predict $\overline{CPI}^{P'}$ as shown below:

$$\overline{CPI}^{P'} = f^{P'}(CPI^{P'}_{steady}, \overline{miss}^{P'}_X, latency^{P'}_X)$$

Power Modeling

We now describe our modeling technique to estimate power on asymmetric multi-cores through cross-core predictions. Unlike performance modeling, which required a combination of mechanistic and empirical modeling, power can be modeled purely based on regression analysis. We use a linear regression model to estimate the power consumption.

In *ARM big.LITTLE platform*, the little cores are superscalar in-order, power-efficient *Cortex-A7* cores. We observe that the power consumption of the little core remains the same across all the benchmarks. The minimum and maximum power consumption measured across training benchmarks are 1.385 and 1.506 W, respectively. Thus, we decide that there is no need to model power for the little cores and we can simply use a constant power value.

While power consumption on the little *A7* core is stable across benchmarks, the big core (power-hungry, out-of-order *A15*) shows significant variation in power consumption within (due to program phases) and across benchmarks. The observed minimum and maximum power consumption on *A15* across training benchmarks are 4.535 and 5.155 W, respectively. This is because complex out-of-order cores exhibit different access profiles of various micro-architectural components across the benchmarks, which contribute to different power consumption by the components. Thus, it is imperative to model task-specific power consumption on *A15*. The power consumption of *A15* critically depends on the pipeline behavior and the memory behavior of the task. In particular, the instruction mix of a task is expected to influence the access profile of different architectural components such as floating point unit, branch predictor, etc., which in turn determines the power consumed in the pipeline. The power consumption in the memory hierarchy is determined by the number of L1 instructions, L1 data, combined L2, and memory accesses. So we define the function h^P in Fig. 6.6 that models the power consumption

$$Power^P = h^P(N_X, miss_X^P, CPI^P)$$

where N_X is the fraction of instructions of type X in the instruction mix.

Given a set of training benchmarks, we first collect the performance counter values on *A15* that capture the instruction mix and the access at different levels of the memory hierarchy. We also measure the power consumption on *A15*. Next, we employ correlation analysis to identify the important performance counters that contribute significantly to the power consumption. The total power consumption of the *A15* core can be expressed in terms of the following linear regression model:

$$\begin{aligned} Power = \beta_1 + \beta_2 \cdot \frac{N_{int}}{N} + \beta_3 \cdot \frac{N_{fp}}{N} + \beta_4 \cdot \frac{1}{CPI} \\ + \beta_5 \cdot \frac{access_{L1D}}{N} + \beta_6 \cdot \frac{access_{L2}}{N} + \beta_7 \cdot \frac{dmiss_{L2}}{N} \end{aligned} \quad (6.1)$$

The first three terms capture the power consumption in the pipeline, which is influenced by the proportion of integer ($\frac{N_{int}}{N}$) and floating point instructions ($\frac{N_{fp}}{N}$), and CPI. N is the total number of instructions. The power consumption is also linearly related to the rate of access to the various memory hierarchy levels, which is captured using the next three terms. We do not include L1 instruction cache access here because it is already included in terms of CPI. The higher the CPI, the lower the rate of access to L1 instruction cache.

The major challenge in estimating the power consumption of a task on the big core while running it on the little core is that we have to predict the access profile. In Eq. (6.1), the total number of instructions and the instruction mix (N, N_{int} and N_{fp}) remain unchanged across cores. The inter-core miss event prediction model discussed earlier estimates $\overline{CPI}$, $\overline{dmiss_{L2}}$, $\overline{miss_{L1D}}$, and $\overline{miss_{L1I}}$ on the big core from the corresponding values on the little core (see also in Fig. 6.6). We can then define

$$\overline{access(L2)} = \overline{miss_{L1D}} + \overline{miss_{L1I}}$$

These estimated values can be plugged into Eq. (6.1) to estimate the power consumption on the big core given the execution profile on the little core.

Accuracy

Across a range of benchmarks from *SD-VBS* [34], *SPEC- CPU2000*, and *SPEC CPU2006* [10] suites, we observe an average error of 13.4 % in predicting little core CPI from big core. The error in predicting big core CPI from little core is 16.7 %. The error for power estimation from little to big core is only 3.9 %.

In the context of predictive/proactive power management approach, our framework needs to continuously monitor the execution profile on one core and estimate the power, performance on the other core. This will allow the scheduler to migrate the task back and forth between the cores depending on the current program phase and the appropriate core type for that phase. We conduct a case study experiment with *astar* benchmark to evaluate the accuracy and robustness of our model in detecting phase changes and accurately predicting the behavior on the target core for each phase. Figure 6.8a, b show the estimated power, performance on *A15* predicted from executing the task on *A7*. For references, we also show the measured power, performance on *A15*. The X-axis on both figures shows the number of committed instructions as time progresses. We set our sampling interval at 500 ms, which roughly corresponds to 500 million instructions on *A7* at 1 GHz. The task demonstrates clear phase behavior as evidenced by the CPI and power values on the Y-axis. Our estimations are fairly close to the measured values. Thus we can track the phase changes accurately to predict performance speedup and energy-efficiency in moving from one core type to another.

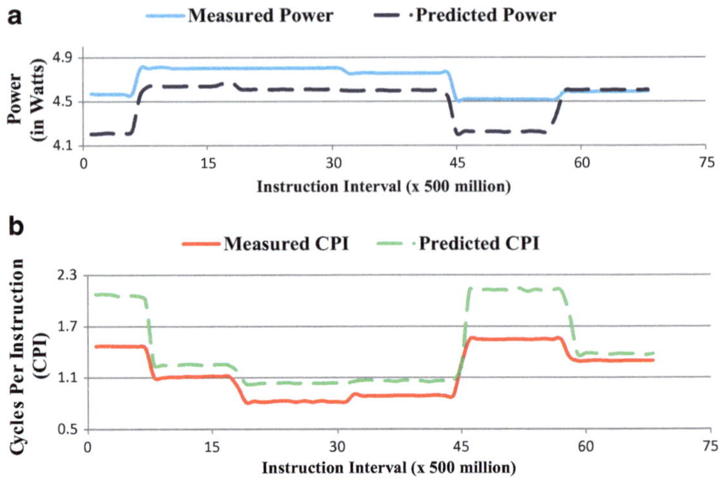

Fig. 6.8 Continuous power and CPI estimation from *A7* to *A15* for *astar* benchmark. (**a**) Power estimation; (**b**) CPI estimation

Proactive Power Management

Given the power-performance models, our Proactive Power Management (PPM) framework [22] is designed on the foundation of the *Price Theory* [16]. The resource allocation, DVFS, task mapping and migration are all controlled through a virtual market place, where the commodity being traded is the processing power.

The framework is realized as a collection of autonomous entities called *agents*. Each agent represents a transactional body in the market [5]. An agent can perform various functionalities such as bidding, purchasing, and distribution of computational resources. Furthermore, the actions of an agent are prompted by the goals of the client(s) it represents. For example, an agent representing a task is motivated to meet the demand imposed by the task. Similarly, the agent representing the entire chip ensures that the TDP is not violated, while meeting system-wide demands.

The efficient functioning of the market is ensured through the regulations imposed on the actions of the agents. The regulations introduced are reflections of the power management goals. According to the price theory, in a competitive market, the quantity and price at which a commodity is traded will be determined by the point of intersection of its supply and demand curve, also known as price equilibrium. Furthermore, if we increase (or decrease) the money in circulation while keeping the supply and demand unchanged, we will observe inflation (or deflation) in the equilibrium market price. Interested readers may refer to [16] for more details on price equilibrium, inflation, and deflation concepts.

Our power management framework consists of two main components: *Supply-Demand (SD) module* and *Load Balancing plus Task migration (LBT) module*.

Given a task-to-core mapping, the SD module attempts to satisfy the demands of all the tasks with minimal power consumption under the TDP constraint. It relies on the concept of regulating inflation–deflation. The LBT module aims to reach a more power-efficient task-to-core mapping through load balancing and task migrations, and employs the concept of reduced spending. Both the modules work in tandem to achieve the final design goal of power management in asymmetric multi-cores.

Overview of the Models and the Agents

The target system is comprised of a set of cores C grouped in a set of V-F clusters V, with each cluster having a separate voltage and frequency regulator. Each cluster v can operate at several discrete V-F levels and consists of a set of cores $C_v \subseteq C$. All the cores within a cluster are symmetric in terms of micro-architecture and are constrained by hardware design to run at the same V-F level, decided by the cluster.

A task t is a computational entity that can execute on a core. Each task t is assigned a priority R_t by the user, where higher value means higher priority. T represents the set of all tasks. Our framework dynamically maps the tasks to the core. A task t is mapped to a core c_t. $T_c \subseteq T$ represents the set of tasks mapped to core c and $T_v = \cup_{c \in C_v} T_c$ denotes the set of tasks mapped to the cores in cluster v. The idle task t_{idle} executes on an empty core. If there are no active tasks in an entire cluster, then we can power down that cluster. We define R_c, R_v, R as the sum of the priorities of all the tasks mapped to core c, cluster v, and the entire system, respectively.

Each core c can supply certain amount of computational resources S_c, which is constrained by the maximum supply $\hat{S}_c$. The computational resource is defined in terms of *Processing Units* (PUs), where one PU is equivalent to one million processor cycles per second. The higher the frequency of a core c, the more it can supply PUs (i.e., higher the value of S_c) and the maximum supply of PUs $\hat{S}_c$ is determined by the maximum possible frequency of the core. For example, a core running at 1000 MHz produces a supply of 1000 PUs. Note that the amount of work (instruction processing) that can be achieved with one PU on a little core is generally less than the amount of work that can be done with one PU on a big core; that is, one PU on a big core is more valuable than one PU on a little core.

The supply of a cluster S_v is the same as the supply of any of the constituent cores, which have identical S_c values; while the supply of the entire chip S is the summation of the cluster supply values. The current supply of PUs to task t on core c_t is represented by S_t ($\equiv S_t^{c_t}$). The supply of PUs to a task t has to be less than the supply produced on the core c_t it is mapped to, that is, $S_t \leq S_{c_t}$.

Each task t demands a certain amount of computational resources (PUs), which can vary dynamically during the course of the execution. In asymmetric multi-cores, a task demands different amount of PUs across different core types. For example, a task would demand more PUs on a little core compared to a big core to achieve the same level of performance. The differing demands for computational resources on different core types model the asymmetry of the architecture. The current demand of task t on core c_t is represented by D_t ($\equiv D_t^{c_t}$).

Let D_c represent the sum of demands of all the tasks mapped to a core c. The core with the highest demand in a cluster is called the constrained core of the cluster. Let $\tilde{c}_v \in C_v$ represent the constrained core of the cluster v. Then the demand of the cluster $D_v = D_{\tilde{c}_v}$ is defined as the demand of its constrained core $\tilde{c}_v$. The demand of the entire chip D is the summation of the demands of the clusters.

The power consumption of a core c represented by W_c depends on the core type, V-F level, and the workload. The power consumption of a cluster v is represented by W_v, while the entire chip power consumption is represented by W. The quality of the cooling solution determines the value of the TDP constraint W_{TDP}. The goal of the framework is to keep the total chip power consumption below the TDP ($W < W_{TDP}$), while meeting the task demands at minimal energy.

Figure 6.9 shows the interaction among the agents. Each task is represented by a *task agent*, who is a buyer in the market. A task agent can receive, spend, or save money to purchase the computational resources (PUs). An agent corresponding to a task gets an allowance (virtual money) that it uses to bid for the resources according to the demands of the task.

Each core is represented by a *core agent*, which determines the price of the computational resources produced by the core. The price for PUs in a core emerges from the bids submitted by the task agents and the current supply of the core. The core agent then distributes the resources among the task agents according to the

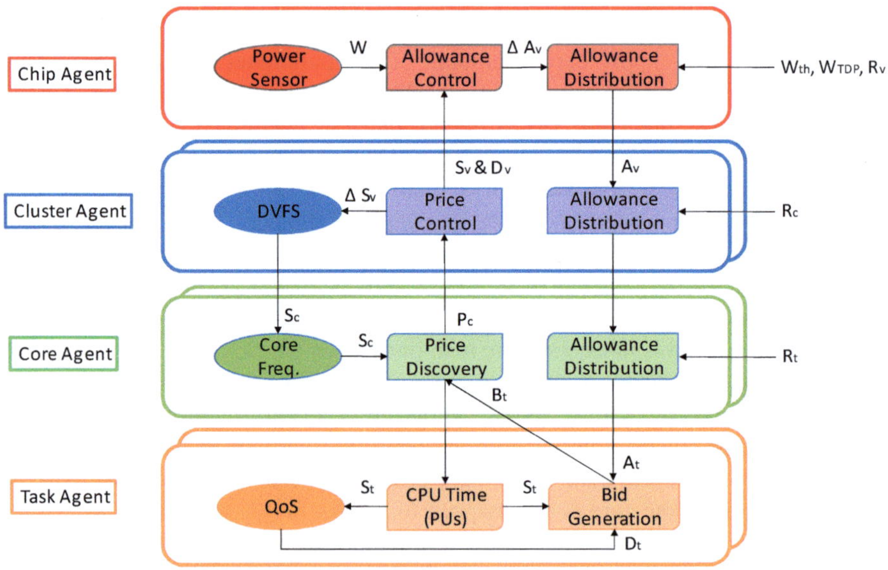

Fig. 6.9 Overview of agent interactions under *PPM*

bids. A core agent also distributes virtual money among the task agents to make their bids.

Each cluster is represented by a *cluster agent*, which controls the price of the resources by manipulating the supply of PUs in the cores under its purview. The increase (or decrease) of the supply is achieved by varying the V-F levels of the cluster. A cluster agent also distributes virtual money among the core agents.

The entire chip is represented by a *chip agent*. The chip agent controls the amount of money in circulation by manipulating the allowances of cluster agents directly and rest of the agents indirectly to ensure that the total power does not exceed the TDP.

Supply-Demand (SD) Module

We first explain the mechanisms employed by the SD module in manipulating the V-F levels to meet the task demands at minimal power consumption. The SD module requires all the task-, core-, cluster-, and chip agents to work in synergy. In terms of price theory, the demands of all the tasks are satisfied only in an economic market without inflation (or deflation). Thus, controlling inflation (or deflation) is equivalent to providing enough supply to satisfy the current demand. This is the basic regulatory principle employed in the SD module.

Task and Core Dynamics

The main goal of a task agent is to sustain the demand of the task it represents. This objective is achieved through an iterative process consisting of three steps per round: bidding by the task agents, price discovery by the core agent, and purchase of the resources. The iterative process continues till a price equilibrium is attained.

Each task agent is given an allowance A_t, which is virtual money, by the core agent according to the priority of the task. The task agent bids an amount B_t to buy resources based on the current demand of the task D_t. If the bid is less than the allowance, then the difference $M_t = A_t - B_t$ is saved for future use. The bid cannot exceed the sum of allowance and savings. We also require the bid to be higher than the predefined minimum bid B_{min}. That is, $B_{min} \leq B_t \leq A_t + M_t$.

For every round, each task agent submits a bid amount B_t based on the experience in the previous round. The task agent increases (or decreases) the bid amount if the supply received was less (or more) than the demand in the previous round. The agents keep the bid unchanged when the demand is satisfied.

Given the bids from all the tasks mapped to core c, the core agent representing c discovers the price per PU P_c as follows:

$$P_c = \frac{\sum_{t \in T_c} B_t}{S_c} \tag{6.2}$$

Each task agent now purchases the resources at the value determined by the core agent and obtains its current supply *St*.

$$S_t = \frac{B_t}{P_c}\bigg|_{t \in T_c} \tag{6.3}$$

The bids in the $(N+1)$th round by the task agents depend on the supply, demand, and prices observed in the Nth round. As mentioned earlier, the bidding amount is capped by the summation of allowance A_t and savings M_t for the task.

$$B_t^{N+1} = \max\left(A_t + M_t,\ \ B_t^N + (D_t - S_t) \times P_c\right)\bigg|_{t \in T_c} \tag{6.4}$$

Cluster Dynamics

The cluster agents are responsible to control the price and prevent both price inflation and deflation in the cores. When a core is undersupplied (oversupplied), we observe price inflation (deflation). The cluster agent adjusts the supply using DVFS to avoid either oversupply or undersupply in the cores, which in turn is reflected as stable price of the PU.

In our architecture, all the cores within a cluster have to run at the same V-F level. Thus, the supply can be modified only at the cluster- level and not at the core level. So the cluster agent observes and responds to price inflation (or deflation) of only the constrained core; the constrained core represents the highest demand among the cores within the cluster. Thus, the supply of the cluster is controlled by the most constrained core. Note that given a task mapping, a non-constrained core may suffer from deflation, while the constrained core suffers from inflation. The cluster agent takes care of the inflation in the constrained core, which can further magnify the deflation in the non-constrained core. The LBT module is responsible for fixing the deflation in the non-constrained core through load balancing.

In order to identify inflation/deflation, we need a base price from which the relative changes can be observed. Every time there is a change in V-F level, we reset the base price to the new price observed in the market. While the V-F level is changing, we do not allow the task agents to change their bids until they have observed the effect of the new supply on their existing bids.

A user supplied parameter called tolerance factor δ defines the rate of inflation (or deflation) that the cluster agent can tolerate before increasing (or decreasing) supply. Let P_c and $PBase_c$ represent the current and base price of resources in a constrained core c, respectively. The cluster agent increases the supply when the current price $P_c \geq PBase_c + PBase_c * \delta$. Similarly, the supply is decreased when $P_c \leq PBase_c - PBase_c * \delta$. The tolerance factor δ determines the response sensitivity of the cluster agents. The lower the value of δ, the faster is the response of the cluster agent. The faster response results in frequent V-F level transitions, which can be detrimental to both the performance and the reliability of the chip due to thermal cycling [26]. Thus, it is important to carefully select the value of δ.

Chip Dynamics

While the cluster agent attempts to set the V-F level at the minimum value so as to meet the demand of the tasks, the chip-level agent is responsible to ensure that the overall chip power does not exceed the TDP budget. The chip agent indirectly controls the power consumption of the chip by manipulating the allowances given to the clusters, cores, and the tasks. It decides on the global allowance value A for the current round. The allowance A is distributed hierarchically throughout the system using the different cluster and core agents.

The global allowance is distributed as cluster allowances (A_v) to the cluster agents and the distribution is inversely proportional to power consumption. The cluster that is consuming more power is given less allowance.

$$A_v = A \cdot \frac{W - W_v}{W} \tag{6.5}$$

The cluster allowance is distributed as core allowance (A_c) to the core agents of the cluster based on the priorities of task agents running on the cluster.

$$A_c = A_v \cdot \frac{R_c}{R_v} \tag{6.6}$$

Finally, the core allowance is further distributed as task allowances (A_t) to the task agents in the core proportional to their priorities.

$$A_t = A_c \cdot \frac{R_t}{R_c} \tag{6.7}$$

When the chip agent increases the global allowance A, the task agents receive additional virtual money to generate higher bids for the resources. The task agents with unsatisfied demands increase their bids with the additional money. This causes inflation in the clusters that are undersupplied, triggering the respective cluster agents to control the inflation by increasing the supply (increase V-F levels). This increased supply in the cluster results in increased power consumption. On the other hand, when the chip agent decreases the global allowance A, all the task agents have less money at their disposal and hence are forced to bid lower values. This causes deflation in the clusters, prompting the cluster agents to decrease the supply (decrease V-F levels) to control the deflation, resulting in reduced power consumption.

When the chip agent decides to keep the allowance A constant, all the cores will reach a stable equilibrium price. With stable prices, neither inflation or deflation will be observed by the cluster agents resulting in a *steady-state* with no changes in V-F levels. The global allowance for the $(N+1)$th round is set as follows

$$A^{N+1} = A^N + \Delta \tag{6.8}$$

where A^{N+1} and A^N are the current and previous round allowances, respectively, and Δ is the change in the allowance. The key question is how to dynamically set the Δ value. The Δ value is set according to the current total power consumption.

When the chip power consumption W is below the TDP, the primary goal of the chip agent is to meet the demands of the tasks. But if the chip power exceeds TDP, then the chip agent is responsible to bring the power below the TDP. In case the system has a demand that is unsatisfiable within the TDP, due to the discrete nature of V-F levels the system will oscillate along the TDP. To stabilize the system when overloaded, we introduce a buffer zone near TDP where the system is ought to stabilize. The size of buffer zone is decided by the parameter W_{th}. Thus, the spectrum of power consumption is divided into three regions.

In the *normal state*, the power consumption of the entire chip is less than the predefined threshold $W < W_{th}$. In this state, the chip agent manipulates the Δ value based on the current total supply S and total demand D. When the demand is not satisfied in at least one of the clusters, the chip is under-utilized and the task agents need extra money to buy more resources. Therefore, the allowance is increased proportional to the difference between the supply and the demand.

$$\Delta = A^N \cdot \frac{D - S}{D} \tag{6.9}$$

In the *threshold state*, the power consumption lies between the threshold W_{th} and TDP W_{TDP}. Ideally, it is desirable for the power consumption to stabilize in threshold state when system is overloaded. The stability is attained by keeping the allowance constant through $\Delta = 0$. With larger buffer zone $(W_{TDP} - W_{th})$, the number of oscillations around the TDP reduces and the stable state is reached quickly, but the chip might be severely under-utilized. On the contrary, a smaller buffer zone leads to frequent oscillations around the TDP, but achieves higher utilization. The idea of achieving stability using a buffer here is similar to the concept of hysteresis in control systems.

In the *emergency state*, the power consumption is above W_{TDP} and must be brought down quickly. The allowances of the task agents have to be curbed to reduce the power consumption. In emergency state, the allowance is reduced proportional to the deviation from the TDP.

$$\Delta = A^N \cdot \frac{W_{TDP} - W}{W_{TDP}} \tag{6.10}$$

Thus, our system can achieve stability in either normal state (supply meets demand) or threshold state (when overloaded) but never in emergency state.

Load Balancing and Task Migration (LBT) module

The SD module achieves a steady-state with permissible power consumption for any given task-to-core mapping by manipulating the V-F levels. But the mapping itself may not be efficient in terms of performance and power leading to a sub-optimal solution. Thus, the goal of the LBT module is to find a task mapping that is superior in terms of both performance and/or power consumption relative to the current mapping. The LBT module first attempts to meet the task demands followed by improving energy-efficiency through load balancing within a cluster and task migration across the clusters. Load balancing within a cluster helps reduce V-F level of the cluster, while task migration exploits the asymmetry of the different clusters to potentially improve power-performance characteristics of the tasks.

If the demands of all the tasks are expected to be satisfied in the steady-state of the current mapping *Map*, then the goal is to reduce the power consumption. In this case, all the task agents in a cluster v estimate the performance and spending if the task migrates to the most oversupplied unconstrained core in each of the other clusters, leading to new mapping. These estimates are obtained using power-performance model presented in Section "Power-Performance Modeling". Each task agent then sends the estimated most power-efficient new mapping Map' to the chip agent. The core agent selects the most power-efficient mapping from the task agents and forwards this mapping to the chip agent through its cluster agent. The chip agent finally selects the most power-efficient mapping from all the clusters.

On the other hand, if some tasks are not expected to meet the demand in the steady-state in the constrained core, then only the clusters with unsatisfied demand consider possible task migration. For each such cluster, the task agents with unsatisfied demand in the constrained core contemplate migration to the other clusters. In particular, at all levels (chip-level, cluster-level, and core-level) we choose the highest priority task that improves its supply-demand ratio through migration without impacting the supply-demand ratio of the higher priority tasks.

The load balancing process is very similar to the task migration. The only difference is that the target core is not in a different cluster but the most oversupplied unconstrained core within the same cluster. Thus only the cluster agent and the subordinate core agents are involved in this process. The chip or the cluster agent approves only one task movement at any given time. The movement of the other tasks, if any, will be performed in future LBT invocation.

Invocation Frequency

The different modules involved in our framework have to be invoked at different rates to reduce the overhead in the system. The bidding process by the task agents takes place in series of rounds. The change in the supply level by the cluster agent occurs asynchronously based on inflation/deflation.

The load balancing within the cluster is invoked much more frequently than the task migrator because task migration across clusters is much more expensive compared to migration within cluster. However, load balancing is still less frequent compared to the bidding rounds. The following equations summarize the periods for LBT, where *Linux* scheduling epoch is 10 ms.

$$task\_migration\_period = 2 \times load\_balancing\_period$$

$$load\_balancing\_period = 3 \times bidding\_rounds\_period$$

$$bidding\_rounds\_period = linux\_sched\_epoch$$

The LBT module is disabled in the emergency state as the immediate goal is to bring the power below TDP through the SD module.

Comparison with *Linux*

We compare our *HPM* framework proposed in section "Reactive Power Management" and *PPM* framework proposed in section "Proactive Power Management" with Heterogeneity aware scheduler in *Linux* kernel (*HL*).

The *HL* scheduler released in *Linux* kernel release 3.8 is aware of the heterogeneity in *ARM big.LITTLE platform*. The activeness of a task (the amount of time spent in the active task run-queue) is used as a proxy for migration decisions. For example, the *HL* scheduler migrates a task to *A15* cluster (*A7* cluster) once the time spent in the active run-queue exceeds (falls below) certain predefined threshold. Furthermore, the *HL* scheduler does not react to the varying demands of the individual tasks. For the *HL* scheduler, we also employ *cpufreq on-demand* governor that changes the frequency value based on processor utilization. In all the experiments related to comparative study, we set all the tasks to run at the same priority because *HPM* and *HL* do not take the priorities into consideration.

We create nine different multiprogrammed workload sets from the *PARSEC* [3] benchmarks based on the following metric

$$intensity = \frac{\sum_{t \in T} D_t^{A7} - S_{A7}^{max\_freq}}{S_{A7}^{max\_freq}} \tag{6.11}$$

where $\sum_{t \in T} D_t^{A7}$ is the total demand of all the tasks in the given workload and $S_{A7}^{max\_freq}$ is the supply at the maximum frequency in the *A7* cluster. The metric *intensity* shows whether the demand of the entire task set in a workload can be accommodated in the *A7* cluster at the highest frequency. If $intensity \leq 0$, the supply exceeds the demand and hence the demand from all the tasks can be satisfied in *A7* cluster at the highest frequency. On the other hand, if $intensity > 0$, some tasks will not meet their demand on *A7* cluster and need to move to the more powerful

A15 cluster. Therefore, based on the *intensity* metric, we classify the workload sets into three types: (a) *light* (*metric* $\leq$ 0), (b) *medium* (0 $<$ *metric* $\leq$ 0.30), and (c) *heavy* (*metric* $>$ 0.30). Table 6.2 summarizes the workload sets in detail.

For the first comparative study, we assume that the system does not have any TDP constraint and hence can consume arbitrarily high power. Figure 6.10 plots the percentage of time the reference QoS range of any task in the workload is not met for the three approaches, i.e., the percentage of time the observed QoS was smaller than the minimum prescribed QoS for any of the task in the workload. It is evident that the *HL* performs better under light workloads (*s1, s2, s3*). This is expected as the *HL* scheduler migrates the tasks to the powerful *A15* cluster at the first opportunity, while *HPM* and *PPM* both take a more judicious approach. The impact is shown

Table 6.2 Workload sets

Light	s1	texture_v, tracking_v, h264_s swaptions_l, x264_l, blackscholes_l
	s2	texture_v, multicnt_v, h264_b swaptions_l, bodytrack_l, blackscholes_l
	s3	tracking_v, multicnt_v, h264_s x264_l, bodytrack_1, blackscholes_l
Medium	m1	swaptions_l, bodytrack_l, blackscholes_l texture_v, tracking_v, h264_b
	m2	texture_v, tracking_v, h262_s swaptions_n, bodytrack_n, x264_n
	m3	tracking_v, multicnt_v, blackscholes_n bodytrack_n, texture_f, h264_fo
Heavy	h1	h264_fo, x264_n, blackscholes_n texture_f, swaptions_n, multicnt_f
	h2	blackscholes_n, x264_n, tracking_f bodytrack_n, texture_f, h264_s
	h3	h264_b, h264_fo, x264_n swaptions_n, bodytrack_n, tracking_f

v vga, *f* fullhd, *n* native, *l* large, *s* soccer, *b* bluesky, *fo* foreman

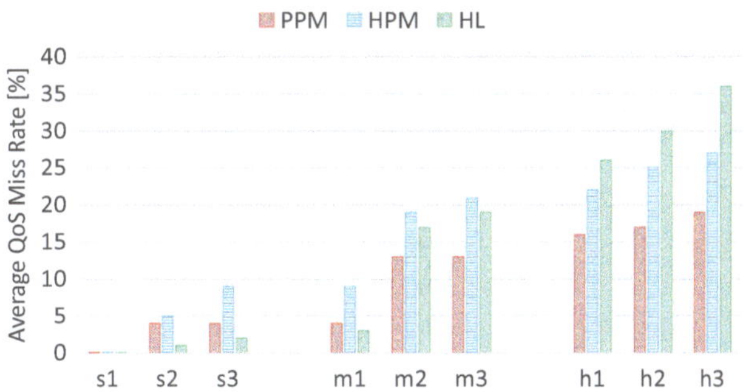

Fig. 6.10 Comparison of percentage of time the tasks do not meet the reference QoS under no TDP constraint

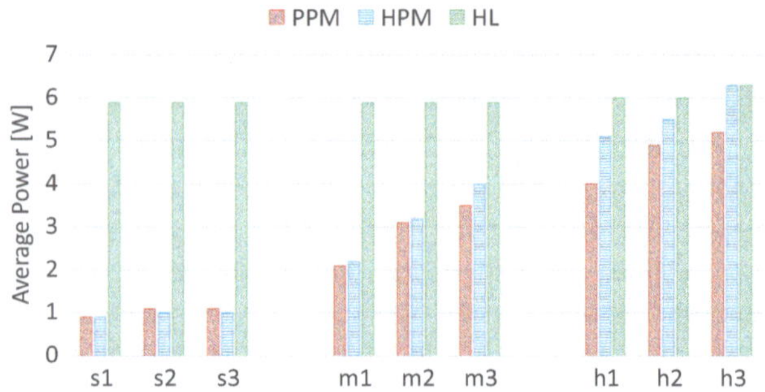

Fig. 6.11 Comparison of power consumption (no TDP constraint)

as significantly higher average power consumption for *HL* compared to *HPM* and *PPM* as shown in Fig. 6.11.

On the contrary, the *PPM* outperforms both *HPM* and *HL* for medium (*m1, m2, m3*) and heavy (*h1, h2, h3*) workloads. The *HPM* scheduler implements a relatively simple and non-speculative load balancer and task migrator that is oblivious to the utilizations in the other clusters. As the *HL* scheduler migrates all the tasks to the A15 cluster, it results in inefficient usage of the resources.

Figure 6.11 plots the average power consumption for the different techniques with no TDP constraint. *HPM* and *PPM* have comparable average power consumption across all types of workloads. The *HL* scheduler with on-demand governor results in an average power consumption of 5.99 W, which is much higher than that of *HPM* (3.43 W) and *PPM* (2.96 W) across all the workloads.

Next, we study how the different techniques cope with strict TDP constraints. We observed through a series of experiments that the in-built thermal management mechanism is invoked when the total power consumption of the platform exceeds 8 W. To emulate a power-constrained environment, we artificially cap the power budget to 4 W. For the *HL* scheduler, we switch off the *A15* cluster once the power exceeds the TDP. This is because the observed maximum power in *A7* cluster and *A15* cluster are 2 and 6 W, respectively. Powering the *A15* cluster down guarantees that the total power will be below the TDP constraint of 4 W.

Figure 6.12 plots the percentage of time any task in the workload does not meet their reference QoS under TDP constraint of 4 W. The tasks are able to meet their reference QoS more often with *PPM* approach compared to *HPM* and *HL*. The improvements are 34 and 44 % compared to *HPM* and *HL*, respectively.

Figure 6.13 summarizes the impact of the runtime management layer that is aware of the thermal design constraints imposed by the dark silicon era. The figure shows QoS miss rate of the workload with *HL* and *PPM* for three different workloads (*W1, W2, W3*). We experiment with no thermal constraint and TDP constraint of 4 and 6 W, respectively. When the total power exceeds the TDP constraint, the system automatically powers down the cores to keep the power within the constraint. Clearly, the stricter the TDP constraint, the higher the QoS miss rates for both policies. But our TDP-aware runtime management fares much better in meeting the demand compared to the stock *Linux* that is agnostic to the TDP

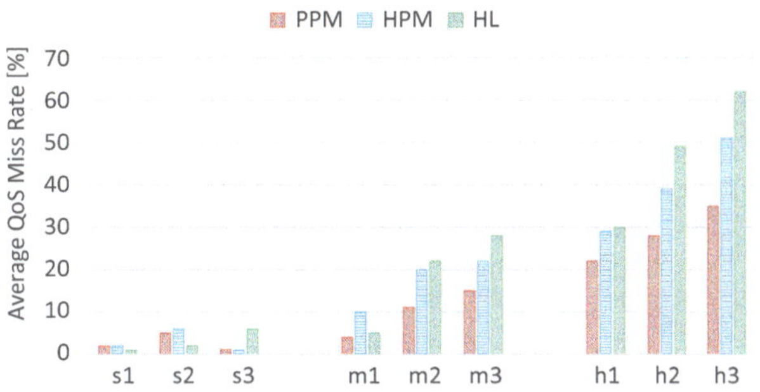

Fig. 6.12 Comparison of the percentage of time the tasks do not meet the reference QoS under TDP constraint of 4 W

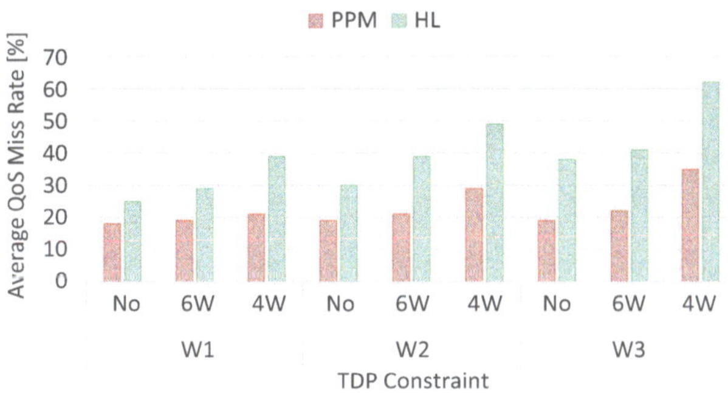

Fig. 6.13 Impact of TDP constraint on QoS for different approaches

constraint. This demonstrates the urgent need for sophisticated runtime management policies for the processors under dark silicon constraints.

Conclusions

Runtime management of asymmetric many-cores, especially in the dark silicon era, is notably more complex compared to the symmetric many-core architectures simply because of the spectrum of choices available for power-performance manipulation as well as the thermal constraints. This chapter presented some of the early efforts in power management of asymmetric multi-cores. The research in this domain is still in its infancy attempting to solve the individual problems. The ultimate objective of a solution that would work for a generic asymmetric multi-core under dark silicon constraints presents myriads of exciting research opportunities.

Acknowledgements This work was partially supported by CSR research funding and Singapore Ministry of Education Academic Research Fund Tier 2 MOE2012-T2-1-115.

References

1. K. Agarwal, K. Nowka, H. Deogun, D. Sylvester, Power gating with multiple sleep modes, in *International Symposium on Quality Electronic Design (ISQED)* (2006)
2. AnandTech, The Samsung Exynos 7420 Deep Dive - Inside A Modern 14nm SoC. http://www.anandtech.com/show/9330/exynos-7420-deep-dive. Accessed 2015
3. C. Bienia, S. Kumar, J.P. Singh, K. Li, The PARSEC benchmark suite: characterization and architectural implications, in *Parallel Architectures and Compilation Techniques (PACT)* (2008)
4. J. Cong, B. Yuan, Energy-efficient scheduling on heterogeneous multi-core architectures, in *International symposium on Low power electronics and design (ISLPED)* (2012)
5. T. Ebi, M Faruque, J. Henkel, TAPE: thermal-aware agent-based power economy for multi/many-core architectures, in *International Conference on Computer-Aided Design (ICCAD)* (2009)
6. H. Esmaeilzadeh, E. Blem, R.S. Amant, K. Sankaralingam, D. Burger, Dark silicon and the end of multicore scaling, in *International Symposium on Computer Architecture (ISCA)* (2011)
7. P. Greenhalgh, Big.LITTLE Processing with ARM Cortex-A15 & Cortex-A7. *ARM Whitepaper* (2011)
8. J. Henkel, M.U.K. Khan, M. Shafique, Energy-efficient multimedia systems for high efficiency video coding, in *International Symposium on Circuits and Systems (ISCAS)* (2015)
9. J. Henkel, H. Khdr, S. Pagani, M. Shafique, New trends in dark silicon, in *Design Automation Conference (DAC)* (2015)
10. J.L. Henning, SPEC CPU2006 benchmark descriptions. *Computer Architecture News (CAN)* (2006)
11. H. Hoffmann, J. Eastep, M.D. Santambrogio, J.E. Miller, A. Agarwal, Application heartbeats: a generic interface for specifying program performance and goals in autonomous computing environments, in *International Conference on Autonomic Computing (ICAC)* (2010)
12. Ineda Systems, http://inedasystems.com/hierarchical-computing.html. Accessed 2016
13. T.S. Karkhanis, J.E. Smith, A first-order superscalar processor model, in *Computer Architecture News (CAN)* (2004)
14. D. Koufaty, D. Reddy, S. Hahn, Bias Scheduling in Heterogeneous Multi-Core Architectures, in *European Conference on Computer Systems (EuroSys)* (2010)
15. R. Kumar, K.I. Farkas, N.P. Jouppi, P. Ranganathan, D.M. Tullsen, Single-ISA heterogeneous multi-core architectures: the potential for processor power reduction, in *International Symposium on Microarchitecture (MICRO)* (2003)
16. S. Landsburg, *Price Theory and Applications*, Cengage Learning, USA 2014
17. T. Li, D. Baumberger, D.A. Koufaty, S. Hahn, Efficient operating system scheduling for performance-asymmetric multi-core architectures, in *International conference on Supercomputing (ICS)* (2007)
18. T. Li, P. Brett, R. Knauerhase, D. Koufaty, D. Reddy, S. Hahn, Operating system support for overlapping-ISA heterogeneous multi-core architectures, in *International Symposium on High Performance Computer Architecture (HPCA)* (2010)
19. R. Merritt, ARM CTO: power surge could create dark silicon. *EE Times* (2009)
20. T. Mitra, Heterogeneous multi-core architectures. *Transactions on System LSI Design Methodology (T-SLDM)* (2015)
21. T.S. Muthukaruppan, M. Pricopi, V. Venkataramani, T. Mitra, S. Vishin, Hierarchical power management for asymmetric multi-core in dark silicon era, in *Design Automation Conference (DAC)* (2013)
22. T.S. Muthukaruppan, A. Pathania, T. Mitra, Price theory based power management for heterogeneous multi-cores, in *Architectural Support for Programming Languages and Operating Systems (ASPLOS)* (2014)
23. nVidia, A multi-core CPU architecture for low power and high performance. *nVidia Whitepaper* (2011)

24. C.S. Pabla, Completely fair scheduler. Linux J. **2009**(184), 4(2009)
25. M. Pricopi, T.S. Muthukaruppan, V. Venkataramani, T. Mitra, S. Vishin, Power-performance modeling on asymmetric multi-cores, in *Compilers, Architecture and Synthesis for Embedded Systems (CASES)* (2013)
26. T.S. Rosing, K. Mihic, G. De Micheli, Power and reliability management of SoCs. Transactions on Very Large Scale Integration Systems (TVLSI) (2007)
27. J.C. Saez, M. Prieto, A. Fedorova, S. Blagodurov, A comprehensive scheduler for asymmetric multicore systems, in *European conference on Computer systems (EuroSys)* (2010)
28. G. Semeraro, G. Magklis, R. Balasubramonian, D.H. Albonesi, S. Dwarkadas, M.L. Scott, Energy-efficient processor design using multiple clock domains with dynamic voltage and frequency scaling, in *International Symposium on High Performance Computer Architecture (HPCA)* (2002)
29. M. Shafique, S. Garg, T. Mitra, S. Parameswaran, J. Henkel, Dark silicon as a challenge for hardware/software co-design, in *Conference on Hardware/Software Codesign and System Synthesis (CODES)* (2014)
30. A.K. Singh, M. Shafique, A. Kumar, J. Henkel, Mapping on multi/many-core systems: survey of current and emerging trends, in *Design Automation Conference (DAC)* (2013)
31. B. Su, J. Gu, L. Shen, W. Huang, J.L. Greathouse, Z. Wang, PPEP: Online performance, power, and energy prediction framework and DVFS space exploration, in *International Symposium on Microarchitecture (MICRO)* (2014)
32. M.B. Taylor, Is dark silicon useful? Harnessing the four horsemen of the coming dark silicon apocalypse, in *Design Automation Conference (DAC)* (2012)
33. K. Van Craeynest, A. Jaleel, L. Eeckhout, P. Narvaez, J. Emer, Scheduling heterogeneous multi-cores through performance impact estimation (PIE), in *International Symposium on Computer Architecture (ISCA)* (2012)
34. S.K. Venkata, I. Ahn, D. Jeon, A. Gupta, C. Louie, S. Garcia, S. Belongie, M.B. Taylor, SD-VBS: the San Diego vision benchmark suite, in *International Symposium on Workload Characterization (IISWC)* (2009)
35. Versatile Express Board from ARM Ltd., http://www.arm.com/products/tools/development-boards/versatile-express/index.php. Accessed 2016

Chapter 7
Multi-Objective Power Management for CMPs in the Dark Silicon Age

Amir M. Rahmani, Mohammad-Hashem Haghbayan, Pasi Liljeberg, Axel Jantsch, and Hannu Tenhunen

Introduction

Dim Silicon concept is a promising approach to increase the overall throughput of chip multiprocessors (CMPs), at the expense of much lower operating frequency [1]. It is considered as one of the most effective methods to mitigate the dark silicon phenomenon. However, implementing an efficient Dim Silicon based approach necessities a comprehensive multi-objective power management mechanism having access to a rich set of on-chip sensors and actuators to utilize several Observe–Decide–Act (ODA) loops (i.e., feedback control) for controlling different aspects of the system. Such a multi-objective power management activity becomes even more challenging when considering near future many-core systems accommodating tens to hundreds of cores interconnected via Network-on-Chip (NoC). On top of that, many-core systems often need to handle an extremely dynamic workload with an unpredictable sequence of different applications entering and leaving the system at a runtime. In addition, due to the need to honor an upper limit on power consumption, i.e., fixed thermal design power (TDP) or dynamic thermal safe power (TSP) [2], in the dark silicon era, a power capping mechanism is required to monitor the instantaneous total system power consumption and manage the power–performance requirements of the system.

A.M. Rahmani (✉) • M.-H. Haghbayan • P. Liljeberg
University of Turku, Turku, Finland
e-mail: amirah@utu.fi; mohhag@utu.fi; pakrli@utu.fi

A. Jantsch
TU Wien, Vienna, Austria
e-mail: axel.jantsch@tuwien.ac.at

H. Tenhunen
KTH Royal Institute of Technology, Stockholm, Sweden
e-mail: hannu@kth.se

A.M. Rahmani et al. (eds.), *The Dark Side of Silicon*,
DOI 10.1007/978-3-319-31596-6_7

The related work on closed-loop dynamic power management for CMPs can be classified into two main categories:

1. **NoC-centric techniques** that utilize different communication related information such as queue length and injection rate as feedback to adjust voltage and frequency of processing elements, routers, or voltage–frequency islands (VFIs) accordingly (e.g., [3, 4]).
2. **Power capping techniques** proposed for bus-based multiprocessor systems which utilize chip/per-core/per-cluster power measurement and per-core performance as sensory data to optimize system power–performance characteristics within a fixed power cap where there is no concern regarding network congestion and saturation (e.g., [5, 6]).

Even though all the techniques in these categories efficiently control the power consumption for their target platforms, they are not comprehensive enough to consider several factors affecting the performance in many-core systems. Therefore, we first characterize different key parameters which should be taken into consideration to devise a proper power management approach for the dark silicon era. In the following, we list the parameters and discuss their significance:

- **Power Budget:** Due to thermal issues in the dark silicon era, there exists an upper limit on power consumption which is called thermal design power (TDP) if it is a fixed value or thermal safe power (TSP) [2] if it can change dynamically at a runtime depending on the number of active cores in a system. To guarantee the safety of the chip this limit should be strictly honored by the power manager.
- **Application Performance:** In order to monitor the impact of DVFS on application performance, a virtual or physical sensor to measure processors' utilization such as performance counters is needed. The main idea is to monitor how much impact voltage–frequency (VF) upscaling has had in the last epoch to increasing performance, and similarly how much VF downscaling has had negative impact on performance in the previous monitoring time-window.
- **NoC Congestion:** Congestion in communication medium can easily lead to a poor efficiency of DVFS process. Assume a pair of producer and consumer processing elements (PEs) where there is a congestion in one or multiple routers in their communication path. VF upscaling of such PEs will result in either zero or marginal performance gain, while a considerable amount of energy can be wasted due to a long waiting time of data transactions. Therefore, utilizing congestion meters in NoC routers can provide to the power manager a beneficial source of information.
- **Application's Network Intensity:** A source-throttling congestion control mechanism will impact in a limited way performance if it is done only based on network load [7]. Such a mechanism is not application aware, but rather throttles all applications equally regardless of applications' sensitivity to latency. Different applications impose different injection rates to the network and suffer differently from network congestion. As DVFS on PEs has also affect on application throttling, i.e., VF upscaling (downscaling) of a PE may result in increasing

(decreasing) packet injection rate by the PE to the network, applications' characteristics in terms of their network-sensitivity should be also monitored and considered in power management.

- **Applications' Priorities:** There are different types of applications, for instance, non-realtime, soft realtime, and hard realtime, where they demand different quality of service at a runtime. These requirements including the minimum required power budget for each application need to be considered in the prioritization phase in the controller.
- **Disturbances Caused by Runtime Mapping:** Whenever a new application is mapped onto the system, it is likely to cause a sudden change in overall power consumption that shoots above the TSP/TDP. Such sporadic rises in power consumption should be also considered and proactively managed.

Network centric techniques in the first category do not consider TDP or TSP, and therefore, they are not closed-loop power budgeting techniques. Power capping techniques from the second category are on the other hand unable to address communication related issues in NoC-based many-core systems. In addition, both categories consider scenarios where the impact of dark silicon phenomena is not yet that significant, e.g., in 45 nm CMOS technology. Furthermore, they often perform off-line application analysis and mapping, without support for Runtime Application Mapping (RTM) and disturbance rejection.

We argue, in the power management context, *dark silicon awareness necessitates an efficient multi-objective ODA control approach which considers workload characteristics, per-core power and performance measurements, network load, disturbances caused by runtime mapping, and total chip power measurement all together*.

In this chapter, we present a comprehensive multi-objective dark silicon aware power management platform for NoC-based many-core systems which is based on our contribution presented in [8] and considers all the discussed parameters.

The rest of the chapter is organized as follows: In section "Related Work", related work is presented. Our proposed multi-objective power management platform is presented in section "Power Management Platform". Experimental results are provided in section "Experimental Results", while section "Conclusions" concludes the chapter.

Related Work

Over the past recent years, researchers have to mitigate the impact of dark silicon to some extent. In general, the major contributions can be classified into three main categories: (1) heterogeneous computing including asymmetric multicore and 3D architectures, (2) Dim Silicon and near-threshold computing including DVFS techniques, and (3) variable symmetric multiprocessing (vSMP) technology (i.e., device-level heterogeneity).

Goulding-Hotta et al. [9] present the GreenDroid architecture which uses specialized energy-efficient processors to execute frequently used portions of the application code. The authors claim that in their Android-based workload, the specialized cores cover about 95 % of the execution time. However, the target of their proposed architecture and approach is general-purpose smartphone applications. Therefore, their specialized cores can be utilized to execute only popular frequently used applications (e.g., browser, gallery, and maps). Esmaeilzadeh et al. [10] propose a neural network based approach to the acceleration of approximate programs. In their approach, the compiler finds code sections which can be replaced by an invocation of a low-power accelerator called a neural processing unit using a learning phase and interchanges these parts accordingly. This approach can be used for specific application domains where approximate computation can be accepted such as signal processing, data mining, and robotics.

In [5], a hierarchical power management framework for asymmetric multicore architectures is demonstrated for ARM big.LITTLE [11] mobile platforms. In this architecture, cores have different size and processing power while having the same instruction-set-architecture (ISA). Ma et al. [6] have done a similar attempt to exploit power gating and DVFS for power capping in symmetric multicore processors. Their technique is demonstrated on the AMD Opteron 6168 processor and is called *PGCapping*. These platforms are energy efficient, yet they suffer from the lack of scalability as both the ARM big.LITTLE and AMD Opteron platforms are bus-based and are limited to a fewer number of cores (i.e., multicore). In contrast, our approach is applied to NoC-based general-purpose many-core systems in excess of hundred cores. The NoC-based communication structure of many-core systems necessitates more advanced controllers capable of considering workload characteristics and network congestion, in addition to power–performance feedbacks. In addition, both platforms [5, 6] do not consider RTM making it trivial to control as no disturbance occurs when an application enters or leaves the system.

Wang et al. [1] attack dark silicon by using near-threshold computing capable of increasing the number of simultaneously active cores, at the expense of lower operating frequency (i.e., Dim Silicon). Even though the approach presents promising speedup with the same power budget, they do not present any solution or control mechanism on how it can be used to manage the power consumption of many-core systems. vSMP [12] is another energy-efficient methodology presented by NVIDIA where cores with the same architecture but fabricated by different silicon processes are integrated, some using a low-power silicon process and others using a standard silicon process. This approach was tested for bus-based embedded mobile processors including five cores and optimized for key mobile use cases. However, the approach does not consider dark silicon related limitations where, for example, TDP should be taken into account.

There are several works dealing with management of dynamic workload in many-core systems. Early works in the supercomputer domain map applications only onto convex set of nodes [13]. While recent works focus on efficient processor allocation methods for many-core systems [14–16]. However, none of these works is dark silicon aware. More precisely, they do not control the system power consumption with respect to TDP.

In [17], a power management technique is presented for on-chip communication network. In this work the voltage and frequency of the interconnection network are adjusted to gain power efficiency when the network operates just below the saturation point. The approach is based on a feedback controller but focuses on the network. Its objective is to minimize power consumption while delivering all the data requested by the application. In contrast, we focus on the processing cores, which have considerably higher potential to save power, and we deal with a hard upper constraint on power consumption. In [3], a control based approach is proposed to minimize dynamic power in MPSoC made of multiple, VFIs. Their goal is to determine optimal operating frequencies for both PEs and routers. This work is not dark silicon aware either, as they do not utilize feedback from power sensors to avoid violating the TSP/TDP. On the other hand, their approach is for VFI-based NoCs where VFIs are formed in design time and hence general-purpose RTM is not properly supported.

Haghbayan et al. [18] present a power management technique for many-core systems using power feedback from the system to meet the TDP bound. This technique is categorized to single objective control approach as it lacks feedbacks from workload characteristics and per-core performance measurements from the system during DVFS process. Lack of information regarding performance and packet injection rate of PEs can easily lead to inefficient core selection for DVFS purpose, as applying DVFS to an under-utilized PE results in a totally different power–performance behavior compared to when it is applied to a busy PE. In addition, the technique presented in [18] is designed for fixed TDP and does not benefit from a dedicated disturbance rejector to handle sudden overshoots when new applications commence execution.

In [19], Chen et al. present a power allocation technique for many-core system performance improvement under power constraints. They formulate a performance optimization problem and apply an optimal power allocation method using on-line distributed reinforcement learning. This contribution does not consider on-chip communication in the problem formulation. However, it should be noted that this work is orthogonal to our multi-objective control management and can complement and enhance our method by integrating the concept of on-line learning towards more efficient global power budget reallocation.

There have been some efforts to minimize the power consumption of on-chip communication network in the dark silicon era [20, 21]. However, we focus on the processing cores, which have considerably higher potential to save power. However, these techniques can also complement our platform to manage the power consumption of the interconnection network to further optimize the power.

Although such efforts have greatly expanded in recent years, there is relatively small improvement in mitigating the dark silicon issue due to highly dynamic nature of general-purpose target workloads. The other important aspect is the dynamicity of the dark/dim area as it grows and shrinks at runtime. This motivates us to provide a general framework that can handle any number of cores in an NoC-based environment running highly dynamic workloads, minimize energy by entering even near-threshold operation, and satisfy QoS and thermal constraints by considering network congestion and application characteristics.

Preliminaries

Each application in the system is represented by a directed graph denoted as a task graph $Ap = TG(T, E)$. Each vertex $t_i \in T$ represents one task of the application, while the edge $e_{i,j} \in E$ stands for a communication between the source task t_i and the destination task t_j [22]. Task graph of an application extracted using TGG [23] is shown in Fig. 7.1.

An architecture graph $AG(N, L)$ describes the communication infrastructure of the processing elements. We consider a 2D mesh NoC (Fig. 7.1) with XY deterministic wormhole routing. The AG graph contains a set of nodes $n_{w,h} \in N$, connected together through unidirectional links $l_k \in L$. Each node is the combination of a PE connected to a router.

We define a non-realtime task as 3 tuples $tnr_i = \langle id_i, ex_i, pr_i \rangle$, and a realtime task as 5 tuples $tr_i = \langle p_i, id_i, ex_i, d_i, pr_i \rangle$, where id_i stands for the task identification, p_i represents period of task i, ex_i represents the task execution time, d_i is its deadline, and pr_i denotes the task priority. We define an abstract time unit, called as tick (e.g., 1 ms) [22].

We define an application as a set of tasks having inter-dependencies. Therefore, application mapping is a one-to-many function. We use a simple mathematical model for representing applications running on the system. Hence, no multi-tasking is assumed in any node. We denote by Application Matrix (AM) the matrix whose entry $(i, j) \in [M] \times [N]$ corresponds to the task's application ID running on the tile located in row i and column j in a mesh-based NoC topology. For example, the following application matrix shows how four applications with IDs from 1 to 4 are mapped onto a 4×4 mesh-based NoC.

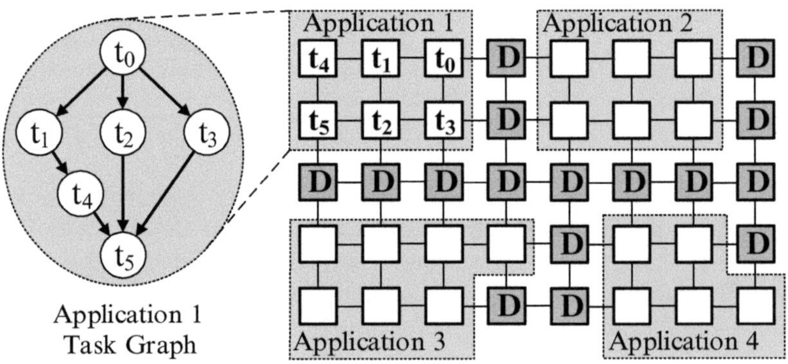

Fig. 7.1 Mesh-based platform with an application mapped onto it (*the highlighted region*) where some cores are *dark* (D)

$$AM = \begin{bmatrix} 1 & 1 & 4 & 4 \\ 1 & 1 & 3 & 4 \\ 1 & 2 & 3 & 3 \\ 2 & 2 & 2 & 3 \end{bmatrix} \quad (7.1)$$

Power Management Platform

Structure of the proposed dark silicon aware power management platform is shown in Fig. 7.2. As can be seen, a feedback controller that makes use of system power measurement is incorporated. Similar to every other control systems, the controller compares the system output with a target value. The system output in our framework is the overall power consumption of the system and the target value can be TDP or TSP. After comparison, it manipulates the system actuators to minimize the error. The controller policy to tune the actuators strongly depends on the dynamic model of the target system and the system robustness against error disturbance. The dynamic model defines how the system reacts to the inputs including actuations and other inputs. The system robustness is defined as the system stability against overshooting of the output values from the target intended output.

In our framework, per-core DVFS, per-core power gating, and application termination are used as actuators. It should be noted that the power manager does not scale the voltage and frequency of the interconnection network components (e.g., routers, links), to ensure that there is no waiting time and gainless static power consumption of the consumer PEs.

Details of the multi-objective controller (MOC) are presented in Fig. 7.3. The framework represents a general controlling strategy for many-core systems enabled with runtime mapping that can easily be applied to any NoC topologies such as 3D architectures. Runtime Mapping Unit (RMU) allocates cores in the NoC-based system to tasks of applications commenced for execution. Some information regarding the mapped applications is also provided by this unit to be passed to other controlling units. This is what we call Runtime Application Information (RAI). The priority of an application is proportional to the amount of expected QoS for that

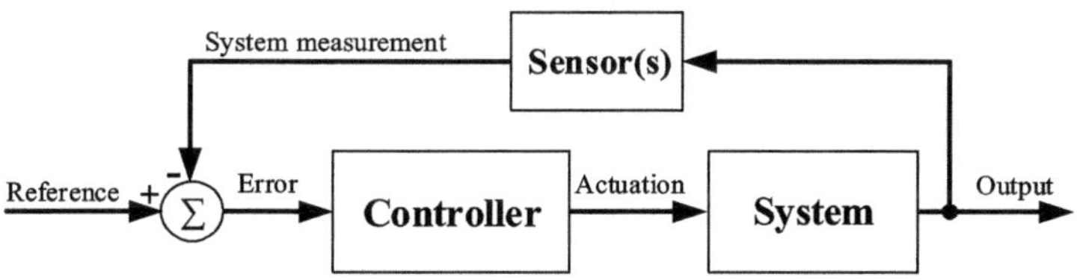

Fig. 7.2 Structure of the feedback controller

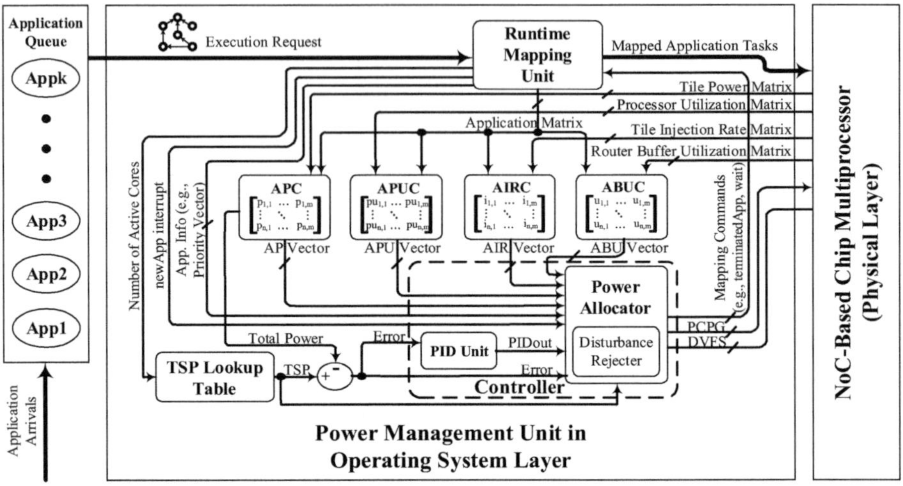

Fig. 7.3 Overview of the multi-objective dark silicon aware power management system (AIRC: Application Injection Rate Calculator, ABUC: Application Buffer Utilization Calculator, APUC: Application Processor Utilization Calculator, and APC: Application Power Calculator)

application. On a system, there might be different types of applications running with different priorities. For example, soft realtime and non-realtime application might have different levels of priority in such systems.

As discussed before, an efficient ODA-based management strategy in this context requires several observation units to monitor different system characteristics at runtime. In the following, we list the observation units integrated in our platform. It should be mentioned that the observation can be enhanced by including other system characteristics such as thermal profile, aging profile, etc.

Application Power Calculator

Each tile in our platform is assumed to be equipped with a power sensor. Power sensors transfer the information about instantaneous power consumption of cores to the central manager which forms Tile Power Matrix. It is worth to mention that many of today's platforms support such power meter equipments, e.g., Versatile Express Development Platform [11] which supports per-cluster power meters.

There are techniques presented in the literature to measure per-core power consumption by reading out current and voltage. For instance, Bakker et al. [24] propose a power measurement algorithm for Intel SCC [25] and Ma et al. [6] in the *PGCapping* approach propose a hybrid technique to monitor power consumption of individual cores. The importance of power meters is getting more evident. For instance, Esmaeilzadeh et al. [26] state that "Just as hardware event counters provide a quantitative grounding for performance innovations, power meters are necessary for optimizing energy."

The power consumption of individual routers also varies dynamically due to primarily uneven traffic distribution in a network. When a set of system resources are allocated to a task, part of the network become active with packets flowing in different directions. The associated routers regulating the packet flow thus dissipate proportional power in order to manage such a traffic.

To accurately measure power dissipation in routers, a power meter is designed within the router micro-architecture [27]. The power meter reads the rate of packet flow at link level and sends its aggregate value to the central control. There are four directional links (South, West, East, and North) and a local link connecting to a processing element. If there is no packet flow in all links, then only the leakage power is consumed. If every link is passing a packet per cycle, then the router is consuming 100 % its dynamic power actively. This happens when the network traffic is congested. However, under optimal conditions for unsaturated network, the router level power reading is less than 100 %.

The aggregate value of both core and router power consumption is sent to the central controller. Application Power Calculator (*APC*) unit in our platform calculates the current power consumption of each application based on the per tile power consumption values received from power meters and mapping information from DMU. This unit calculates Application Power Vector (*APV*), which contains the current power consumption value for each application. This is done by masking Application Matrix on the Tile Power Matrix.

Application Processor Utilization Calculator

To monitor the impact of power capping on performance, we equipped processors with performance counter through which the utilization of the processing element, in a specified time interval, is calculated and reported to the central controller. In the same manner as *APC*, Application Processor Utilization Calculator (*APUC*) unit calculates the aggregate processor utilization for each application based on the Processor Utilization Matrix generating the Application Processor Utilization Vector (*APUV*).

Application Buffer Utilization Calculator

An ideal network topology is the one with a scalable network configuration and traffic distribution where every packet is transmitted and received without delay and bandwidth limitations. For example, a network running a highly localized traffic where every node sends packets only to its immediate neighboring node can be considered as an ideal one because it exploits its maximum performance. There is no traffic congestion in the network and each packet reaches its destination within a predictable latency. Nevertheless, in practice, network traffic distribution is non-uniform and due to interconnection complexity and intrinsic wire delays such an ideal topology is not feasible. Instead more practical configurations, such as generic 2D mesh or 3D cube topologies, are used. However, the scalability of such practical topologies is limited as the capacity of the networks does not grow proportionally to accommodate traffic generated with increasing number of cores [28].

For each added core, the network traffic gets more easily congested and the overall throughput per-core decreases. Hence the total network performance gives a diminishing return due to increased communication distance. This leads to a network performance gap where every core is not able to send or receive packets in every cycle. In such cases, there is no need for a core to operate at high frequencies or voltages. Thus, we find it imperative to take the network performance gap into account when designing dynamic power management for many-core systems.

In our platform, each router is equipped with a buffer utilization meter that measures router congestion level in a specified time interval. More precisely, it measures the traffic dynamically by calculating the moving average of packet flow in every link of a router as follows:

$$C_{Total} = \frac{1}{W} \sum_{cycle=i}^{i+w} (\theta_{South,i} + \theta_{North,i} + \theta_{East,i} + \theta_{West,i} + \theta_{Local,i}) \quad (7.2)$$

where $\theta = \{0, 1\}$ is the presence or absence of a packet in a link at any given cycle, W is the width of the moving window, and C_{Total} is the moving average congestion level. This buffer utilization of each router (Router Buffer Utilization Matrix in Fig. 7.3) is sent to the Application Buffer Utilization Calculator (*ABUC*). Then, by masking the Application Matrix (provided by the RMU) on the Router Buffer Utilization Matrix, *ABUC* calculates the average buffer utilization vector, i.e., *ABUV*, which is exploited in the controller unit.

Application Injection Rate Calculator

In the power management process, applications are categorized and managed based on their computation intensiveness as well as communication intensiveness. In [7], it is shown that a source-throttling congestion control mechanism will have a limited performance improvement if it is done only based on the network load. Such a mechanism is not application aware, but rather throttles all applications equally regardless of applications' sensitivity to latency. Inspired from [7], we also consider applications' network intensity in order to classify them into intensive and non-intensive categories in the power management process. We use application injection rate as a metric that closely correlates to network intensity. It should be noted that

DVFS has a throttling effect on the system as voltage and frequency (VF) upscaling results in increasing packet injection rate, and likewise, VF downscaling leads to decreasing packet injection rate.

The injection rate of each task running on a tile is measured at the tile's network interface for the last epoch and transferred to the Application Injection Rate Calculator (*AIRC*). By masking the Application Matrix (provided by the RMU) on the Tile Injection Rate Matrix, *AIRC* calculates the average injection rate for each application and puts it on the use at the controller unit.

TSP Lookup

TDP calculation is performed at a design time regardless of runtime thermal distribution across the silicon area. Therefore, using a constant value as an upper bound for power (i.e., TDP) can result in large performance losses [2]. To optimize the power budget calculation, a new power budget concept called Thermal Safe Power (TSP) has been proposed which is a function of the number of active, i.e., non-dark, cores in a system determined at runtime. In our platform, TSP_{worst} is used which is calculated for the worst-case mapping through which it is assumed that all active cores are physically packed and influencing the temperature of their adjacent cores. Other parameters needed to calculate TSP, e.g., floorplan, power consumption of an inactive core, etc., are generally available at a design time. The worst-case calculation function returns a uniform value of TSP per-core for all active cores.

In our system, TSP_{worst} values for different number of active cores are pre-calculated and stored in a small lookup table (a one-dimensional array). The overhead of such a lookup table is negligible as it needs H entries where H is the number of cores in the system. The *number of active cores* can be used as the index for the array to avoid any search function. For example, $P_{TSP}^{worst}(9) = 12.5\,\text{W}$ indicates that the safe power budget is 12.5 W when there are 9 active cores in the system. This function is called whenever an application enters or leaves the system at runtime. It should be noted that if a fixed TDP value is desired, the lookup table can be simply replaced with the fixed value.

Power Management Policy

In the previous sections, the multi-objective ODA loop was discussed. As shown in Fig. 7.3, within the Controller Unit, a Proportional–Integral–Derivative (PID) controller monitors the error between the actual power consumed by the system and the reference thermal power (i.e., TDP or TSP). Then, the downstream Power Allocator Unit decides how to handle the power management features based on the output of PID controller and other system parameters. In this section, we present the decision-making policy (i.e., decide state comprising different controllers) in detail.

PID Controller Unit

The general expression for a PID controller is formulated as follows:

$$PID_{out}(t) = K_p e(t) + K_i \int e(t)\,\mathrm{d}t + K_d \frac{de(t)}{dt} \tag{7.3}$$

where $PID_{out}(t)$, $e(t)$, K_p, K_i, and K_d are the controller output, error, proportional gain, integral gain, and derivative gain, respectively.

Several Matlab simulations are performed to adjust the gains of the PID controller. The system stability and robustness are two essential aspects that need to be carefully considered when adjusting the gains. In a PID controller, each gain magnifies the importance of a specific system behavior. The proportional gain directs a change in the controller output by magnifying the change of the error value. The integral gain responds to the accumulated errors that magnify the effect of history on controller's output value to eliminate the residual steady-state error. The derivative gain predicts the system behavior by magnifying the most recent changes in error value.

A high proportional gain leads to a large change in the output for a given change in the error value. Therefore, the proportional gain should have the foremost contribution when determining the controller's output. On the other hand, the integral term accelerates the movement of the process towards the target value. However, as the integral term responds to the accumulated errors from the past, it can affect the present value by overshooting the target value. Derivative action predicts system behavior and thus improves settling time and stability of the system.

In multiprocessor systems utilizing runtime task mapping, there are often three influential bearings in the total power trace curve: (1) when an application enters the system, (2) when an application leaves the system, and (3) when there is no incoming or outgoing application to/from the system yet power consumption changes due to different changes in intra-application task behaviors such as task dependencies and varying switching activities. These three behaviors make the system workload often highly dynamic and unpredictable, resulting in sudden steep slopes in the power trace curve, i.e., disturbances. We address all the three behaviors. The PID controller can efficiently handle the second and third behavior. However, when an application enters the system (the first behavior), a high overshoot may happen especially if the application size is large and demands several dark cores to be activated. This situation is separately handled by the disturbance rejector unit in a proactive way, discussed in later sections. In summary, *Power Allocator handles the second and third behaviors using PID Controller unit while it manages the first behavior with the help of Disturbance Rejector unit.*

Power Allocator

The main task of the power allocator unit is to manipulate voltage and frequency of the processing elements, i.e., (V_{PEs}, $Freq_{PEs}$), by utilizing the information obtained from the observation units and the directions provided by the PID controller. Algorithm 1 shows the process to obtain V_{PEs} and $Freq_{PEs}$. At the first step, each active core's power limit is determined by dividing the overall power budget i.e., TDP or TSP, by the number of active cores. After that, based on applications' injection rate (IR) information obtained from Application Injection Rate Vector (AIRV), all the applications are classified into two categories, intensive (I_{set}) and non-intensive (NI_{set}), by making use of *IRClassifier* function. Similarly, through *BUClassifier* function the applications are also classified into two categories, congested (C_{set}) and non-congested (NC_{set}), based on their buffer utilization (BU) information obtained from Application Buffer Utilization Vector (ABUV). An application is considered to be congested if its corresponding routers' buffer utilization value is larger than a predefined threshold, for example, 75 %. Figure 7.4 shows four possible application types after classification by *IRClassifier* and *BUClassifier* functions. In this way, every application is tagged at runtime with a 2-bit label which can get one of these values: *NI_NC* (non-intensive, non-congested), *NI_C* (non-intensive, congested), *I_NC* (intensive, non-congested), and *I_C* (intensive, congested). These tags are variable and updated in every iteration. This provides appropriate target set of applications that can be upscaled or downscaled to maximize network throughput.

After classification, based on the applications' type and the PID_{out} (the output of the PID controller), voltage/frequency of each processing elements is downscaled or upscaled. There can be two possible scenarios to deal with: overshoot (i.e., power consumption exceeding TSP/TDP) and undershoot (i.e., power consumption beneath the TSP/TDP). An overshoot violates TSP/TDP constraint, while an undershoot represents resource under-utilization. $VF_{downscaler}$ and $VF_{upscaler}$ functions are used to scale the voltage and frequency of target applications. When a new application to be mapped arrives, Power Allocator receives a *newApp interrupt* from the mapping unit. This interrupt is serviced by the *proactiveDistRej* function implemented in the Disturbance Rejecter module which proactively scales currently running applications. The power monitoring continues normally when there is no new application arrival. Pruning the application space in case of an overshoot and undershoot is explained in the following subsections.

Voltage–Frequency Downscaler

VF downscaling process of PEs is explained in Algorithm 2. We consider the entire application space ($C_{set} \cup NC_{set}$) to choose the target application set to be downscaled. When there is an overshoot, applications with the lowest priority are chosen by the LP_{apps} function, letting the high priority applications run at a higher QoS level. *RAI* includes application priorities known from the application priority

Fig. 7.4 Four possible applications types classified by *IRClassifier* and *BUClassifier* functions

Application Types	
Non-Congested (*NC*) **Non-Intensive (*NI*)**	**Non-Congested (NC)** **Intensive (I)**
Congested (C) **Non-Intensive (NI)**	**Congested (C)** **Intensive (I)**

Algorithm 1: Power allocation algorithm

Inputs: PID_{out}, RAI, $ABUV$, $AIRV$, APV, $APUV$, TSP, $newAppinterrupt$, $Error$
Output: V_{PEs}, $Freq_{PEs}$, $terminatedApp$
Global Variables: $DVFSList$, I_{set}, NI_{set}, C_{set}, NC_{set}, $PCPowerLimit$
Constant values: $bufferUtilizationLimit$
Body:

$PCPowerLimit \leftarrow \frac{TSP}{\#activeCores}$; // calculating per-core power limit
$(I_{set}, NI_{set}) \leftarrow IRClassifier\ (AIRV, RAI)$; // classify I and NI
$(C_{set}, NC_{set}) \leftarrow BUCassifier\ (ABUV, RAI)$; // classify C and NC
if $newAppinterrupt$ **then** {// interrupt - new application to be mapped}
  $(V_{PEs}, Freq_{PEs}, terminatedApp) \leftarrow proactiveDistRej\ (Error, RAI, ABUV, APV, APUV)$;
else
  if $PID_{out} < 0$ **then**
    $(V_{PEs}, Freq_{PEs}, terminatedApp) \leftarrow VF_{downscaler}\ (RAI, ABUV, APV, APUV, PID_{out}, PCPowerLimit)$;
  else
    $(V_{PEs}, Freq_{PEs}, terminatedApp) \leftarrow VF_{upscaler}\ (RAI, ABUV, APV, APUV, PID_{out}, PCPowerLimit)$;
  end if
end if

vector. Among them, applications that are tagged as congested (C_{set}) are chosen to minimize congestion and improve network throughput. PEs residing in a congested area can dissipate unnecessary power (particularly static) due to a low network throughput. As VF downscaling also affects throttling of packet injection, it can alleviate the network congestion for such applications and save power. In case of unavailability, congested set is replaced by non-congested set (NC_{set}). These are further narrowed down to the application with the lowest performance loss to power reduction ratio by the $lowD_{prf-pwr}$ function which is presented in detail in the following. Finalized target application (*targetApp*) is then downscaled by the *DVFS* function as per PID_{out} and *PCPowerLimit*.

The DVFS function fails when it cannot throttle the target application any further according to the application type, which occurs when VF level cannot be reduced anymore. When the *failedDVFS* flag is asserted, the application will be removed from the *availableApp* and the algorithm will keep on searching until an alternative application is found.

Algorithm 2: Voltage and frequency downscaling function

```
Inputs: RAI, ABUV, APV, APUV, PID_out, PCPowerLimit
Outputs: V_PEs, Freq_PEs, terminatedApp
Variables: availableApps, targetApp, failedDVFS, appSet
Body:
availableApps ← C_set ∪ NC_set; // application space
while true do
    targetApp ← ∅; // the application targeted for DVFS
    appSet ← LP_apps (availableApps, RAI); // low priority apps
    appSet ← appSet ∩ C_set;
    if appSet = ∅ then {// consider congested apps}
        appSet ← appSet ∩ NC_set; // consider non-congested apps
    end if
    targetApp ← lowD_prf−pwr (appSet, APV, APUV, PID_out);
    (V_PEs,Freq_PEs,failedDVFS) ← DVFS(targetApp, PID_out, PCPowerLimit);
    if failedDVFS then
        remove targetApp from availableApps; continue;
        if availableApps is empty then
            terminatedApp ← targetApp; break;
        end if
    else
        DVFSList ← targetApp; break;
    end if
end while
```

Voltage–Frequency Upscaler

Voltage/Frequency upscaling of processing elements is presented in Algorithm 3. When there is an undershoot, first, the set of applications that are already downscaled and applications that are non-intensive and non-congested ($availableApps \cap NI_{set} \cap NC_{set}$) are chosen. The ground for this selection is that upscaling voltage and frequency of a PE residing in a congested area and having a high injection rate may result in zero performance gain if on-chip communication network is the bottleneck. That is the reason why in VF upscaling process, in contrast to downscaling, a higher priority is given to congestion than application priority in the algorithm. If there is no *NI_NC* application in the system, *I_NC* applications will be the next candidates set ($DVFSList \cap I_{set} \cap NC_{set}$). Among these, applications with the highest priority are picked by the HP_{apps} function to meet system's QoS demands. These are further narrowed down to the application with the highest performance gain to power increase ratio ($HighD_{prf-pwr}$). The chosen target application is then upscaled by the *DVFS* function as per PID_{out} and *PCPowerLimit*.

The DVFS function considers both normal and near-threshold operations. Voltage-to-frequency scalings are modeled by interpolating empirical results from circuit simulations [29]. Transistor switching speed scales exponentially with the threshold voltage while operating at near-threshold voltage. As a result, near-threshold operation region is highly sensitive to the threshold voltage [29]. More

Algorithm 3: Voltage and frequency upscaling function

```
Inputs: RAI, ABUV, APV, APUV, PID_out, PCPowerLimit
Outputs: V_PEs, Freq_PEs, terminatedApp
Variables: availableApps, targetApp, failedDVFS, appSet
Body:
targetApp ← Ø;
availableApps ← DVFSList;
while targetApp = Ø do
    appSet ← availableApps ∩ NC_set ∩ NI_set; // non-congested/non-intensive apps
    if appSet = Ø then
        appSet ← availableApps ∩ NC_set ∩ I_set; // non-congested/intensive apps
        if appSet = Ø then
            appSet ← availableApps;
        end if
    end if
    appSet ← HP_apps(appSet, RAI); // high priority apps
    targetApp ← highD_prf−pwr (appSet, APV, APUV, PID_out);
    (V_PEs, Freq_PEs, failedDVFS) ← DVFS (targetApp, PID_out, PCPowerLimit);
    if failedDVFS then
        remove targetApp from availableApps;
        targetApp ← Ø; continue;
    end if
end while
remove targetApp from DVFSList;
```

details regarding near-threshold frequency and voltage modeling can be found in [29]. As there are limited amount of voltage and frequency levels, the *DVFS* function ignores marginal deviations of PID_{out} from its previous value for the sake of stability.

Functions $highD_{prf-pwr}$ and $lowD_{prf-pwr}$

Functions $highD_{prf-pwr}$ and $lowD_{prf-pwr}$ search for an application with the highest or lowest performance-power ratio (i.e., $D_{prf-pwr}$) in a given set to be the target of VF upscaling or downscaling. In [6], product of core utilization (*Util*) and aggregated frequency (*Freq*) is used as a high-level computational capacity metric. In this metric, the frequency is weighted to deduct the idling cycles. We extend this metric by aggregating core utilization in an application (*appUtil*), provided by *APUC*, to calculate the performance of an application as:

$$Perf_{current} = appUtil \times Freq_{current} \tag{7.4}$$

Then, the performance-power ratio is calculated as the following:

$$D_{prf-pwr} = \frac{Perf_{next} - Perf_{current}}{Power_{next} - Power_{current}} \tag{7.5}$$

$Power_{current}$ is the power consumption of the current application provided by the *APC* unit. $Power_{next}$ and $Perf_{next}$ are the estimated power consumption and performance of the application after the DVFS process. The next level of voltage and frequency (V_{ddnext} and $Freq_{next}$) is estimated for the candidate applications based on the magnitude of PID_{out} and application size. The $Perf_{next}$ and $Power_{next}$ are calculated as follows:

$$Perf_{next} = Perf_{current} \times \frac{Freq_{next}}{Freq_{current}} \tag{7.6}$$

$$Power_{next} = Power_{current} \times \frac{Freq_{next}}{Freq_{current}} \times (\frac{V_{ddnext}}{V_{ddcurrent}})^2 \tag{7.7}$$

After calculating $D_{prf-pwr}$ for all the applications in *appSet*, $lowD_{prf-pwr}$ and $highD_{prf-pwr}$ functions use a simple *quicksearch* algorithm to find the application with the lowest and highest $D_{prf-pwr}$ value as the target application for DVFS, respectively.

Proactive Disturbance Rejection

Whenever a new application is mapped onto the system, it is likely to cause a sudden change in overall power consumption that shoots above the TSP/TDP. Such sporadic rises in power consumption can be minimized by proactively scaling down applications that are currently running on the system. Algorithm 4 details the PDR function. If *Error* is positive, indicating that new application can be accommodated, the predicted power consumption (*appPredictedPower*) is calculated based on the number of tasks (*N* extracted from *RAI*) of the new application and average power consumed by actively running cores (P_{avg}). The difference between *Error* and *appPredictedPower* is the *proactiveError*, which is fed back to a proportional controller with gain K'_p. Here, the integral and derivative terms are removed because when such sporadic rises occur, history-based (i.e., integral term) or prediction-based (i.e., derivative term) decision-making will most likely affect the controller's response. Output of the controller (P_{out}) determines the extent by which currently running applications are to be scaled so that the new application can be mapped without violating TSP/TDP. If the ($Error > 0$) and ($proactiveError > 0$), indicating availability of power budget that can be allocated to new application, it is mapped as it is without any further scaling. If ($Error > 0$) and ($proactiveError < 0$), indicating that power allocation to new application would violate TSP/TDP, currently running applications are downscaled by $VF_{downscaler}$ based on P_{out}. The new application is pushed onto the stack, annexed to the list of applications running with DVFS.

Algorithm 4: Proactive disturbance rejection (*proactiveDistRej()*).

Inputs: *Error*, *RAI*, *ABUV*, *APV*, *APUV*, *PCPowerLimit*
Outputs: V_{PEs}, $Freq_{PEs}$, *terminatedApp*
Variables: *failedDVFS*, *appPredictedPower*, *proactiveError*, P_{out}, P_{avg}
Constant values: K'_p
Body:

appPredictedPower $\leftarrow N \times P_{avg}$;
proactiveError $\leftarrow$ *Error* - *appPredictedPower*;
if *proactiveError* < 0 **then**
  $P_{out} \leftarrow K'_p \times$ *proactiveError*;
  ($V(PEs)$, *Freq*(PEs), *terminatedApp*) $\leftarrow VF_{downscaler}$ (*RAI*, *ABUV*, *APV*, *APUV*, P_{out}, *PCPowerLimit*);
end if

Experimental Results

In this section, we present the experimental results with 16 nm Complementary metal "oxide" semiconductor (CMOS) technology node for evaluating our proposed multi-objective dark silicon aware power management platform.

Experiment Setup

To experimentally evaluate the proposed approach, we implemented a system-level simulation platform for the described many-core architecture together with accompanying runtime management layer and testing procedures in SystemC on the basis of Noxim NoC simulator [30]. The basic core has been characterized by using the Niagara2-like in-order core specifications obtained from McPAT [31]. Physical scaling parameters were extracted from the Lumos framework (by Wang and Skadron) [29]. Lumos is a framework to analytically quantify the power–performance characteristics of many-core systems especially in near-threshold operation. Lumos is open source and publicly available [32]. The physical scaling parameters have been calibrated by circuit simulations with a modified Predictive Technology Model [33]. Moreover, we have imported other models and specifications such as power modeling, voltage–frequency scaling, TDP calculation, and near-threshold computing parameters from the Lumos framework. Our many-core platform was reinforced to support RTM by implementing a central manager (CM) residing in the node $n_{0,0}$. The network size is 12×12 and the chip area is $138\,\text{mm}^2$.

We model two application categories—non-realtime (the lowest priority) and soft realtime (the highest priority). Several sets of non-realtime applications with 4–35 tasks are generated using TGG [23] where the communication and computation volumes are randomly distributed. We model MPEG4 and VOPD multimedia applications as soft realtime applications. The realtime requirements of these applications require the system to respond within certain deadlines for different priority levels. We pre-calculate the minimum VF level for soft realtime tasks for their worst-case contiguous mapping. Soft realtime here means that these applications can provide different quality of services, for example, by processing different frame rates per second depending on the availability of system resources.

In our multi-application many-core system, a random sequence of applications enter the scheduler FIFO. This sequence is kept fixed in all experiments for the sake of fair comparison. The probabilities of selecting soft realtime and non-realtime applications from the application repository are 30 % and 70 %, respectively. CM selects the *first node* using SHiC [34] method, and maps the application based on its realtime attributes. The soft realtime applications are mapped contiguously. In addition to the RMU, our multi-objective power management platform (including the controller, AIRC, ABUC, etc.) is also implemented in software (i.e., soft coded) as a part of the CM. This makes the area overhead of the proposed method so negligible. In other words, the congestion meters embedded in the NoC routers and the power sensors are the only extra hardware components needed to implement our idea. CM receives feedbacks from the whole network and sends actuation commands to each tile. These short control packets are synchronously sent along with the ordinary traffic using the same on-chip network. At first glance, the centralized approach seems unscalable and the control traffic overhead looks considerable. As the control interval can be long (i.e., millisecond scale) compared to the system clock period (i.e., nanosecond scale), the control traffic overhead is negligible and control packets have enough time to reach the destination even in large networks. For example, an NoC system running at 750 MHz can be as large as 75,000 cores, while the time for control packet collection is <1 % of sampling interval of 10 ms.

In our platform, we assumed that the chip is equipped with power sensors. Therefore, we need to model the power sensors to estimate the power consumption in our simulations. A PE can be in three different states in our simulations: (1) "*Busy*" when all the required input packets have been received and the corresponding task is being executed, (2) "*Waiting*" when a PE is clock-gated as it is waiting (due to data dependencies) for upcoming packets to be completely stored in the input FIFO, and (3) "*Dark*" when a PE is idle and power-gated. We estimate the power consumption for these different states according to the following equations:

$$P_{Busy} = P_{static} + P_{dynamic} \tag{7.8}$$

$$P_{Waiting} = P_{static} \tag{7.9}$$

$$P_{Dark} \simeq 0 \tag{7.10}$$

where P_{static} and $P_{dynamic}$ are static power and dynamic power of each core in the system, respectively. The static and dynamic power consumption are calculated using the following equations [29]:

$$P_{dynamic} = \alpha.C_{eff}.V_{dd}^2.f \tag{7.11}$$

$$P_{static} = V_{dd}.N.k_{design}.\hat{I}_{leak} \tag{7.12}$$

where α denotes the switching activity factor of a PE while running a task, C_{eff} is the effective capacitance, V_{dd} is the supply voltage, f is the core's frequency, N is the number of transistors, k_{design} is a device-specific constant, and $\hat{I}_{leak}$ is the normalized per-transistor leakage current. To be consistent with the parameters used in the Lumos framework, we assume a constant activity factor and effective capacitance when the core frequency is scaled with various supply voltages. More details regarding the above equations can be found in [29].

For the DVFS purpose, we use 15 VF levels (similar to Intel SCC) including near-threshold operation extracted from the Lumos framework. The minimum and maximum VF levels are set to (0.456 V, 300 MHz) and (0.908 V, 5.2 GHz), respectively. Lumos models voltage–frequency scalings in an *optimistic way*—without considering reliabilities and also in *pessimistic way*—considering the process variations through an analytical model with rigid voltage downscaling. We chose the pessimistic option in our simulations to ensure a higher degree of reliability for same architecture, sacrificing possible energy gains.

We define different minimum voltage–frequency levels depending on the application type. For example, we use all the 15 levels to scale voltages and frequencies of PEs running non-realtime applications. The frequency of the on-chip communication network (e.g., routers) is set to the maximum level (i.e., 5.2 GHz) to demonstrate that even at the maximum NoC speed, the network can get congested and should be taken into account in power management along with the other parameters. For the TSP calculation, we follow the same floorplan style, chip thickness, silicon thermal conductivity, and heat sink model as [2]. We set ambient temperature to 45 °C, a threshold temperature that triggers thermal management to 80 °C, maximum chip power consumption from the power supply to 300 W, and the power consumption of an inactive core to 0.3 W.

We compare different characteristics of the many-core system under four different management scenarios: (1) our MOC with proactive disturbance rejection (PDR), (2) PGCapping [6] in which the power management technique only considers core's power–performance ratio as the feedback for the PCPG and per-core DVFS actuation, (3) DSAPM [18] in which PID controller is used to control the system power consumption, however no information regarding performance and packet injection rate of PEs is considered in power allocation policy, and (4) without TSP/TDP constraint in which there is not any policy to control the power. Without TSP/TDP constraint is the scenario where the system is not limited in terms of maximum power consumption. This is the situation when, in reality, the chip is damaged due to overheating. We consider 10 s warm up phase for the results. To perform a fair comparison, we run PGCapping and DSAPM techniques with the same 15 VF levels for per-core DVFS actuation.

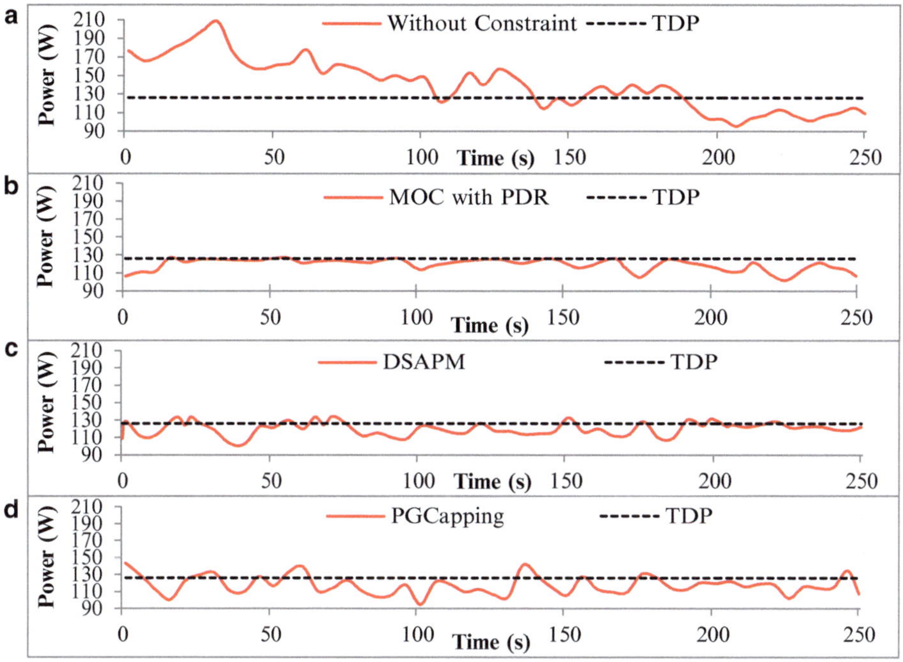

Fig. 7.5 The power consumption of the system using (**a**) without TSP/TDP constraint, (**b**) MOC with PDR, (**c**) DSAPM, and (**d**) PGCapping power management policies to honor TDP

Results

Power consumption of the system under the aforementioned power management scenarios to honor constant TDP is presented in Fig. 7.5. The dashed black curve represents maximum power budget for the system (i.e., TDP). The TDP value is set to 126 W which is calculated based on the chip power density. Deviation of power consumption from the baseline reflects either violation or under-utilization of power budget. Power consumption in case of the PGCapping, DSAPM, and without-constraint power managements mostly tends to overshoot or undershoot from TDP. The without-constraint power management does not consider any upper bound on power consumption, subsequently it violates the TDP constraint right through.

PGCapping benefits from the cores' power–performance values, fed back by the controller and thus increases the system throughput to some extent. However, it suffers from the under-utilization issue as it does not consider the network congestion and applications injection rates. DSAPM considers network congestion, however it also suffers from the under-utilization issue as it is agnostic of cores' performance value and applications' injection rates. Moreover, both PGCapping and DSAPM techniques refuse to properly handle occasional overshoots due to new application arrivals. Evidently, MOC with PDR stays in close proximity with

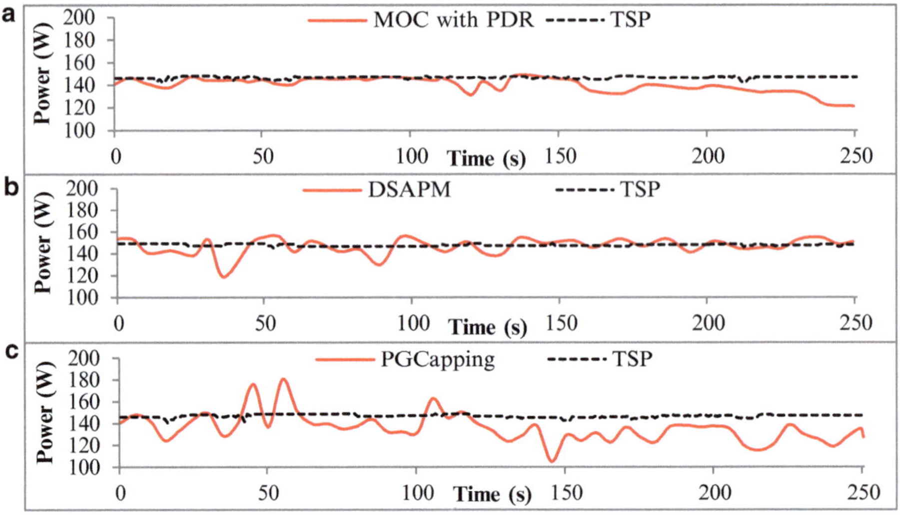

Fig. 7.6 The power consumption of the system using (**a**) MOC with PDR, (**b**) DSAPM, and (**c**) PGCapping power management policies to honor TSP

TDP and hence has the best power management mechanism in comparison with the others. In cases where power consumption exceeds TDP, the MOC controller rapidly reduces the power consumption by a proper voltage and frequency scaling. The control system is stable even for large fluctuations in power consumption that occur with arrival of intense applications. Figure 7.6 demonstrates the aforementioned power management scenarios to honor dynamic TSP values. As can be observed from the figure, the conclusions we made for Fig. 7.5 are also valid for dynamic TSP, the MOC-based system is stable even when budget is changed at runtime. In the figure, TSP does not radically change (often between 141 and 149 W) as the system is mostly busy and the majority of cores are active.

To assess the efficiency of our platform, we compare the normalized throughput for the set of applications under MOC (with PDR), PGCapping, and DSAPM policies, as shown in Fig. 7.7. The results reveal that our proposed method can significantly improve the overall system throughput for different power budget types (up to 29 % compared with PGCapping and up to 15 % compared with DSAPM). The results reveal the advantage of our proposed MOC which considers both the computation and communication aspects in power management. Figure 7.8 shows TDP/TSP violation for different power management policies over time. We measure violation as the ratio of time for which power consumption exceeded TDP/TSP (resulting in a violation) to the entire simulation time. It can be observed that the proposed disturbance rejection technique honors the TDP/TSP constraints for more than 99 % of the simulation time. System utilization is another important factor to be analyzed for the aforementioned policies. As shown in Fig. 7.9, the MOC policy considerably increases the system utilization compared to the DSAPM and

Fig. 7.7 Normalized throughput of MOC vs. PGCapping vs. DSAPM

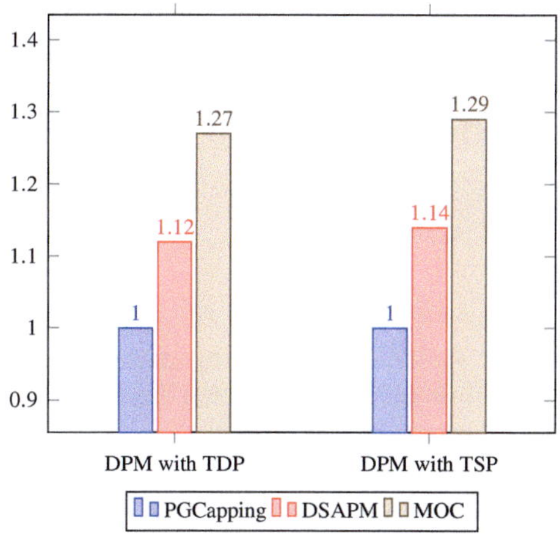

Fig. 7.8 TDP/TSP violation for different dynamic power managements

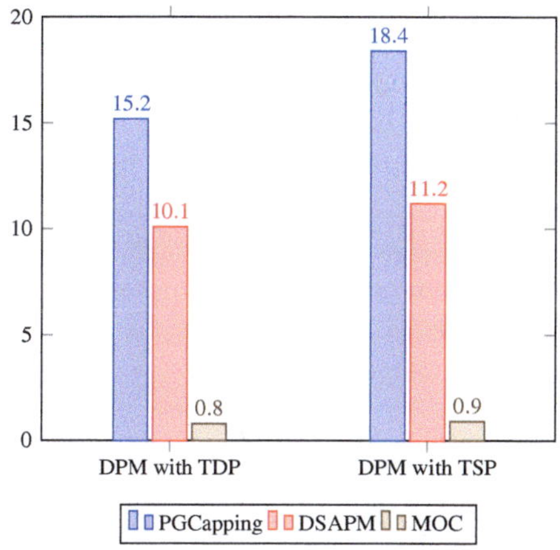

Fig. 7.9 Average of the core utilization (%) for different dynamic power managements

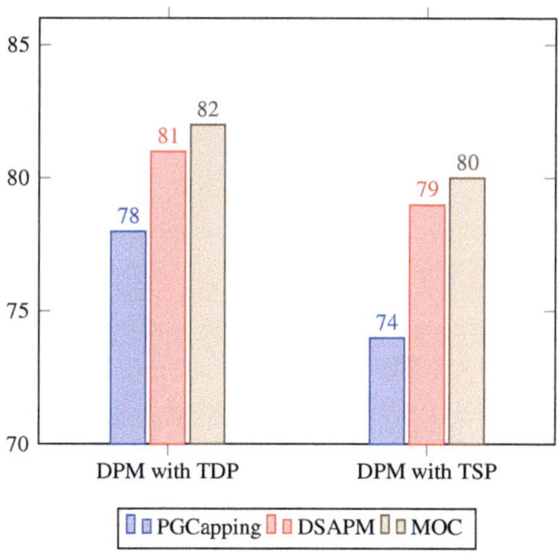

PGCapping policies. The main reason for the improvements in terms of system utilization is that MOC exploits the power saved from the scenarios where network congestion is the bottleneck.

Conclusions

The need to utilize controlling mechanisms in management of complex multiprocessor systems is becoming more evident particularly when the number of cores in a chip increases. In this chapter, we introduced a multi-objective feedback based controller approach to protect many-core systems against exceeding the power consumption from a certain limit while maximizing system utilization and throughput. The target system architecture was an NoC-based multiprocessor system using dynamic application mapping where applications enter and leave the system at runtime. We utilized a variety of feedbacks such as processing elements' power–performance measurements, application workloads, and network congestion to monitor the system. Two different algorithms for downscaling and upscaling the voltage and frequencies of the cores, and a proactive strategy to avoid power consumption violations when the system encounters rapid power increases were discussed. It was shown that the controller efficiently changes voltage and frequency of appropriate processing elements, down to near-threshold operation when needed. The results show improvements in system throughput as well as reductions in TDP/TSP violations, for the proposed platform when compared to state-of-the-art power management policies.

References

1. W. Liang, K. Skadron, Implications of the power wall: dim cores and reconfigurable logic. IEEE Micro **33**(5), 40–48 (2013)
2. S. Pagani, H. Khdr, W. Munawar, J. Chen, M. Shafique, M. Li, J. Henkel, TSP: thermal safe power: efficient power budgeting for many-core systems in dark silicon, in *Proceedings of the 2014 International Conference on Hardware/Software Codesign and System Synthesis*, CODES '14 (2014)
3. P. Bogdan, R. Marculescu, S. Jain, Dynamic power management for multidomain system-on-chip platforms: an optimal control approach. ACM Trans. Des. Autom. Electron. Syst. **18**(4), 46:1–46:20 (2013)
4. R. David, P. Bogdan, R. Marculescu, U. Ogras, Dynamic power management of voltage-frequency island partitioned networks-on-chip using intel sing-chip cloud computer, in *Proceedings of the Fifth ACM/IEEE International Symposium on Networks-on-Chip*, NOCS '11 (2011), pp. 257–258
5. T.S. Muthukaruppan, M. Pricopi, V. Venkataramani, T. Mitra, S. Vishin, Hierarchical power management for asymmetric multi-core in dark silicon era, in *Proceedings of Design Automation Conference* (2013), pp. 1–9

6. K. Ma, X. Wang, PGCapping: exploiting power gating for power capping and core lifetime balancing in CMPs, in *Proceedings of the 21st International Conference on Parallel Architectures and Compilation Techniques*, PACT '12 (2012), pp. 13–22
7. K.K. Chang, R. Ausavarungnirun, C. Fallin, O. Mutlu, HAT: heterogeneous adaptive throttling for on-chip networks, in *Proceedings of IEEE 24th International Symposium on Computer Architecture and High Performance Computing (SBAC-PAD)* (2012), pp. 9–18
8. A.-M. Rahmani, M.-H. Haghbayan, A. Kanduri, A.Y. Weldezion, P. Liljeberg, J. Plosila, A. Jantsch, H. Tenhunen, Dynamic power management for many-core platforms in the dark silicon era: a multi-objective control approach, in *Proceedings of International Symposium on Low Power Electronics and Design (ISLPED)* (2015), pp. 1–6
9. N. Goulding-Hotta, J. Sampson, G. Venkatesh, S. Garcia, J. Auricchio, P. Huang, M. Arora, S. Nath, V. Bhatt, J. Babb, S. Swanson, M.B. Taylor, The GreenDroid mobile application processor: an architecture for silicon's dark future. IEEE Micro **31**(2), 86–95 (2011)
10. H. Esmaeilzadeh, A. Sampson, L. Ceze, D. Burger, Neural acceleration for general-purpose approximate programs, in *Proceedings of IEEE/ACM International Symposium on Microarchitecture* (2012), pp. 449–460
11. ARM Ltd. (2011), http://www.arm.com/products/tools/development-boards/versatile-express/index.php
12. Variable SMP: a multi-core CPU architecture for low power and high performance, in *White paper, NVIDIA* (2011)
13. M.A. Bender, D.P. Bunde, E.D. Demaine, S.P. Fekete, V.J. Leung, H. Meijer, C.A. Phillips, Communication-aware processor allocation for supercomputers: finding point sets of small average distance, in *Proceedings of Algorithmica* (2008), pp. 279–298
14. E. Carvalho, N. Calazans, F. Moraes, Heuristics for dynamic task mapping in NoC-based heterogeneous MPSoCs, in *Proceedings of 18th IEEE/IFIP International Workshop on Rapid System Prototyping* (2007), pp. 34–40
15. M. Fattah, P. Liljeberg, J. Plosila, H. Tenhunen, Adjustable contiguity of run-time task allocation in networked many-core systems, in *Proceedings of Asia and South Pacific Design Automation Conference (ASP-DAC)* (2014), pp. 349–354
16. M.-H. Haghbayan, A. Kanduri, A.-M. Rahmani, P. Liljeberg, A. Jantsch, H. Tenhunen, MapPro: proactive runtime mapping for dynamic workloads by quantifying ripple effect of applications on networks-on-chip, in *Proceedings of International Symposium on Networks-on-Chip (NOCS)* (2015), pp. 1–8
17. G. Liang, A. Jantsch, Adaptive power management for the on-chip communication network, in *Proceedings of the EUROMICRO Conference on Digital System Design* (2006), pp. 649–656
18. M.-H. Haghbayan, A.-M. Rahmani, A.Y. Weldezion, P. Liljeberg, J. Plosila, A. Jantsch, H. Tenhunen, Dark silicon aware power management for manycore systems under dynamic workloads, in *Proceedings of International Conference on Computer Design (ICCD)* (2014), pp. 509–512
19. Z. Chen, D. Marculescu, Distributed reinforcement learning for power limited many-core system performance optimization, in *Proceedings of the 2015 Design, Automation & Test in Europe Conference & Exhibition* (2015), pp. 1521–1526
20. H. Bokhari, H. Javaid, M. Shafique, J. Henkel, S. Parameswaran, darkNoC: designing energy-efficient network-on-chip with multi-Vt cells for dark silicon, in *Proceedings of 51st ACM/EDAC/IEEE Design Automation Conference (DAC)* (2014), pp. 1–6
21. J. Zhan, Y. Xie, G. Sun, NoC-sprinting: Interconnect for fine-grained sprinting in the dark silicon era, in *Proceedings of 51st ACM/EDAC/IEEE Design Automation Conference (DAC)* (2014), pp. 1–6
22. M. Fattah, A.-M. Rahmani, T.C. Xu, A. Kanduri, P. Liljeberg, J. Plosila, H. Tenhunen, Mixed-criticality run-time task mapping for NoC-based many-core systems, in *Proceedings of International Conference on Parallel, Distributed and Network-Based Processing* (2014), pp. 458–465
23. TGG: Task Graph Generator (2010), http://sourceforge.net/projects/taskgraphgen/

24. R. Bakker, M.W. van Tol, A.D. Pimentel, Emulating asymmetric MPSoCs on the intel SCC many-core processor, in *Proceedings of the Euromicro International Conference on Parallel, Distributed, and Network-Based Processing* (2014), pp. 520–527
25. J. Howard, S. Dighe, Y. Hoskote, Vangal, A 48-Core IA-32 message-passing processor with DVFS in 45nm CMOS, in *Proceedings of Solid-State Circuits Conference Digest of Technical Papers (ISSCC)* (2010), pp. 108–109
26. H. Esmaeilzadeh, T. Cao, Y. Xi, S.M. Blackburn, K.S. McKinley, Looking back on the language and hardware revolutions: measured power, performance, and scaling, in *Proceedings of the Sixteenth International Conference on Architectural Support for Programming Languages and Operating Systems*, ASPLOS XVI (2011), pp. 319–332
27. A.Y. Weldezion, M. Grange, D. Pamunuwa, A. Jantsch, H. Tenhunen, A scalable multi-dimensional NoC simulation model for diverse spatio-temporal traffic patterns, in *Proceedings of the IEEE International 3D Systems Integration Conference* (2013), pp. 1–5
28. A.Y. Weldezion, M. Grange, D. Pamunuwa, Z. Lu, A. Jantsch, R. Weerasekera, H. Tenhunen, Scalability of network-on-chip communication architecture for 3-D meshes, in *Proceedings of the International Symposium on Networks-on-Chip* (2009), pp. 114–123
29. L. Wang, K. Skadron, Dark vs. Dim Silicon and near-threshold computing extended results. University of Virginia Department of Computer Science, Technical Report TR-2013-01 (2012)
30. F. Fazzino, M. Palesi, D. Patti, Noxim: Network-on-chip simulator (2008), http://sourceforge.net/projects/noxim
31. S. Li, J.H. Ahn, R.D. Strong, J.B. Brockman, D.M. Tullsen, N.P. Jouppi, McPAT: an integrated power, area, and timing modeling framework for multicore and manycore architectures, in *Proceedings of the 42Nd Annual IEEE/ACM International Symposium on Microarchitecture* (2009), pp. 469–480
32. Lumos Framework (2014), http://liangwang.github.io/lumos/. Accessed 20 May 2014
33. B.H. Calhoun, S. Khanna, R. Mann, J. Wang, Sub-threshold circuit design with shrinking CMOS devices, in *Proceedings of the International Symposium on Circuits and Systems* (2009), pp. 2541–2544
34. M. Fattah, M. Daneshtalab, P. Liljeberg, J. Plosila, Smart hill climbing for agile dynamic mapping in many-core systems, in *Proceedings of Design Automation Conference (DAC)* (2013), pp. 1–6

Chapter 8
Robust Application Scheduling with Adaptive Parallelism in Dark-Silicon Constrained Multicore Systems

Nishit Kapadia and Sudeep Pasricha

Introduction

With increasing transistor miniaturization, circuit densities have drastically increased, and the critical charge, which is the minimum charge capable of a bit-flip in a memory- or a logic-cell, has significantly decreased [1–3]. This phenomenon has caused newer process technologies to be more susceptible to transient faults due to the effects of radiation, e.g., alpha-particle and neutron strikes. The rate of such transient faults at run-time has thus been increasing with technology scaling [4]. Studies have also shown that hardened flip-flops are only 30–50 % more resilient than unprotected ones, at sub-40 nm nodes [5], i.e., circuit-level hardening may not sufficiently suppress soft-errors. Thus, system-level run-time approaches to cope with transient faults and complement circuit-level techniques will be increasingly essential in newer process technologies.

Simultaneously, unpredictability in leakage power and circuit-delay due to variability in modern fabrication processes have become a serious concern. In emerging CMPs, spatially correlated systematic within-die (WID) variations manifest across multiple cores, creating core-to-core (C2C) variations [6]. At the same time, die-to-die (D2D) variations remain quite significant [7]. Both WID and D2D variations have random and systematic components. Although several prior works such as [8, 9] have proposed variation-aware design-time application-mapping frameworks, it is generally quite difficult to predict the variation profile of a fabricated chip at design time [10]. It is however possible to extract the variation map of a chip at

N. Kapadia (✉) • S. Pasricha
Department of Electrical and Computer Engineering, Colorado State University, Fort Collins, CO, USA
e-mail: nkapadia@colostate.edu; sudeep@colostate.edu

A.M. Rahmani et al. (eds.), *The Dark Side of Silicon*,
DOI 10.1007/978-3-319-31596-6_8

run-time from the chip-frequency-profile obtained using ring-oscillator based delay sensors [11, 12]. Thus, run-time approaches to mitigate adverse effects of process variations become possible, and will be vital for scaled technologies.

The slowdown of power scaling with technology scaling, due to leakage and reliability concerns [13, 14], has led to further rise in chip power-densities, exacerbating the dark-silicon phenomenon. With the extent of dark silicon increasing every technology-generation (30–50 % for 22 nm) [15, 16], designs are becoming increasingly power-limited rather than area-limited. Run-time power-saving techniques such as dynamic voltage scaling (DVS) are thus becoming increasingly important.

Given these multiple daunting design challenges, there is a critical need for a system-level solution that can simultaneously and adaptively manage the constraints imposed by dark silicon, process variations, and soft-error reliability, while executing applications. In this chapter, we address this need by proposing a novel run-time variation- and reliability-aware application scheduling framework (VARSHA) that employs dynamically adaptable application degrees of parallelism (DoPs) to minimize average application service times and energy, while meeting a chip-wide dark-silicon power constraint, i.e., the thermal design power (TDP) and application performance and reliability constraints, in the presence of process variations. Our novel contributions are summarized below:

- We design a novel run-time application-mapping methodology for the emerging dark-silicon-constrained-design-regime, improving over traditional mapping approaches optimized for the area-constrained-design-regime;
- Our framework simultaneously manages all dynamically arriving applications while adapting application-DoPs (app-DoPs) to optimally utilize the system-power-slack (difference between TDP and current system-power dissipation);
- We design a novel heuristic to integrate within the application-mapping process a DVS mechanism that is constrained not just by performance but also by application-reliability requirements;
- Our combined mapping and DVS approach is also enhanced to take advantage of both D2D and WID variations, performing WID variation-aware mapping on to cores with the optimal power/performance characteristics, and D2D variation-aware chip-wide DVS where faster (leakier) chips would need lower V_{dd}-levels and slower chips may run at higher V_{dd}-levels.

Related Work

A few recent works have explored dark-silicon aware CMP design methodologies. Allred et al. [15] and Turakhia et al. [17] propose design-time frameworks for synthesis of heterogeneous CMPs to extract better energy efficiency and performance in the dark-silicon regime. However, these works do not consider the effects of reliability and impact of process variations on CMP performance and power profiles. Raghunathan et al. [16] exploit the variation profile for run-time application mapping on to a homogeneous CMP die, to maximize performance and reduce leakage power given a fixed dark-silicon power budget. But the authors do not consider overheads of inter-core communication and application-reliability requirements. Chou et al. [18] and Fattah et al. [19] consider run-time mapping of a queue of incoming applications, and propose mapping techniques to fit the maximum number of applications possible on a homogeneous CMP die, while minimizing inter-core communication distances. However, these approaches do not consider reliability and system power, and are geared towards traditional area-constrained designs.

Very recently, a few research efforts have proposed dark-silicon aware run-time application-mapping frameworks for homogeneous multicore processors. A. Kanduri et al. [20] advocate a mapping approach to evenly distribute power density across the CMP chip to provide a higher power budget within a safe operating peak temperature. The authors attempt to maximize the sparsity between existing applications in order to increase the available power-slack, thereby improving the application throughput. On the other hand, authors in [21] propose a combined run-time application mapping and online core-testing (for the effects of aging) framework, given the chip power budget. This work defines a test-criticality parameter based on the utilization of each core in the past, and prioritizes online-testing of cores with higher test-criticality while taking advantage of the run-time power-slack on the CMP die. Although these are significant contributions, these works do not consider either process variations, soft-error reliability, or DoP-adaptation while making run-time mapping decisions.

Several other efforts (e.g., [22–24]) have proposed design-time fault-tolerant scheduling frameworks that assume fault-detection mechanisms implemented on the multicore platform. Hardening techniques such as task-re-execution and replication are utilized in these works to probabilistically meet task-completion deadlines for real-time tasks of varying criticalities. However, these efforts do not consider dark silicon and process variation challenges, run-time adaptation support, or a dynamically parallel application workload.

Some recent works advocate varying the DoP of multithreaded applications at run-time, to adapt to changes in the execution environment (power/performance profiles, core-availability, etc.) of a CMP while optimizing metrics such as power and energy-delay product (EDP). Given enough parallelism within the application, [13] showed that increasing DoP is a more energy-efficient way of boosting performance, compared to hiking the core-frequency/voltage values. Variable DoPs can be implemented by saving multiple versions of application-code at compile-time, and choosing the DoP at run-time that is most appropriate for the execution environment; or via more sophisticated techniques [25]. The variable-DoP run-time scheduling frameworks in [26, 27] report improvements over scheduling techniques that employ statically fixed DoPs, and search for the best combination of voltage, frequency, number of cores, and number of threads, to optimize power and performance of a single application on a CMP. The VARSHA framework is different from these efforts in that we assume such information to be pre-profiled at design time and the focus is on run-time management of app-DoPs in a multi-application power-constrained system. Besides, unlike VARSHA, the frameworks in [26, 27] consider neither the effects of process variations nor the design implications of system-reliability.

Motivational Example

In this section, we illustrate the advantages of some of the key aspects of our VARSHA framework with the help of a small example. We consider a scenario in which applications arrive at run-time at a service-queue to be served. The example assumes applications App-1 and App-2 arriving and being mapped on the CMP at time $t = 0$ s, and a third application App-3 arriving at $t = 1$ s.

Prior works such as [18, 19] search for square or near-convex regions on the CMP die to map arriving application at run-time, so that a maximum number of applications can fit on the die-area, while also minimizing inter-core communication distances. In this chapter, we note that in the modern dark-silicon era, performance is generally not limited by the chip area but by the dark-silicon power constraint (TDP) of the chip. Therefore, while frameworks proposed in prior works would map App-1 and App-2 at $t = 0$ as shown in Fig. 8.1a with a nominal DoP of 8, our framework, which adapts app-DoPs to extract maximum system performance (i.e., minimum application service times) for a given TDP, increases the DoP of App-1–16 (as shown in Fig. 8.1b). When App-3 arrives at $t = 1$ s, our framework maps it with a reduced DoP of 4 with a zero-wait time while meeting the TDP of 60 W (Fig. 8.1b), whereas [18, 19] map App-3 at its nominal DoP of 8 (not shown), and require waiting until App-1 finishes at $t = 4.1$ s, so that the TDP is not violated. Thus in our framework, app-DoPs are hiked from their nominal values opportunistically, to minimize run-times, and app-DoPs are reduced to cut-down on wait times thereby minimizing average application service times (service time = run-time + wait time).

We define reliability of the ith application as $R_i = \{$1-Probability of one or more soft-errors during the execution of App-i$\}$. In this chapter, we assume applications with different minimum reliability constraints running simultaneously on a CMP. Rc_i represents the reliability constraint of the ith application. The application reliability (R_i) primarily depends on V_{dd} and frequency of the cores (as shown in Eqs. (8.1), (8.2), and (8.3) in the section “Problem Formulation”). Additionally, R_i depends upon the app-DoP—although application execution time typically reduces with higher DoP (as long as the DoP is below an application-specific performance saturation point), the chip area susceptible to soft-errors increases. Therefore, for fixed values of voltage and frequency, soft-error probability increases (i.e., application reliability decreases) with increasing app-DoP. Our framework exhibits reliability awareness by reducing the DoP of App-2 (with a stringent reliability constraint) from the nominal value of 8 to 4, thereby meeting Rc_2 (Fig. 8.1b). However, Rc_2 is violated when the reliability-unaware frameworks [18, 19] are employed (Fig. 8.1a).

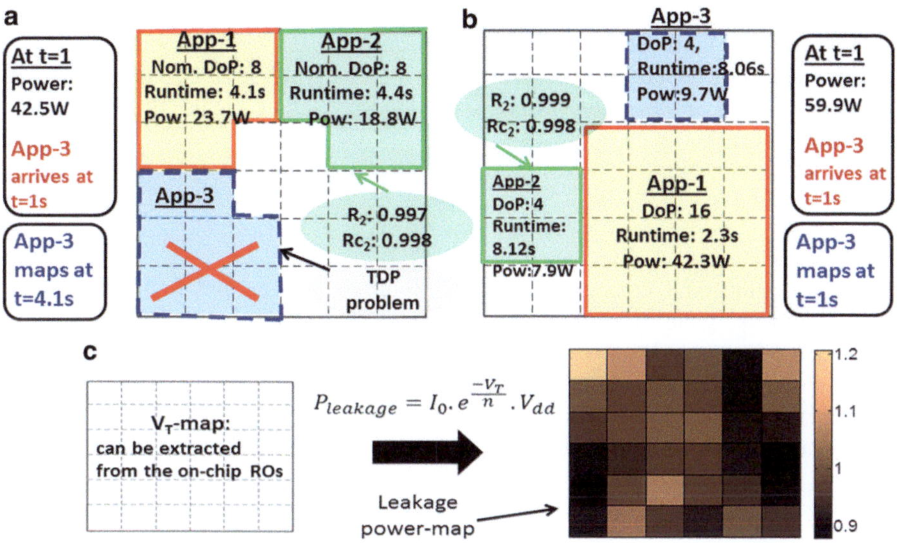

$$P_{leakage} = I_0 . e^{\frac{-V_T}{n}} . V_{dd}$$

Fig. 8.1 Motivational example using a 6 × 6 chip multiprocessor (CMP) where App-1 and App-2 are simultaneously mapped at time t = 0 and App-3 arrives at t = 1 s. A thermal design power (TDP) = 60 W is assumed: (**a**) application-scheduling representative of prior works [18, 19], (**b**) application scheduling obtained by our variation- and reliability-aware application scheduling framework (VARSHA), and (**c**) extracted V_T-map and corresponding estimated leakage-power profile (in Watts) of the chip

Furthermore, due to the variation-awareness of our framework, applications are mapped to regions with minimum core-leakage powers (see Fig. 8.1c). Therefore, the average leakage power per core is 1 W for the variation-unaware frameworks (Fig. 8.1a), and 0.94 W for our framework (Fig. 8.1b). Even though the initial allocation's power dissipation is higher with our framework (at t = 0, in Fig. 8.1), our lower service time and variation-aware mapping result in less overall energy consumption for the applications—184 J compared to 213 J for [18, 19]. Our framework also employs an energy-saving mechanism to opportunistically scale V_{dd}-levels while meeting performance and reliability constraints of all applications. This feature is omitted from the above simple example for brevity.

Problem Formulation

Reliability Modeling

Computer systems are susceptible to both transient and permanent faults, the latter being caused by fabrication defects or wear-out. We only consider the impact of transient faults on reliability and do not consider permanent faults in this work. It is assumed that permanent faults could either be detected during the testing phase or are mitigated by hardware-redundancy techniques.

To model the dependence of raw soft-error rate, raw-SER (λ), in a hardware component (core or router) on voltage and frequency values, we use the relationship proposed in [4]:

$$\lambda\left(\omega_j\right) = \lambda_0 \cdot 10^{\frac{d\left(1-\omega_j\right)}{1-\omega_{min}}} \tag{8.1}$$

where λ_0 is the SER corresponding to the highest voltage and frequency values ($\omega_{\max}$), and ω_j is the average of the normalized values of the jth combination of voltage and frequency (such that $\omega_{\max} = 1$). For any compute core, we use a value of 10^{-6} errors/s for λ_0 and assume $d = 3$ as in [22, 23]. However, we use $\lambda_0 = 10^{-6}/3$ for network-on-chip (NoC) routers, given that our router-area is roughly a third of the area of a compute core. The reliability of a core or a router is given by:

$$\mathbb{R}\left(\omega_j\right) = e^{-\lambda\left(\omega_j\right).\tau} \tag{8.2}$$

where τ is the execution time of the component (time-duration that the component stays active). Assuming $n =$ app-DoP, reliability of the ith application running on a CMP can be given as the product of all $2n$ (n routers and n compute cores) component-reliabilities:

$$R_i = \prod_{2n} \mathbb{R}\left(\omega_j\right) \tag{8.3}$$

Prior works [1, 2] have shown that at technology nodes of 32 nm and below, process variations have almost no effect on SER. Therefore, in this chapter, we assume no dependence of V_T-variations on SER, instead exploiting variations for speed/power benefits only.

Inputs, Assumptions, and Problem Objective

We assume the following *inputs* to our problem:

- A CMP with a regular mesh-based 2D NoC, with T tiles: $T = (d^2)$, where d is the mesh dimension, and each tile consists of a compute core and an NoC router;
- A set S of candidate supply voltage (V_{dd}) levels for the chip;
- A set of N V_T-maps (N test-chips) incorporating the effects of WID and D2D variations with continuous distribution over the die; the estimated leakage power profile for a given value of V_{dd} can be obtained from a V_T-map (using the relation shown in Fig. 8.1);
- Application sequence s of length l, made up of η different applications, with arbitrary application inter-arrival times;

- Application-task-graphs for the set $P = \{P_1, P_2, \ldots P_\eta\}$ of DoPs for all applications; an application i possesses $|P_i|$ viable DoPs; an application has a maximum DoP value beyond which performance does not improve (or gets worse)—such DoP values are ignored;
- Vertices of each task-graph with execution-times of compute cores and edges with inter-task communication volumes; execution time and volume values are assumed available from offline profiling;
- Energy-optimal frequency constraints $\{f_1, f_2, \ldots, f_\eta\}$ and minimum reliability constraints $\{Rc_1, Rc_2, \ldots, Rc_\eta\}$ for all η applications;
- A chip-wide dark-silicon power dissipation constraint (TDP).

We make the following *assumptions* in our work:

- There exists one-to-one mapping between tasks and cores;
- Applications are mapped contiguously on non-overlapping rectangular regions of the CMP die, for inter-application isolation;
- All cores executing an application run at the same frequency, to avoid imbalances during multithreaded execution [16];
- Variation-map data for a chip is available at run-time, in terms of the threshold-voltage (V_T) distribution, from the chip-frequency-profile obtained using ring-oscillator based delay sensors [11];
- A chip-wide supply voltage that can be scaled using DVS at run-time, as the overheads of implementing DVS at a per-core granularity may incur significant overheads [28].

Problem Objective Given the above inputs and assumptions, our objective is to perform run-time application scheduling and DVS on a given CMP platform such that the average application service time and average energy (across all *N* test-chips) are minimized, while all application-specific operating frequency and reliability constraints, as well as CMP platform-specific TDP, are satisfied.

VARSHA Framework: Overview

Figure 8.2 illustrates the key aspects of our VARSHA framework. The knowledge of the chip-variation profile is continuously utilized in the scheduling and DVS steps. Assuming equal priority for all incoming applications, the application-scheduling step consists of: (1) determining the DoP (out of the $|P_i|$ DoPs) for each waiting application in the service-queue, and (2) mapping the appropriate task-graphs on to the tiles of the CMP. For a given V_{dd}, scaling-up of *app-DoPs* is constrained by the available power-slack (difference between TDP and current system power), application-reliability constraints, and the available tiles meeting the application-frequency constraints. At any given time, the scaling-down of V_{dd} (to save power/energy) is constrained by the frequency and reliability constraints

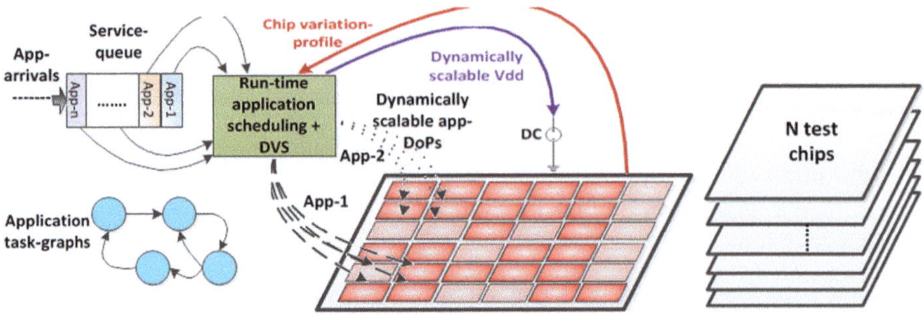

Fig. 8.2 Overview of our application scheduling + dynamic voltage scaling (DVS) framework

of the applications running on the CMP, whereas scaling-up of V_{dd} (to boost *app-DoPs*) is constrained by the TDP for the CMP.

The VARSHA framework is effectively executed in two nested procedures: (1) V_{dd}-level selection (outer loop), triggered on an arrival or a departure of any application; and (2) determination of application schedule for the current V_{dd}-level (inner loop). These procedures are discussed in detail in sections "V_{dd}-Level Selection" and "Determination of Application Schedule," respectively, and the corresponding design-flows for the procedures are shown in Fig. 8.3a, b, respectively.

V_{dd}-Level Selection

To extract maximum performance from the applications being considered for mapping at any time instant, the first-order objective in VARSHA is to maximize overall DoP of the system (sum-total of all *app-DoPs*). Recall that an application typically has a maximum viable DoP, and higher DoPs can cause performance to degrade (due to high synchronization overheads)—such higher DoP configurations are ignored (section "Inputs, Assumptions, and Problem Objective"). Our V_{dd}-selection heuristic (Fig. 8.3a) selects the V_{dd}-level that yields the maximum overall DoP. As a second-order power/energy-saving objective, on completion of any application, our framework reduces V_{dd} to the lowest allowable level that would not introduce any violations in frequency and reliability constraints of existing (already running) applications.

We assume that all incoming applications are buffered in a service-queue. On arrival or completion of any application, the V_{dd}-selection heuristic is triggered, which processes the entire service-queue. The V_{dd}-selection heuristic iteratively invokes the application-schedule determination procedure (discussed in the section "Determination of Application Schedule"), which produces the mapping solution with the highest overall DoP corresponding to the current V_{dd}-level. The V_{dd}-level

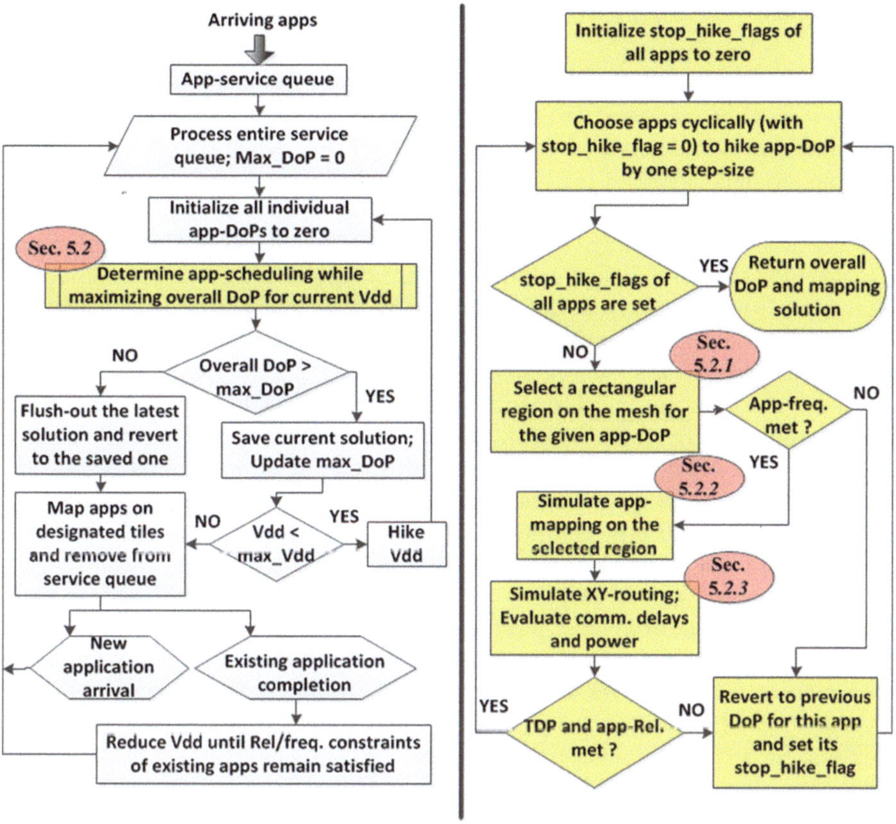

Fig. 8.3 Design-flows for the VARSHA framework: (**a**) V_{dd}-level selection (discussed in the section "V_{dd}-Level Selection"), (**b**) determination of application schedule for the current V_{dd}-level (discussed in the section "Determination of Application Schedule")

is hiked (in increments of 0.1 V) until either the overall DoP reduces, in which case the immediately preceding solution with the highest overall DoP is reverted to, or the maximum allowable V_{dd}-level (max_V_{dd}) is reached. Note that the overall DoP may increase with increasing V_{dd}-levels, as more applications satisfy frequency and reliability constraints for higher DoPs; at the same time, the chip power will reach the TDP quicker at higher V_{dd}-levels, thereby limiting overall DoP. Therefore, our search for the optimal V_{dd}-level culminates when the increase in overall DoP is limited by the TDP. Finally, the best application-mapping solution with the highest overall DoP is mapped to the CMP. The voltage supply is hiked to the selected V_{dd}-level, and the mapped applications are then removed from the service-queue.

Determination of Application Schedule

Given a specific execution environment for the CMP (including the V_{dd}-level, available power-slack, and variation profile), the objective of the application-schedule determination heuristic is to maximize the overall DoP, while simultaneously considering all applications in the service-queue and satisfying application-frequency and -reliability constraints. Figure 8.3b shows the design-flow of this heuristic. Starting at the least possible DoP value of zero (DoP of zero leaves the application unmapped at the current time) applications are considered cyclically for hiking of DoP to their next higher valid DoP-level. Here, to extract maximum performance from the CMP, we choose applications for hiking of DoP in order of their compute-intensiveness, because of the relatively smaller communication delay and power overheads for compute-intensive applications at higher DoPs. Also, we hike *app-DoPs* symmetrically across all applications because execution of an application is generally more energy efficient at lower DoPs (due to lower parallelization- and communication-overheads). For instance, running four applications simultaneously, each with DoP = 4, is typically more energy efficient than running them one after the other with DoP = 16 each.

To produce an optimal mapping for the application under consideration (with a specific DoP), the following three steps are performed: (1) rectangular-region selection on the CMP for the current DoP (section "Rectangular-Region Selection"); (2) mapping of the corresponding task-graph to the selected region (section "Application Mapping"); and (3) communication-flow routing and delay/power analysis (section "Communication Delay and Power Estimation"). After the above steps, the mapping is evaluated for overall power footprint and reliability of the applications. Satisfaction of the application-frequency constraints are checked during the rectangular-region-selection step. As shown in Fig. 8.3b, DoP-hike of any application could fail due to potential violation(s) in application-frequency and -reliability constraints or TDP. When an attempted DoP-hike is stalled for any application, its *stop_hike_flag* is set to preclude it from future DoP-hike consideration, and the feasible mapping with the preceding DoP is finalized for this application.

Rectangular-Region Selection

We consider application mapping on rectangular regions, of pre-determined dimensions corresponding to each possible DoP value in set *P*. Therefore, all intra-application communication is contained within a rectangular region. This provides inter-application isolation, and eliminates communication cross-interference overhead. Given the V_T-map, the maximum frequency that each core can be reliably clocked at depends upon the V_T and V_{dd} values, given by:

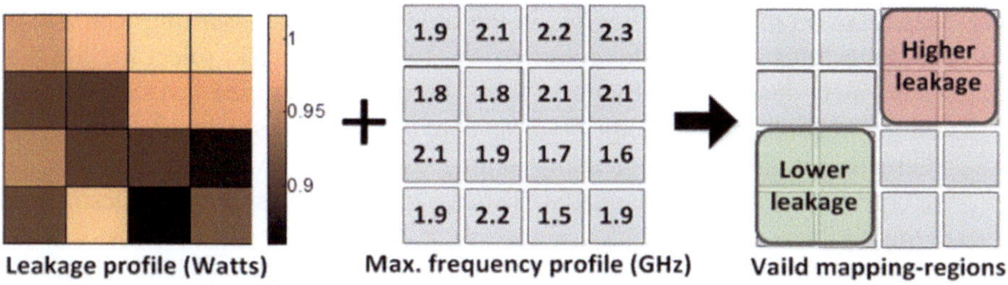

Fig. 8.4 Two feasible regions are shown for the application-frequency constraint of 1.9 GHz. Our method chooses the one with lower leakage

$$f_{max} = \frac{\mu(V_{dd} - V_T)^{\alpha}}{C_0 \cdot V_{dd}} \tag{8.4}$$

where α and μ are technology-dependent constants, and C_0 is switching capacitance of the critical path [29]. In our region-selection method, we utilize the knowledge of both frequency and leakage-power profiles of the chip. Note that dynamic power remains unaffected by V_T-variations and is thus not considered in this step. Our objective is to find the region on the mesh (of pre-defined dimensions) that dissipates the least total leakage power and all cores within which satisfy the frequency constraint of the application being mapped. To this end, we perform a simple exhaustive search over all tiles on the mesh as the left-upper corner of the given rectangular region. If the rectangle is not a square, then both of its orientations need to be checked to find the optimal rectangular region. Figure 8.4 shows an illustration of our region-selection method for a 4 × 4 mesh.

Theoretical Time-Complexity Analysis At most T tiles (total tiles on the chip) are considered for the prospective rectangular region. For non-squares, e.g., a 2 × 4 rectangle, rectangles of both orientations need to be evaluated at each tile. Note that *app-DoP* (relatively small integer *c*—treated as constant) number of tiles are to be evaluated for frequency and leakage power at each of these iterations. This gives a linear-time complexity for the region-selection step: O(2*cT*).

Application Mapping

After the rectangular region (of size equal to app-DoP) on the mesh has been selected, our mapping heuristic maps the appropriate application-task-graph on to the CMP tiles. We employ an incremental-mapping approach that is carried out in two steps:

(1) Starting with the task with the highest communication-volume, list tasks in decreasing order of inter-task communication volumes with all preceding tasks in the list;

(2) Map tasks on the selected region, in the order of this list—starting with one of the center tiles on the rectangular region. To map any task, select the tile that would incur least traffic footprint with the already mapped tasks. We define communication-traffic footprint of a core (mapped task) as: $\sum MD_j \times vol_j$, where vol_j is the communication-volume of the jth flow entering or leaving the core and MD_j is the Manhattan distance between the source and destination of the jth flow.

Theoretical time-complexity analysis: Step (1) sorts *app-DoP* tasks—the time complexity could be bounded by O((*app-DoP*)^2). In step (2), *app-DoP* number of tasks are mapped and each could consider a maximum of *app-DoP* number of tile-locations. Thus this step could be loosely bounded by O((*app-DoP*)^2). As *app-DoP* is a relatively small integer, it can be considered a constant (between 4 and 16 in our experiments). Thus, our application mapping has constant time complexity. Note that mapping-information can potentially even be saved with the task-graph information at compile-time, thus saving valuable run-time resources and overhead.

Communication Delay and Power Estimation

Similar to numerous prior works such as [19], we use the XY-routing scheme to route the communication-flows of applications. The communication-delays with congestion-overheads are calculated from the application frequency (routers and links run at the rated application frequency) and link BWs. Profiling of compute- and communication-delays could potentially be performed at design time [30–32]. Dynamic powers of routers and links (corresponding to different voltages, frequencies, communication loads, and router-sizes), are assumed to be saved in the read-only (or non-volatile) memory on the CMP die. Router leakage powers, estimated from the chip-variation profile, are added to the appropriate dynamic-power values to produce the total communication-power for running each application. Based on the active-times (execution-times) of routers and compute cores, the application reliability is computed using Eqs. (8.1), (8.2), and (8.3). For our analyses, the energies and run-times of applications are calculated from component powers and active-times.

Complexity Analysis of VARSHA Framework

Our application-schedule determination heuristic (discussed in the section "Determination of Application Schedule"), which maximizes the DoP of all applications being processed at the current V_{dd}-level, attempts to find mapping solutions for the *w* waiting applications for up to $|P_i|$ DoP-levels. This heuristic could be required to execute for at most $|S|$ (total number of candidate V_{dd}-levels) times. $|S|$ and $|P_i|$ are small constant integers in our experiments. Also, at any given time, only a small number of applications (up to w) are generally expected to be processed given the TDP. Therefore, whenever the service-queue is processed in our VARSHA framework, the number of times that our region-selection heuristic would need to be invoked is bounded by a relatively small constant integer. As the region-selection step, which has the highest theoretical time complexity in the VARSHA design-flow, finishes in linear time (as discussed in the section "Rectangular-Region Selection"), the time complexity of our framework is linear, with respect to the CMP mesh-size, T.

Experiments

Our experiments were conducted using $\eta = 14$ different parallel application benchmarks: seven from the SPLASH-2 benchmark suite [33] (*cholesky*, *fft*, *lu*, *ocean*, *radix*, *radiosity*, *and raytrace*), and seven from the PARSEC benchmark suite [34] (*vips*, *swaptions*, *fluidanimate*, *dedup*, *streamcluster*, *canneal*, *and blackscholes*). We consider DoPs that are multiples of 4, up to 16, where a DoP of 8 is considered the nominal-DoP value, as a reasonable trade-off between speed and energy. As discussed earlier, every application has a unique maximum viable DoP beyond which further performance gains cannot be achieved. Our task-graphs for all applications at different DoPs are modeled based on the system-traces generated by running topology-agnostic gem5 simulations [35]. The reliability constraints of different applications are set in the range: 0.99–0.999. We assume the ARM Cortex-A9 processors [36] as the baseline CMP compute cores, which support five operating voltage levels ($|S| = 5$): 0.8 V, 0.9 V, 1.0 V, 1.1 V, and 1.2 V. The application-specific energy-optimal core frequencies range from 1300 to 1900 MHz, based on the level of compute intensity of the tasks assigned to cores. We use a 100-core mesh for the CMP platform with dimensions 10×10. The dark-silicon power constraint (TDP) is set at 100 W. The delay-overhead for V_{dd}-scaling (PLL lock time) is estimated to be less than 5 μs, similar to [37], which is negligible compared to the granularity of application run-times (2–9 s each) in our experiments.

To investigate the applicability of our approach to CMP dies with diverse variation profiles, we use 1000 test-chips ($N = 1000$), in our experiments. The 1000 V_T-maps are generated using the open-source tool [38] (based on systematic and random WID-variation model in [29]). The values of 0.3 and 0.09 are used for the statistical mean and a standard deviation of the parameter V_T, respectively, and a correlation range (ϕ) of 0.5 is used (as recommended in [29]). A normally distributed V_T bias, representing the D2D variation component, is superimposed onto these V_T-maps; the standard deviation of the D2D V_T is assumed to be 6 %, as in [39]. Each V_T-map represents a 30×30 grid of points corresponding to nine ring-oscillator test-sites for each core on the CMP. For a given V_T-map, the maximum core frequencies are calculated by using the V_T-max values and the core-leakage powers are calculated using the V_T-avg values (out of the nine V_T values per core). The power values of routers and links (32-bit wide) for different voltages and frequencies at varying communication loads for the 32 nm technology node are obtained from ORION 2.0 [40]. Note that the router power values obtained are for nominal V_T, and are scaled for varying V_T values.

Results

We compare the results obtained from our VARSHA framework with those obtained from using run-time application-mapping frameworks proposed in recent prior works [16, 19]. A variation- and dark-silicon-aware mapping technique is proposed in [16], whereas [19] advocates for a traditional area-constrained design approach. We implemented these prior works to the best of our understanding. Our experiments considered two unique application sequences (Seq-A and Seq-B) that represent an ordering of arriving application instances, with instances randomly chosen from among the 14 applications considered. For each sequence, we vary the inter-arrival times of application instances randomly within the following ranges: 0–1 s (Seq-1A and Seq-1B), 0–2 s (Seq-2A and Seq-2B), 0–4 s (Seq-3A and Seq-3B), and 0–8 s (Seq-4A and Seq-4B). We assume $l = 100$ application instances in any application sequence. Also, our results (as shown in Fig. 8.5 and Table 8.1) show the mean values across a 1000 test-chips.

The prior works [16, 19] assume fixed nominal app-DoPs. Our framework adapts app-DoPs in accordance with the application inter-arrival rates to minimize the application service times. Observe in Table 8.1 that for both sequences, the average app-DoP reduces with increasing inter-arrival-rates for VARSHA. At higher inter-arrival rates when applications with nominal DoPs cannot be quickly serviced due to the TDP constraint, our VARSHA framework cuts down application wait times significantly by reducing DoPs (as shown in Fig. 8.5a—Seq-1A,B and Seq-2A,B), although the application run-times tend to increase due to the reduction in DoPs. On the other hand, at lower inter-arrival rates, with on average fewer applications to be serviced simultaneously, our framework opportunistically hikes the app-DoPs to minimize run-times (as shown in Fig. 8.5a—Seq-3A,B and Seq-4A,B). In comparison with [19], we obtain 27–43 % savings in average service times with our VARSHA framework. Note that maximum savings are obtained when the inter-arrival-rates are most stringent, as shown for Seq-1A and Seq-1B in Fig. 8.5a. The communication-unaware framework in [16] maps applications on to large rectangular regions of non-contiguous tiles, resulting in longer run-times due to longer communication-latencies. Compared to [16], we obtain 70–87 % savings in average service times with our VARSHA framework.

Our framework also opportunistically lowers V_{dd}-levels while ensuring that frequency and reliability constraints remain satisfied for the set of existing applications. The weighted-average V_{dd} metric ($V_{w\text{-}avg}$) represents the average value of

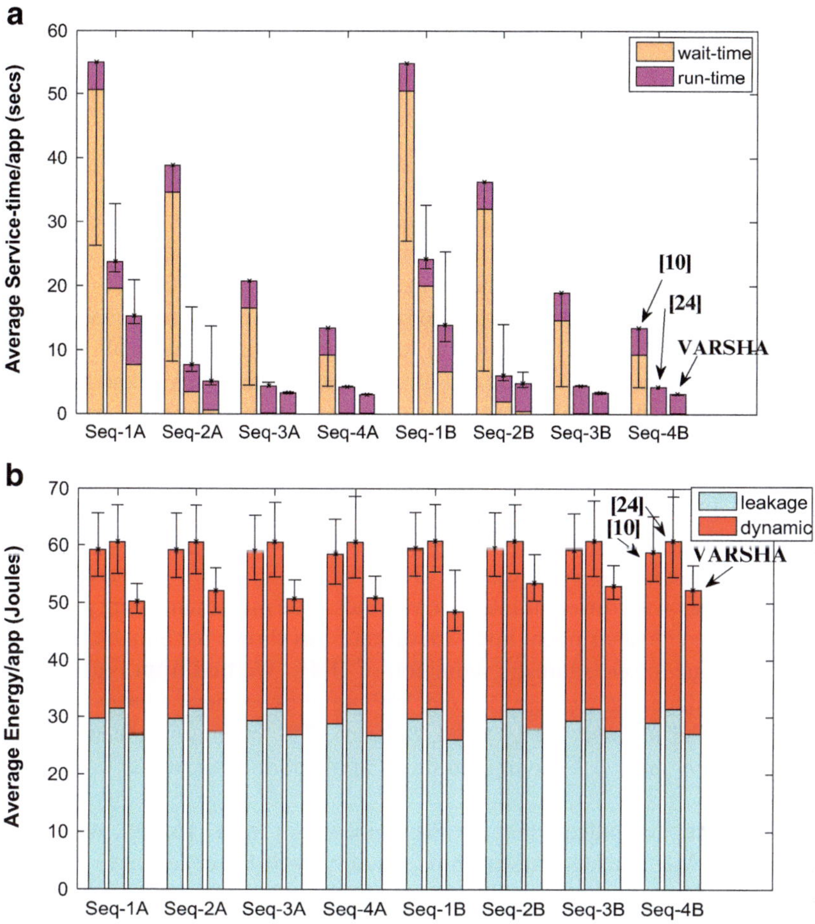

Fig. 8.5 Results of VARSHA framework vs. [10, 24]: (**a**) average service time per application instance (wait time + run-time), (**b**) average energy per application instance (leakage + dynamic). The *bars* represent mean values of service times and energies across 1000 test-chips, while variation in service times and energies is shown by confidence intervals

Table 8.1 Mean values across all 1000 test-chips for average degree of parallelism (DoP) per application instance and weighted-average V_{dd} for VARSHA framework

	Seq-1A	Seq-2A	Seq-3A	Seq-4A	Seq-1B	Seq-2B	Seq-3B	Seq-4B
DoP	4.5	10.3	14.1	14.5	5.1	10.7	13.7	14.2
$V_{w\text{-}avg}$	1.09	1.10	1.07	1.07	1.07	1.11	1.09	1.07

V_{dd} throughout the execution of the entire application sequence for a given chip, and is defined as

$$V_{w-avg} = \frac{\left((V_{\Delta t1} \times \Delta t1) + (V_{\Delta t2} \times \Delta t2) + \cdots + \left(V_{\Delta tp} \times \Delta tp\right)\right)}{\sum_{i=0}^{i=p} \Delta ti} \tag{8.5}$$

where Δti is the ith time-interval during which the V_{dd}-level stays constant at a value of $V_{\Delta ti}$, and the execution of the entire application sequence s takes p such time-intervals. Table 8.1 shows the mean $V_{w\text{-}avg}$ across all 1000 test-chips. In the absence of DVS in [16, 19], in our implementation of these frameworks, we assume the highest voltage of 1.2 V, which is just high enough to meet all application-frequency constraints for the 1000 test-chips. Despite the high V_{dd} and much longer communication paths, [16] saves leakage power by performing variation-aware mapping. In the absence of such variation-awareness, [19] consumes the highest energy (as shown in Fig. 8.5b). In our VARSHA framework, energy efficiency decreases with increasing app-DoPs (due to increased communication and sub-linear increase in computation-performance), while decreasing V_{dd}-levels generally cause an increase in energy efficiency. Therefore, in Fig. 8.5b, we observe variation in energy values with different inter-arrival rates. The maximum energy savings are obtained for the most stringent inter-arrival rates, as shown for Seq-1A and Seq-1B in Fig. 8.5b, due to opportunistic reduction of DoPs. We find from Fig. 8.5a, b that VARSHA produces on average, 13 and 15 % savings in mean energy (8 and 14 % savings in mean leakage energy) and, 80 and 35 % savings in mean application service time (93 and 66 % savings in mean application wait times), over [16, 19], respectively.

Our experiments indicate that for the frameworks from [16, 19], where a fixed high value of V_{dd} is used, slower chips (usually with lesser leakage) could prove to be advantageous as they would be left with greater power-slack to accommodate more applications at any given time (less %dark silicon), thus resulting in better service times and energies. Conversely, in our VARSHA framework, given a TDP, faster chips (that also dissipate higher leakage power) can usually utilize lower V_{dd}-levels to minimize energy, and slower chips (that dissipate lower leakage power) can utilize higher V_{dd}-levels to improve performance. We observed that test-chips with moderate leakage and performance profiles produce desirable service times and energy results, whereas chips that are too leaky or too slow yield worse results, as indicated by the upper-bounds of the confidence intervals in Fig. 8.5a, b for the VARSHA framework. To explain this phenomenon, we compare the average energy per application obtained for all 1000 test-chips with [19] and VARSHA, using the application sequence Seq-1B (as shown in Fig. 8.6). With a high value of V_{dd} set pessimistically for the WC (highest) V_T values, we observe that the energy efficiency obtained using [19] generally steadily reduces as the leakage power of test-chips increases (going from right to left in Fig. 8.6a). On the other hand, the VARSHA framework can intelligently mitigate the adverse effects of D2D variations resulting in much higher energy efficiency (as shown in Fig. 8.6). Figure 8.6b also shows

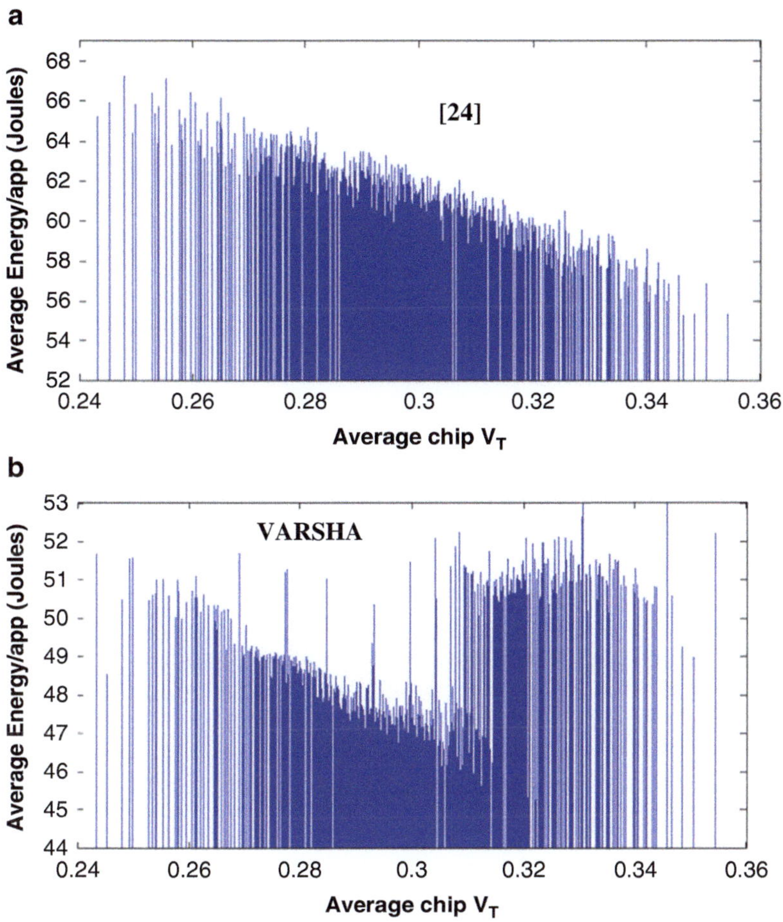

Fig. 8.6 Effects of die-to-die (D2D) variation in terms of average energy/app obtained for all 1000 test-chips using Seq-1B: (**a**) With [24], performance degradation due to D2D variations cannot be mitigated efficiently with a high value of V_{dd}, and faster chips (with higher leakage power) generally result in much worse energy efficiency, (**b**) due to intelligent DVS scheme integrated within the VARSHA framework, best energy efficiency is obtained for test-chips with moderate leakage and performance profiles

that the test-chips with moderate variation profiles result in better energy efficiency compared to the ones that are too leaky or too slow.

For applications with relatively more stringent reliability constraints, it may not be possible to support nominal DoP even at the highest V_{dd}-level (1.2 V). Therefore, when using reliability-unaware frameworks with no DoP-adaptivity, reliability constraints (Rc) of such applications may be violated. Table 8.2 shows the Rc-violations across 1000 test-chips for schedules produced by [16, 19]. With longer routing distances, [16] consumes far more routing resources compared to

Table 8.2 Reliability-constraint (Rc) violations (average per sequence)

	Seq-A	Seq-B
[16]	8	9.5
[19]	6	8

[19], thus even with the same app-DoPs (nominal DoPs), the estimated failure-rates are greater for [16]. Our reliability-aware VARSHA framework results in no Rc-violations because of its ability to dynamically reduce app-DoPs as well as hike V_{dd}-levels in accordance with application-reliability requirements.

Conclusion

VARSHA represents one of the first efforts to integrate reliability and variation-awareness in a run-time variable DoP application-scheduling methodology to enhance performance of multicore systems in the new dark-silicon era. Tailored for deeply scaled technologies, our VARSHA framework generates highly optimized application-mapping solutions, while exhibiting linear run-time complexity. Our experimental results show that VARSHA produces savings of 35–80 % in application service times, 13–15 % in energy, and avoids reliability violations unlike state of the art prior works that suffer from reliability violations in up to 11 % of application instances arriving at run-time.

Acknowledgements This research is supported by grants from SRC, NSF (CCF-1252500, CCF-1302693), and AFOSR (FA9550-13-1-0110).

References

1. A. Kaouache, F. Wrobel, F. Saigne, A.D. Touboul, R.D. Schrimpf, Analytical method to evaluate soft error rate due to alpha contamination. IEEE Trans. Nucl. Sci. **60**(6), 4059–4066 (2013)
2. N. Gaspard, S. Jagannathan, Z. Diggins, A.V. Kauppila, T.D. Loveless, J.S. Kauppila, B.L. Bhuva, L.W. Massengill, W.T. Holman, Effect of threshold voltage implants on single-event error rates of D flip-flops in 28-nm bulk CMOS. *IEEE International Reliability Physics Symposium*, Apr 2013, pp. SE.7.1–SE.7.3
3. S. Abe, Y. Watanabe, N. Shibano, N. Sano, H. Furuta, M. Tsutsui, T. Uemura, T. Arakawa, Neutron-induced soft error analysis in MOSFETs from a 65 to a 25 nm design rule using multi-scale Monte-Carlo simulation method. *IEEE International Reliability Physics Symposium*, Apr 2012, pp. SE.3.1–SE.3.6
4. D. Zhu, R. Melhem, D. Mosse, The effects of energy management on reliability in real-time embedded systems. *IEEE/ACM International Conference on Computer Aided Design (ICCAD)*, San Jose, California, Nov 2004, pp. 35–40

5. T.D. Loveless, S. Jagannathan, T. Reece, J. Chetia, B.L. Bhuva, M.W. McCurdy, L.W. Massengill, S.J. Wen, R. Wong, D. Rennie, Neutron- and proton-induced single event upsets for D- and DICE-flip/flop designs at a 40 nm technology node. IEEE Trans. Nucl. Sci. **58**(3), 1008–1014 (2011)
6. E. Humenay, D. Tarjan, K. Skadron, Impact of process variations on multicore performance symmetry. *ACM/IEEE Design, Automation and Test in Europe Conference (DATE)*, Nice, France, Apr 2007, pp. 1653–1658
7. L. Pang, K. Qian, C.J. Spanos, B. Nikolic, Measurement and analysis of variability in 45 nm strained-Si CMOS technology. IEEE J. Solid State Circuits **44**(8), 2233–2243 (2009)
8. L. Huang, Q. Xu, Performance yield-driven task allocation and scheduling for MPSoCs under process variation. *ACM/IEEE Design Automation Conference (DAC)*, Anaheim, California, June 2010, pp. 326–331
9. N. Kapadia, S. Pasricha, Process variation aware synthesis of application-specific MPSoCs to maximize yield. *IEEE International Conference on VLSI Design (VLSID)*, Mumbai, 2014, pp. 270–275
10. L. Zhang, L.S. Bai, R.P. Dick, L. Shang, R. Joseph, Process variation characterization of chip-level multiprocessors. *ACM/IEEE Design Automation Conference (DAC)*, San Francisco, California, July 2009, pp. 694–697
11. X. Wang, M. Tehranipoor, S. George, D. Tran, L. Winemberg, Design and analysis of a delay sensor applicable to process/environmental variations and aging measurements. IEEE Trans. Very Large Scale Integr. VLSI Syst. **20**(8), 1405–1418 (2012)
12. A.A.M. Bsoul, N. Manjikian, L. Shang, Reliability- and process variation-aware placement for FPGAs. *ACM/IEEE Design, Automation and Test in Europe Conference (DATE)*, Dresden, Germany, Mar 2010, pp. 1809–1814
13. J. Lee, N.S. Kim, Optimizing total power of many-core processors considering voltage scaling limit and process variations. *ACM/IEEE International Symposium on Low Power Electronics and Design (ISLPED)*, San Francisco, California, July 2009, pp. 201–206
14. S. Borkar, Design perspectives on 22nm CMOS and beyond. *IEEE/ACM Design Automation Conference (DAC)*, San Francisco, California, July 2009, pp. 93–94
15. J. Allred, S. Roy, K. Chakraborty, Designing for dark silicon: a methodical perspective on energy efficient systems. *ACM/IEEE International Symposium on Low Power Electronics and Design (ISLPED)*, Redondo Beach, CA, July 2012, pp. 255–260
16. B. Raghunathan, Y. Turakhia, S. Garg, D. Marculescu, Cherry-picking: exploiting process variations in dark-silicon homogeneous chip multi-processors. *ACM/IEEE Design, Automation and Test in Europe Conference (DATE)*, Grenoble, France, Mar 2013, pp. 39–44
17. Y. Turakhia, B. Raghunathan, S. Garg, D. Marculescu, HADES: architectural synthesis for heterogeneous dark silicon chip multi-processors. *ACM/EDAC/IEEE Design Automation Conference (DAC)*, Austin, Texas, June 2013, pp. 1–7
18. C. Chou, U. Ogras, R. Marculescu, Energy- and performance-aware incremental mapping for networks on chip with multiple voltage levels. IEEE Trans. Comput. Aided Des. Integr. Circuits Syst. **27**(10), 1866–1879 (2008)
19. M. Fattah, M. Daneshtalab, P. Liljeberg, J. Plosila, Smart hill climbing for agile dynamic mapping in many-core systems. *ACM/EDAC/IEEE Design Automation Conference (DAC)*, Austin, Texas, June 2013, pp. 1–6
20. A. Kanduri, M.-H. Haghbayan, A.-M. Rahmani, P. Liljeberg, A. Jantsch, H. Tenhunen, Dark silicon aware runtime mapping for many-core systems: a patterning approach. *IEEE Conference on Computer Design (ICCD)*, New York. Oct 2015
21. M.-H. Haghbayan, A.-M. Rahmani, A. Miele, M. Fattah, J. Plosila, P. Liljeberg, H. Tenhunen, A power aware approach for online test scheduling in many-core architectures. IEEE Trans. Comput. **65**(3), 730–743 (2015)
22. A. Das, A. Kumar, B. Veeravalli, C. Bolchini, A. Miele, Combined DVFS and mapping exploration for lifetime and soft-error susceptibility improvement in MPSoCs. *ACM/IEEE Design, Automation and Test in Europe Conference (DATE)*, Dresden, Germany, Mar 2014, pp. 1–6

23. M. Haque, H. Aydin, D. Zhu, Energy-aware task replication to manage reliability for periodic real-time applications on multicore platforms. *International Green Computing Conference (IGCC)*, Arlington, Virginia, June 2013, pp. 1–11
24. S. Kang, H. Yang, S. Kim, I. Bacivarov, S. Ha, L. Thiele, Static mapping of mixed-critical applications for fault-tolerant MPSoCs. *ACM/EDAC/IEEE Design Automation Conference (DAC)*, San Francisco, California, June 2014, pp. 1–6
25. A. Raman, H.K. Taewook, O.J.W. Lee, D.I. August, Parallelism orchestration using DoPE: the degree of parallelism executive. *ACM SIGPLAN Conference on Programming Language Design and Implementation (PLDI)*, San Jose, California, June 2011, pp. 26–37
26. J. Li, J.F. Martinez, Dynamic power-performance adaptation of parallel computation on chip multiprocessors. *International Symposium on High-Performance Computer Architecture (HPCA)*, Austin, Texas, Feb 2006, pp. 77–87
27. Y. Ding, M. Kandemir, P. Raghavan, M.J. Irwin. A helper thread based EDP reduction scheme for adapting application execution in CMPs. *IEEE International Symposium on Parallel and Distributed Processing (IPDPS)*, Miami, Florida, Apr 2008, pp. 1–14
28. S. Dighe, S.R. Vangal, P. Aseron, S. Kumar, T. Jacob, K.A. Bowman, J. Howard, J. Tschanz, V. Erraguntla, N. Borkar, V.K. De, S. Borkar, Within-die variation-aware dynamic-voltage-frequency-scaling with optimal core allocation and thread hopping for the 80-core TeraFLOPS processor. IEEE J. Solid State Circuits **46**(1), 184–193 (2011)
29. S. Sarangi, B. Greskamp, R. Teodorescu, J. Nakano, A. Tiwari, J. Torrellas, VARIUS: a model of process variation and resulting timing errors for microarchitects. IEEE Trans. Semicond. Manuf. **21**(1), 3–13 (2008)
30. N. Kapadia, S. Pasricha, VISION: a framework for voltage island aware synthesis of interconnection networks-on-chip. *ACM Great Lakes Symposium on VLSI (GLSVLSI)*, Switzerland, May 2011, pp. 31–36
31. N. Kapadia, S. Pasricha, A framework for low power synthesis of interconnection networks-on-chip with multiple voltage islands. Integr. VLSI J. **45**(3), 271–281 (2012)
32. N. Kapadia, S. Pasricha, VERVE: a framework for variation-aware energy efficient synthesis of NoC-based MPSoCs with voltage islands. *IEEE International Symposium on Quality Electronic Design (ISQED)*, San Jose, California, Mar 2013, pp. 603–610
33. S.V. Woo, M. Ohara, E. Torriet, The SPLASH-2 programs: characterization and methodological characterization. *IEEE/ACM International Symposium on Computer Architecture (ISCA)*, Santa Margherita Ligure, Italy, May 1995, pp. 24–36
34. C. Bienia, S. Kumar, J.P. Singh, K. Li, The PARSEC benchmark suite: characterization and architectural implications. *ACM International Conference on Parallel Architectures and Compilation Techniques (PACT)*, Princeton University, Oct 2008, pp. 72–81
35. N. Binkert, B. Beckmann, G. Black, S.K. Reinhardt, A. Saidi, A. Basu, J. Hestness, D.R. Hower, T. Krishna, S. Sardashti, R. Sen, K. Sewell, M. Shoaib, N. Vaish, M.D. Hill, D.A. Wood, The gem5 simulator. ACM Special Interest Group on Computer Architecture (SIGARCH) **39**(2), 1–7 (2011)
36. ARM, http://www.arm.com/products/processors/selector.php
37. A. Bashir, J. Li, K. Ivatury, N. Khan, N. Gala, N. Familia, Z. Mohammed, Fast lock scheme for phase-locked loops. *IEEE Custom Integrated Circuit Conference*, San Jose, California, Sept 2009, pp. 319–322
38. VARIUS model, http://www.cse.ohiostate.edu/~teodores/arch/tools/
39. B. Li, L. Peh, P. Patra. Impact of process and temperature variation on network-on-chip design exploration. *ACM/IEEE International Symposium on Networks-on-Chip (NOCS)*, Newcastle, U.K, Apr 2008, pp. 117–126
40. A. Kahng, B. Li, L. Peh, K. Samadi, ORION 2.0: a fast and accurate NoC power and area model for early-stage design space exploration. *ACM/IEEE Design Automation and Test in Europe (DATE)*, Nice, France, Apr 2009, pp. 423–428

Chapter 9
Dark Silicon Patterning: Efficient Power Utilization Through Run-Time Mapping

Anil Kanduri, Mohammad-Hashem Haghbayan, Amir M. Rahmani, Pasi Liljeberg, Axel Jantsch, and Hannu Tenhunen

Introduction

Power density, the power used per unit area, is an increasingly critical issue to be considered in computer design. Working components of a chip consume power, thus dissipate heat that raises the chip's temperature. Increase in temperature beyond a certain level would result in unreliable functionality, faster ageing, and even permanent failure of the chip. Scaling down transistor gate length makes it possible to operate them at lower voltages and frequencies, managing the power density automatically. Since voltage scaling does not keep pace with transistor size scaling ever more power is consumed per unit area driving up power density and heat dissipation. It is hence critical to operate a chip under specific thermal constraints and consequently within a fixed power budget [1]. Chip manufacturers thus are forced to operate working cores at less than their full capacity, and the issue of dark silicon surfaces, limiting performance, resource utilization, and efficiency of the system.

A. Kanduri (✉) • M.-H. Haghbayan • A.M. Rahmani • P. Lijeberg
University of Turku, Turku, Finland
e-mail: spakan@utu.fi; mohhag@utu.fi; amirah@utu.fi; pakrli@utu.fi

A. Jantsch
TU Wien, Vienna, Austria
e-mail: axel.jantsch@tuwien.ac.at

H. Tenhunen
KTH Royal Institute of Technology, Stockholm, Sweden
e-mail: hannu@kth.se

A.M. Rahmani et al. (eds.), *The Dark Side of Silicon*,
DOI 10.1007/978-3-319-31596-6_9

Challenges of Many-Core Design at Dark Silicon Era

Thermal Awareness

Challenge: Thermal Design Power

The thermal profile of a chip would determine safe peak operating temperature and the corresponding power consumption. Since heat dissipation and temperature are resultant phenomena of power consumption, an upper limit on power usage is to be set. This upper limit varies from device to device based on its power-performance characteristics. The upper limit is popularly known as Thermal Design Power (TDP), used as an industrial standard that is determined during the design process. Thus far, common practice of ensuring safe chip functionality has been by estimating TDP [2] in a conservative manner. Safer operation of a chip is guaranteed as long as power consumption and dissipation stays within TDP [2]. It is to be noted that TDP is not the maximum power that can be consumed by a chip, but is only a safer upper bound. TDP is calculated assuming a worst case application running at worst case voltage and frequencies of components. Although, a typical application might consume power that is lesser than TDP, yet runs under TDP constraint. On average, performance is sacrificed for the sake guaranteed safe operation. The penalty is low for applications that are power hungry; however, other applications that may never exceed TDP would still end up running at lower than full compute capacity of the chip. Recent upgrades on AMD and Intel Corporations' CPUs have the option of a configurable TDP; however, they are limited to a maximum of 3 modes, without any fine grained control [3, 4]. In this sense, current systems are rather conservatively, thermal aware.

Solution: Thermal Saturation Power

TDP is single fixed upper bound that is pessimistically estimated assuming that all the cores are active and operating at a worst case voltage and frequency. However, a variable number of cores will be inactive, depending on current set of applications running. Thus, the safe upper bound on power budget varies during run-time, as opposed to the conservative upper bound of TDP which leads to under-utilization of available power budget [5]. A sensible way to avoid the conservative limit of TDP is to use a variable and realistic upper bound on power consumption, Thermal Saturation Power (TSP), as proposed by Pagani et al. [5]. TSP is modeled as a function of simultaneously active cores, their alignment, effect of temperature of a core on its neighbors, and ambient temperature. At any given time instance, the amount of power the chip can consume to safely operate will depend on the alignment of active cores, which is determined by application mapping. It can be deduced that for the same set of applications and the same number of active cores, a certain mapping can result in a higher power budget over other mappings, based

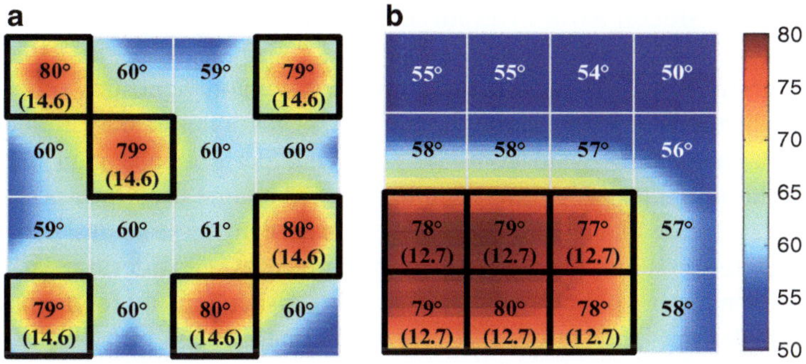

Fig. 9.1 Effect of mapping on power budget [5]. Each block shows temperature of the core and utilized power budget. Sparse mapping (left) has higher power budget utilization than dense mapping (right)

on appropriate alignment of active and dark cores. Subsequently, the surplus budget gained through mapping can be utilized to power up more cores, minimizing dark silicon and thus offering higher performance. This is explained through an example presented in Fig. 9.1.

Resource Allocation

Challenge: Greedy Resource Allocation

Run-time application mapping policy is one of the key factors that can influence performance and energy efficiency of many-core systems [6]. Mapping is the process of choosing a preferable set of cores on the chip to run tasks of an application, minimizing congestion and maximizing performance [7]. With dynamic workload characteristics and an unpredictable sequence and arrival of applications, mapping decisions have to be made at run-time for current and future many-core systems. State-of-the-art dynamic application mapping strategies have targeted benefits in terms of network performance, minimizing congestion, system throughput, power optimization, etc. [7–9], without any fixed upper bound on power consumption. Designers have been using strategies that map applications assuming that all the cores of a chip are active and available. However in a dark silicon scenario this assumption is not valid as not all the cores can be active at a given time and the number of cores that can be active varies depending on activity of other working cores [5]. The main focus of mapping strategies so far is on inter-task communication and their objective is maximizing performance, forcing them to map contiguously [10–12]. Precisely, these algorithms do not consider the issue of dark silicon and the availability of active cores without violating a safe power budget. Subsequently, they will not be able to provide the highest possible performance.

Solution: Dark Silicon Aware Resource Allocation

Existing run-time mapping algorithms are dark silicon agnostic, offering poor energy efficiency and lower throughput. This necessitates a dark silicon aware run-time mapping strategy to continue achieving performance gain and energy efficiency through technology node scaling. Power budget (and thus performance) limitations of future many-core systems emphasize to re-structure the objective of mapping towards improving the power budget by considering dark silicon. Conventional mapping policies advocate avoiding dispersion and fragmentation and do not consider the issue of dark silicon which changes the impact of dispersion and fragmentation on system performance [13]. A non-contiguous mapping through geometrical partitioning of the network is presented in [14], which shows that penalties on performance can be minimized by mapping communicating tasks on nearby cores, but not necessarily contiguously. They have established that non-contiguous mapping not necessarily affects system performance, although they do not exploit this fact to reduce dark silicon. A patterning approach to avoid congestion between packets routed from the same row or column is proposed in [15]. They limit the number of tasks to one per row and one per column and do not consider any power budget.

Re-Thinking Many-Core Design for the Dark Silicon Era

Several attempts have been made recently to address the issues of dark silicon, typically by using power management and thermal management individually. A feedback based power management system is proposed in [1]. However, they are limited to restricting the violation of TDP through a PID controller and they use a conventional dynamic mapping approach [7], which is not dark silicon aware. An online learning approach is presented in [16] that employs various power management techniques whenever there are hot spots identified in the system with changing workloads. It is restricted to multi-core systems and does not consider dark silicon. Thermal aware system analysis and calibration is presented in [17] at a lower level of abstraction, yet they do not propose effective task allocation based on their detailed analysis. Liu et al. [18] present an energy and thermal aware mapping strategy for NoC-based systems. Their approach is limited to design time (static mapping) which is based on a heuristic that estimates temperature, resulting in a near exhaustive search to find optimal nodes. A proactive estimation of potential hot spots through temperature sensors is presented in [19]. They mitigate identified hot spots through thread migration and dynamic voltage scaling. This method suffers from performance degradation by voltage scaling and overhead in migration. Bao et al. [20] present a case for balancing the heat dissipation of the chip evenly through a temperature aware mapping. Their mapping is based on voltage downscaling when needed as per critical temperatures of cores, putting a limit on its performance. Power capping of the cores through dynamic voltage and frequency scaling by identifying hot spots at run-time is presented in [21]. The authors try to minimize the power consumption of a specific section of the chip and thus to balance heat distribution, however, no run-time mapping strategy is presented in their work. On the whole, thermal aware mappings stay within upper bound of power budgets and avoid hot spots, but do not utilize the available budget effectively.

Pagani et al. have proposed an adaptive way of setting the upper bound on power consumption by expressing it as a function of simultaneously active cores [5]. They provide a light weight C library to estimate TSP for a given mapping, and the worst case TSP for a given number of active cores. Despite these, they do not provide any insight on which mapping would offer a better power budget. The importance of spatial alignment of dark cores along with active cores and its effects on the power budget have been presented in [22]. It shows the gain of patterning in two perspectives: better power budgets and lower operating temperatures. Shafique et al. [23] quantitatively showed that different patterns of dark silicon result in different power budgets and temperature profiles of the chip. However, both these works do not propose any method on how to pattern the dark tiles to get such higher power budgets. Taking all these things together, it is necessary for a run-time mapping algorithm to pattern active and dark cores to maximize the utilization of power budget.

Implications of Power Budget and Mapping

A contiguous mapping of an application with 6 tasks is shown in Fig. 9.1a. Since all the active cores are tightly packed, the heat dissipated by every active core affects its neighbors hazardously. As a result, the cores reach their critical temperature of (80 °C) after consuming 12.7 W of power. This configuration of mapping can effectively utilize $12.7 \times 6 = 76.2$ W of power, which would be its power budget. Utilizing power beyond this budget would heat up the cores beyond their critical temperature leading to permanent failure. The mapping shown in Fig. 9.1b is sparse in comparison with the previous mapping. In this case, as the cores are relatively apart, the heating effect of one active core on the others is minimized. Therefore, the cores can now consume 14.6 W before they reach their critical temperature. Effectively, the chip's power budget in this case is $14.6 \times 6 = 87.6$ W, which is 11.2 W (14.9 %) more than in the previous case. This surplus budget can in turn be used to: (1) activate more cores, (2) run current tasks faster, and (3) run more tasks without reaching critical temperatures.

Although it has been shown that optimal mapping can provide a higher power budget [22, 23], to the best of our knowledge, no methodical approach exists on how to pattern the active cores alongside dark cores such that they result in a higher power budget. In this chapter, we present a dark silicon aware run-time mapping approach which aligns active cores along with dark cores that can evenly distribute heat dissipation across the chip. The surplus budget we gain through mapping raises the upper bound on power consumption which is used to activate more cores.

The patterning technique presented in this chapter is based on our recent work [24]. To the best of our knowledge, this is the first work to consider a dark silicon scenario for run-time application mapping and to pattern the active and dark cores to improve the power budget. The key contributions of this work are as follows:

- A dark silicon aware run-time application mapping approach that aligns active cores with dark cores to offer higher power budget.
- A closed-loop power budgeting platform that keeps the maximum power consumption under safe operational power (i.e., TSP) which varies at run-time.

Effect of Mapping on Dark Silicon

The heat dissipated by a core C_i is a 3-tuple (P_i, T_n, T_{amb}), where P_i is the power dissipated by itself, T_n is the temperature of neighboring cores, and T_{amb} is the ambient temperature. When active cores function at full throttle, they dissipate power and heat that is proportional to this power. If heat goes beyond the safer limit, chip's functionality would fail permanently. Generally, dynamic thermal management techniques such as clock and power gating, voltage and frequency scaling, increased fan speed, etc., are triggered to manage any such sudden phases of a chip's overheating. This would reduce activity inside the chip and let the chip back to steady state temperature over a period of time. Although uneven heat distribution can be managed with these, it hampers the performance by a great deal. A better way for even distribution of heat is to pattern the dark cores along with working cores through run-time mapping.

The impact of spatial alignment of active cores on the power budget is explained through a motivational example, presented in Fig. 9.2. Three applications App1, App2, and App3 with 9, 12, and 7 tasks, respectively, are assumed to be running on the system. Conventional mapping approaches offer lower inter-task communication latency by greedily mapping all the applications contiguously. Mapping of the three applications contiguously on an NoC-based many-core system with 144 cores is shown in Fig. 9.2a. The power budget (TSP) of this system as computed by TSP library is 66 W. A non-contiguous and spread-out mapping of the same applications is shown in Fig. 9.2b. This mapping provides a power budget (TSP) of 74.6 W, as calculated by TSP library. An improvement of 8.6 W in the power budget can be observed for the spread-out, patterned mapping as opposed to tightly packed and contiguous mapping. Contiguous mapping avoids dispersion, but it leads to a poor thermal profile of the chip due to prorogation of heat among neighboring applications and tasks and thus resulting in lower power budget. Contrastingly, a spatially distributed mapping of applications offers higher power budget as effect of heat among different applications is negligible. Also, active cores are patterned along with inactive cores such that heat effects of neighboring cores running the same application are minimized. Based on the spatial alignment and a minor compromise on dispersion, we could gain up to 13 % of power budget. It is to be

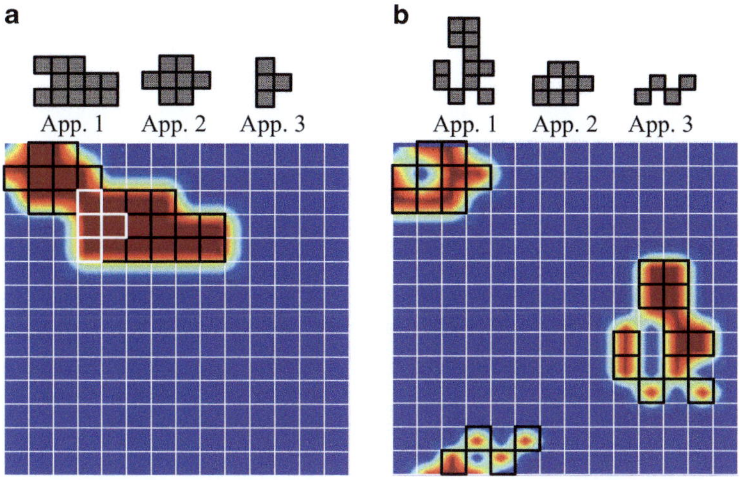

Fig. 9.2 Thermal profiles of contiguous and spatially distributed mappings. (**a**) Contiguous mapping - temperature accumulation is concentrated around the applications increasing power density. (**b**) Spatially distributed mapping - temperature accumulation is distributed out balancing power density

noted that the inactive cores are the inevitable dark cores—instead of leaving them out naively, we use them to balance out heat distribution and gain power budget. In the mapping presented in Fig. 9.2a, the budget of 66 W is used to power 22 cores, giving a per-core budget of 3 W. Assuming a per-core budget of 3 W, the mapping configuration in Fig. 9.2b could power $74.6/3 = 24.7$ cores. The surplus budget we could gain via patterning could thus be used to power up 2.7 more cores, reducing dark silicon by 12.2 %. This benefit could be better realized with increased size of the many-core system, technology node scaling, and more dark silicon which is expected in the near future. The average packet latency in each of the patterned mappings for App1, App2, and App3 is 1, 3.5, and 3.8 % more than that of the contiguous mappings. Although there is a minor penalty in terms of latency, the gain in power budget outweighs the odds against it.

Figure 9.3 shows the surplus power budget that was gained for different mapping configurations over the worst case TSP budget. We ran random mappings with fixed amount of inactive cores ranging from 25 to 90 % of darkness, over mesh sizes of 8×8, 12×12 and 15×15 with 22 nm technology. For the same number of active cores, we collected the worst case TSP budgets. The difference between a random configuration and the worst case budget is presented as percentage gain in power budget. It is evident that the gain in the power budget increases with increase in amount of dark silicon, to an extent where we can get a surplus of up to 114 % when 90 % of core is dark. It is to be noted that this surplus is purely from a random mapping generated to represent the power budget gain that can possibly be achieved, although the realistic gain could be variable.

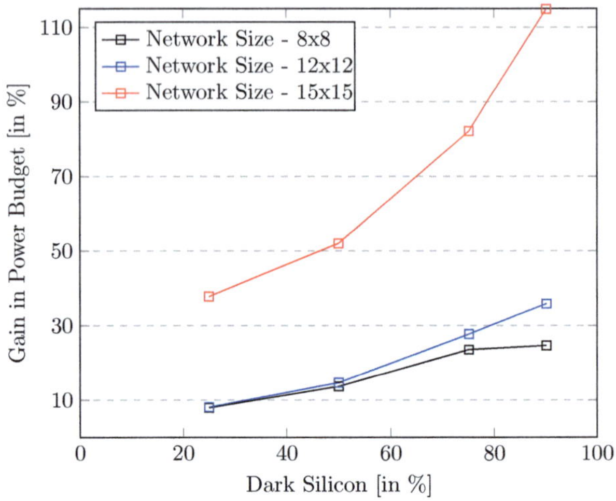

Fig. 9.3 Surplus power budget with increasing dark silicon and network sizes

Dark Silicon Aware Run-Time Mapping

System Architecture

The top level abstraction of the system we implemented in our approach is shown in Fig. 9.4. Application Repository holds the applications modeled as task graphs and are released onto the system for execution over time. The Run-time Mapping Unit (RMU) monitors the power profile of incoming applications issued by the Application Repository, to ensure that the upper bound of the power budget is not violated by mapping a new application. The RMU estimates the power of incoming application and checks if the chip currently has a high enough power budget to run the new application. An example of such an estimation can be found in [1]. The application is forwarded to the system if there is available power. In case of un-availability, the application waits until the system can allocate enough power, perhaps with currently running application(s) leaving the system after finishing their execution. TSP Calculator receives current mapping configuration of the system as input and computes the realistic upper bound on power budget, the thermal safe power. The RMU feeds this new budget value to the chip and updates the maximum power budget of the chip to the TSP provided by the TSP Calculator.

Application mapping is to find a free region for an application on the chip, which is of polynomial time complexity [25]. However, we have constraints on finding an optimal region such as power consumption, communication latency, dispersion, patterning the dark cores, etc., making mapping an NP-hard problem. Surplus budget gained from mapping in Fig. 9.2b can be attributed to the mapping policy. Two major factors that distinguish it from contiguous mapping are spatial distribution of applications and sparsity among tasks of each individual application. In view of these factors, we split our mapping approach into two phases, viz.,

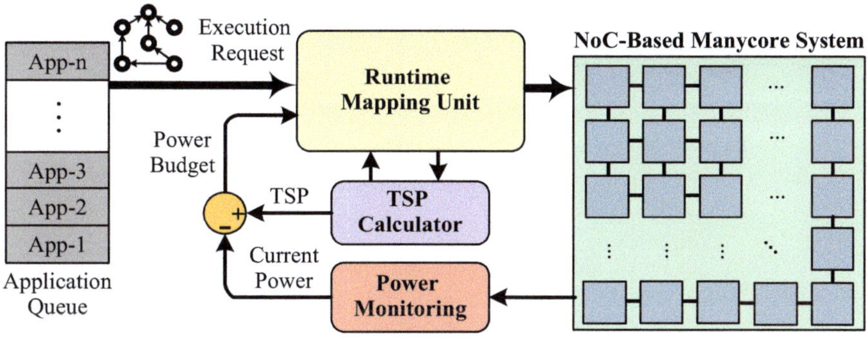

Fig. 9.4 Architecture implementing the proposed mapping strategy

selecting a region that is spatially dispersed from the current set of applications and mapping tasks of the application sparsely such that active and inactive cores are patterned in the selected region.

Finding an optimal region for an application starts with finding an optimal node, the *first node*, around which an application can be mapped. In our approach, we prioritize nodes (thus regions) that are spatially far from currently active cores as *first nodes*. Tightly packed up active cores are the major reason for hop-spots which eventually lead to under-utilization of power budget. Hence, after region selection, we map tasks of the application in a sparse pattern that will reduce the probability of heat accumulating at specific regions that potentially turn out as hot spots. Our approach tries to attain an even distribution of heat at both region selection (through optimal first node selection) and mapping (through patterning).

First Node Selection

Given the unpredictable application sequence, system's throughput is decreased when the first node is not found quickly. To avoid this, we cluster the mesh into square shaped regions of different radii, in order to fit applications of different sizes. *Square*(S_r): A square $S_r \subseteq Mesh$ is a closed square shaped region of radius r, with a set of nodes $n_{ij} \in N$. Radius r of a square region S_r is the distance from the center of the square to any of its edges. Figure 9.5a shows square regions centered at two different nodes, n_{37} and n_{63}, with radii 1 and 2. The node n_{37} is a center for the squares S^1_{37} and S^2_{37} and the node n_{63} is a center for the squares S^1_{63} and S^2_{63}. Note that each node is a center for squares with radius 1, 2, 3, ... up to $(\sqrt{M} - 1)/2$, where M is the total number of nodes in the mesh.

In contrast to reactive strategies, our first node selection method pro-actively calculates an optimal first node for incoming applications of every possible size by assigning the square that can fit the application. This way, we totally eliminate the time spent in first node selection, which automatically reflects in the overall

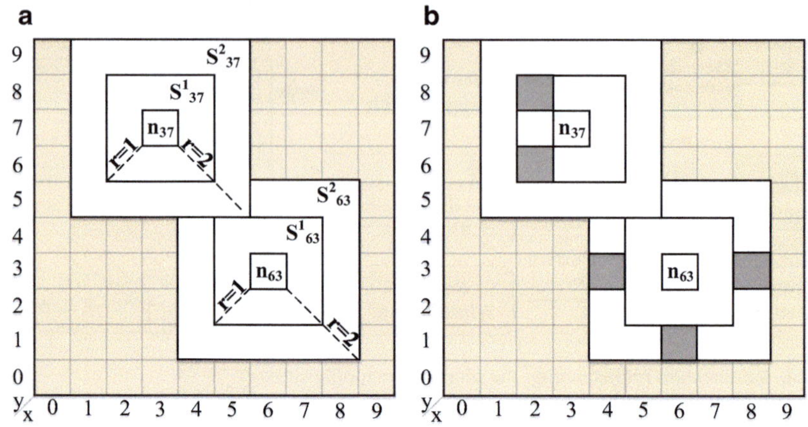

Fig. 9.5 Clustering of mesh into square regions. (**a**) Square regions of radii 1 and 2. (**b**) Affect of occupied nodes on square selection

execution time. The choice of a particular first node and thus a particular square region is based on our objective to find a region that has: (1) free nodes to allocate for incoming application with minimal internal congestion and (2) minimal effect in terms of temperature on other regions. We define two parameters, viz., the Vicinity Counter (*VC*) and the Distance Factor (*DF*) that quantify the above mentioned objectives. Vicinity Counter represents the number of free nodes available around a given node while Distance Factor represents the distance of the selected node (and thus region) from all the other existing regions with active cores.

Definition. *Vicinity Counter,* VC^r_{ij}, *for a node located at (i,j) is the weighted sum of the number of free nodes in the square centered at (i,j) with radius r.* The *VC* for a node located at (i,j) is expressed as

$$VC^r_{i,j} = \sum Wn_{i,j} \times (r - d + 1) \tag{9.1}$$

where Wn_{ij} is the weight of node n_{ij}, r is radius of square, and d is the distance from occupied node to the center. The weight for a node n_{ij} is assigned as

$$Wn_{i,j} = \begin{cases} 1 & \text{if } n_{i,j} \text{ is unoccupied} \\ -1 & \text{if } n_{i,j} \text{ is occupied} \end{cases}$$

VC expresses the number of free nodes, which also represents the number of occupied nodes. We consider the affect of occupied nodes in the region on internal congestion, by quantifying the location of occupied nodes. For instance, a node that is occupied in the innermost square close to the first node has more affect on internal congestion than the ones that are occupied in outer squares, far from the first node. We quantify this by pegging the weight of an occupied node with its distance from the central node, by assigning a higher penalty to occupied nodes closer to central node and relatively lower penalty to the ones that are far. The VC value represents the availability and congestion around a chosen node, and makes it easier to select between nodes with the same VC value, by considering congestion. Figure 9.5 shows the VC values for two different scenarios of occupied nodes. In case of a square centered at tile t_{37}, the number of occupied nodes in the square is 2, with both the nodes being in the innermost square close to the first node. The VC for t_{37} is $25 - (1 \times 2 + 1 \times 2) = 21$. The other scenario presented is the square centered at node t_{63}, with three nodes occupied, all being in outer most square far from the central node. The VC for this square $= 25 - (1 \times 1 + 1 \times 1 + 1 \times 1) = 22$. The square centered at t_{63} has a better VC value and would be the preferred despite having more number of occupied nodes than that of square centered at t_{37}. This can be attributed to the fact that t_{37} has occupied nodes closer which would result in higher congestion and dispersion. The fact that tasks with higher communication volume will be mapped at the central node and its neighbors makes the impact of occupied nodes in inner squares more significant.

When an un-occupied node becomes occupied by a new task, it starts dissipating power and thus heat. Initially, the heat is concentrated on the node itself, over time it starts impacting its neighbors. However, the effect of heat dissipated by this node on neighboring nodes gradually decreases as we move towards farther nodes. Inspired by the surface tension phenomenon [26], where energy distribution of a flat surface turns into a curved surface with applied surface pressure, we model the effect of heat transfer in a similar fashion. The temperature effect of every active core decreases exponentially, but not linearly, with distance from the active (hot) core. In other words, the greater the distance from an active core, the smaller the effect of heat from it. We illustrate this effect in Figure 9.6, for the chip running three applications App1, App2, and App3. The regions where the applications are mapped (deeper zones) have a DF as low as -4, while the regions that are far away (shallow zones) from them have a DF of 1. Also, as we traverse towards the far off shallow zones, the DF value improves indicating the importance of distributing the application spatially across the chip. This effect also depends on mesh size such that a smaller mesh results in greater impact, due to more proximity among its neighboring nodes. In order to minimize this heating effect from active cores of other applications, we prioritize nodes that are as far as possible from such active cores.

Definition. *Distance Factor, DF_{ij}, for a node located at (i,j) is the weighted sum of impact of distance from all the other occupied nodes located at (x,y) such that $(x, y) \in Mesh$.*

$$DF_{i,j} = \sum Wn_{i,j} \times (e^{-\alpha(d_{ij-xy})}) \tag{9.2}$$

where Wn_{ij} is the weight of node n_{ij}, as in Eq. (9.1), d_{ij-xy} is the distance from nodes located at (i,j) and (x,y) and α is the mesh size. Thus the DF of a node represents the effect of heat from concurrently running applications on the chip.

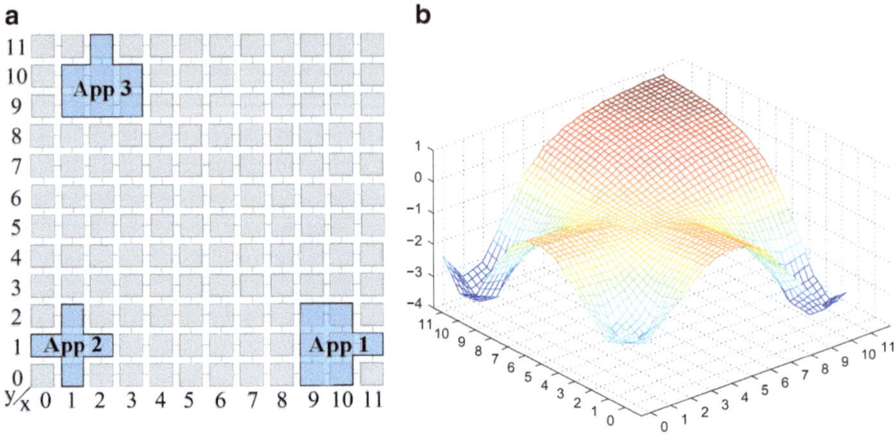

Fig. 9.6 Effect of occupied cores on Distance Factor of un-occupied cores. (**a**) x (**b**) y

This helps in selection of a region that is less probable to generate any potential hot spots. In addition, the issue of two or more nodes having the same VC value can be resolved by examining their corresponding DF values, and the node with the lower (best) DF value is chosen.

The algorithmic flow of first node selection is shown in Algorithm 1. When a new application enters the system, the required power budget is estimated. If the system has enough budget to be allocated for the new application, then the application request is serviced by finding the first node (lines 1–3). The node with maximum VC value, *maxVC*, among all nodes is chosen as the first node, and thus also selecting the square centered at *maxVC*. The chosen first node and the application are passed to the patterning phase (line 4). Mapping the tasks based on patterning is explained in the following section in Algorithm 2. The chip's mapping configuration changes after mapping one application. As a result, the VC and DF values for the remaining un-occupied nodes need to be updated for every newly occupied node. Once the application is mapped, we calculate the new node with maximum value of VC, *maxVC*, for different radii and also update the DF values of un-occupied nodes. This way, we pro-actively calculate the first node for next incoming application and avoid the overhead caused in first node selection. VC values for square of different groups are updated (lines 6–23) to determine the new *maxVC* node. In case of a conflict when more than one node has the same *maxVC*, it is resolved by comparing their DF values (lines 16–18). Thus the VC and DF values are pro-actively updated with the intention of servicing the next incoming application immediately, without any overhead. Nodes are released from the system when an application leaves after finishing its execution. The VC and DF values are once again updated according to the changed mapping configuration of the chip, with the exit of an application.

Algorithm 1: The mapping algorithm

Inputs: *newApp*:New application, *budget*:Available power budget;
Outputs: Q:Mapping;
Constants: M :Size of the mesh, *groups* : Number of square groups = $[(\sqrt{M}-1)/2]$,*maxRadius*:Maximum radius of the square $((\sqrt{M}-1)/2)$, α:Mesh size parameter $\sqrt{M}$,P_{avg}: Average power consumption per node;
Global Variables: VC : Vicinity Count of a node, $maxVC$: Node with the maximum VC, *firstNode* : Selected first node for mapping, DF : Distance factor;

Body:

$appPredictedPower \leftarrow |newApp| \times P_{avg}$;
if $appPredictedPower \leq budget$ **then**
    $firstNode \leftarrow maxVC_{[(\sqrt{appSize}-1)/2]}$
    $Q \leftarrow pattern(firstNode, newApp)$;
    //Updating VC and DF values after mapping
    for each $n_{xy} \in newApp$ **do**
        for each core n_{ij} located in Row i and Column j **do**
            $r' = maximum(|i-x|, |j-y|)$;
            $DF_{ij} -= e^{-\alpha r'}$;
            for $r = 1$ to $maxRadius$ **do**
                if $r - r' \geq 0$ **then**
                    VC^r_{ij} -= r - r';
                    if $VC^r_{ij} > maxVC_r$ **then**
                        $maxVC_r \leftarrow VC^r_{ij}$;
                    else
                        if $VC^r_{ij} = maxVC_r$ and $DF_{ij} > DF_{maxVC_r}$ **then**
                            $maxVC_r \leftarrow VC^r_{ij}$;
                        end if
                    end if
                end if
            end for
        end for
    end for
end if

Dark Silicon Patterning

An optimal region for mapping an incoming application is chosen via first node selection. The mapping receives the first node (*fn*) as an input and builds a polygon P which is the set of all un-occupied nodes around the selected first node. We choose nodes among the region P such that the tasks are run on nodes that are sparsely aligned. Sparsity of a node n_{ij} represents the number of free nodes that are neighboring it in all four cardinal directions (North, East, West, and South). The Sparsity Factor (SF_{ij}) for a node n_{ij} located at (i,j) is expressed as

$$SF_{i,j} = \sum_{i=1}^{4} \sum_{j=1}^{4} F(i+i', j+j') \tag{9.3}$$

where $i' = [0, 1, 1, -1], j' = [-1, 0, 1, 1]$. $F(i,j)$ denotes if a node located at (i,j) is free or not, such that

$$F(i,j) = \begin{cases} 1 & \text{if } n_{i,j} \text{ is unoccupied} \\ 0 & \text{if } n_{i,j} \text{ is occupied} \end{cases}$$

Based on the Sparsity Factor, we prioritize the nodes that are more sparse. Sparse nodes ensure that they have less effect on neighbor's temperature. As a result, the dense nodes are assigned least priority and are patterned in a way that they provide necessary cooling effect needed by their active neighbors. While mapping, we sort the tasks of the incoming application as per communication volume. We choose the task with the highest communication volume and map it on to the first node, the one with the highest sparsity factor among available nodes. We proceed to the task with the next highest communication expense and map it onto the node with the highest *SF* among the nodes in the selected square *S*. We continue similarly, in the order of communication of tasks and sparsity of nodes so that tasks with higher communication get mapped onto nodes with higher sparsity, until all tasks of the application are mapped. This way, the less sparse nodes (i.e., the dense nodes) which are tightly packed with active cores as neighbors get least priority. These denser nodes eventually remain un-occupied among the other occupied nodes of the region, minimizing the probability of creating potential hot spots. For tasks that are mapped at non-minimal distance, there is an increase in energy and latency. However, prioritizing communication weight among tasks makes sure that this latency and energy penalty are minimal.

The patterning is explained through an example presented in Fig. 9.7. Assuming that an application with four tasks, as in Fig. 9.7a, has arrived, the selected first node *fn* and the corresponding square region is shown in Fig. 9.7b. The most communication intensive task of the application is task 3, hence it is mapped onto most sparse node (first node), *fn*. This automatically effects the SF values of nodes surrounding it. The next in order of communication is task 1, and the next node with higher SF is chosen and task 1 is mapped on it. In the similar fashion, the remaining tasks are mapped subsequently, as in Fig. 9.7c–f. It is to be observed that after the application is mapped, one node in the square is left such that it has SF = 0, which would be of least priority for any incoming application. Along the same lines, there are other nodes with SF = 1, which also attained a relatively lower priority for getting mapped. These nodes are the candidates that would potentially be dark, and provide the cooling effect needed by their active neighbors. Thus, we pattern the dark cores among active cores to minimize the probability of heat getting accumulating at any single point on the chip. Conversely, had all the tasks of the application been mapped contiguously, they would have reached their critical temperatures by consuming only lower amount of power and eventually trigger dark cores elsewhere on the chip. However, in the case where we patterned the inevitable dark cores, the cores will reach critical temperatures only after utilizing available power budget to a better extent, or offer more budget to activate more cores.

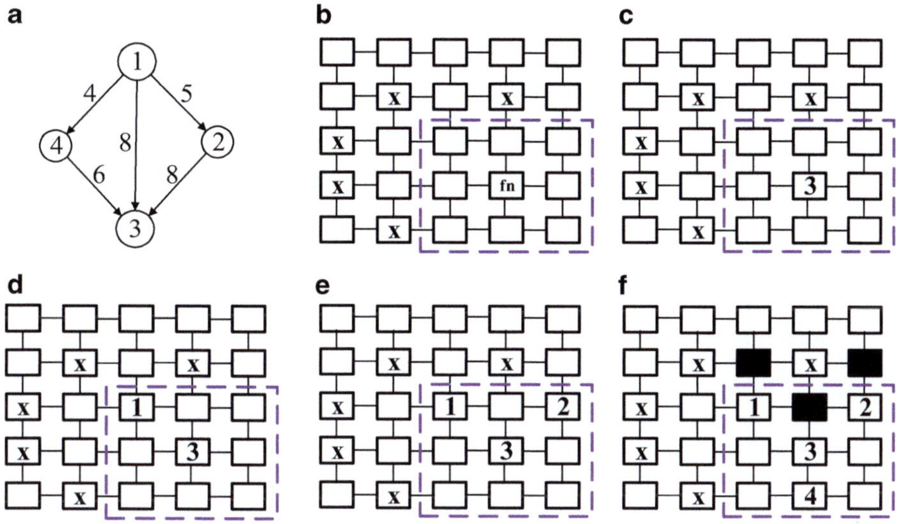

Fig. 9.7 Example for pattern based mapping. X represents already mapped tasks. (**a**) Application task graph. (**b**) First node selection. (**c**)-(**e**) Mapping. (**f**) Patterning

The algorithmic flow of patterning is shown in Algorithm 2. Patterning starts with a selected first node, *firstNode*, and chooses the square region ($Square^{size}_{firstNode}$) that can fit the application (*App*) (line 1). The application is sorted into tasks (*Tasks*) as per their communication volume (line 2). The most expensive task is mapped onto the node with maximum Sparsity Factor, *maxSF* (lines 5–6). The mapped task is removed from the list of tasks to be mapped and the mapped node is removed from the list of available nodes in the square (lines 7–8). This procedure is repeated until all the tasks of the application are mapped. With every node that gets occupied in the selected square, the *SF* value for other nodes changes accordingly. A new *maxSF* is computed for every unmapped task, given that *SF* values of every occupied node varies.

Evaluation

Testing and validating the proposed algorithm require simulation of different traffic patterns running on a many-core platform that supports run-time mapping. Comparing it with other state-of-the-art mapping strategies would provide a deeper insight on gain in power budget, also viewed from the dark silicon perspective.

Algorithm 2: Patterning

Inputs: *App* : Application,*firstNode* : Selected First Node.
Global Variables: *S* Selected square, *SF* : Sparsity Factor for each node in a square, *maxSF* : Node with the maximum SF in a square, *currentNode* : Node onto which current task is being mapped;
Global Constants: *Tasks* : Vector of tasks of the application *App*,*size*: Radius of square close to size of *App*

Body:

$S \leftarrow Square^{size}_{firstNode}$;
$Tasks = sort(App)$;
while $App \neq \emptyset$ **do**
  for each $t_i \in Tasks$ **do**
    $currentNode \leftarrow maxSF$;
    $map\ (t_i) \rightarrow maxSF$;
    $App - t_i$;
    $S - currentNode$;
  end for
end while

Simulation Platform for Mapping Evaluation

Throughout the evaluation, it is assumed that applications are modeled as task graphs of different sizes ranging from 4 to 35 tasks, generated using [27]. Communication volumes among these tasks are distributed randomly using Gaussian distribution. An in-house cycle-accurate many-core platform implemented in SystemC is used to simulate these traffic patterns. The specifications of Niagara-2 like in-order cores obtained from McPAT [28] are used as the baseline for processing elements. The communication network infrastructure between processing elements is provided by a pruned version of Noxim [29] that uses mesh topology and XY routing. Parameters related to technology node scaling are extracted from the Lumos framework [30], an open source framework which quantifies power-performance characteristics of many-core systems with technology node scaling. TSP library [5] is used to calculate the thermal safe power. Proposed dynamic mapping is implemented by a Central Manager (*CM*), which is the node $n_{(0,0)}$ of the mesh in our many-core platform. In the many-core platform we implemented (overview as in Fig. 9.4), a random sequence of applications enter the system and are buffered into a FIFO. Applications are serviced in a first-come-first-serve policy, subject to availability of enough power budget. If a high enough power budget can be allocated, the application is scheduled by the Central Manager (*CM*). A suitable first node for mapping the incoming application is chosen by the *CM*, followed by mapping all the remaining tasks of the application.

Power Budget Gain

References [7] and [13] present state-of-the-art approaches for first node selection and application mapping for NoC-based many-core systems. These two approaches prioritize regions with free nodes and contiguity among nodes selected for mapping, with the primary objective of minimizing communication latency. The proposed dark silicon patterning approach (from here on referred to as *PAT*) is evaluated by comparing it against the combination of *SHiC* [7] and *CoNA* [13] (from here on referred to as *SC*), for first node selection and mapping, respectively. *PAT* relaxes the constraint on contiguity among individual applications, hence comparing it against *SC* would quantify the effect of patterning approach as opposed to contiguity. In view of applications that require a conservative upper bound on power, we also compare *PAT* against worst case power budget that can be offered with a given number of active cores (TSP_{wc}). Given a fixed number and sequence of applications, we compare power budgets offered by different mappings based on different mapping strategies. The entry sequence of applications is maintained the same for different mapping approaches for a fair comparison. In case of TSP_{wc}, we compare the power budget given by Pagani et al. [5] for the number of active cores in the mapping generated by *PAT*. We run our simulations over different network sizes of 16×16 and 20×20. In addition, we emulate a varying dark silicon behavior ranging from 50 % darkness through 90 % by adjusting initial upper bound on power consumption (TDP). We set the TDP to 177.7 and 277.7 W, respectively, for 16×16 and 20×20 network sizes using 22 nm technology, keeping the power density constant [30]. ITRS projections present the fact that by the year 2020, computer systems would face 90 % dark silicon and that many-core platforms would be in upwards of 512 cores to gain maximum peak performance [4]. Hence, we considered the case of the chip being 90 % dark, while also including contemporary projection of 50 % dark area. We limit the number of applications entering the system to be in accordance with dark areas. We evaluate the proposed approach over power budget provided per mapping and corresponding throughput. The power budget is computed using TSP library for every mapping, traced with entry of a new application. The ambient temperature is set to 45 °C and the safe operating temperature beyond which chip's functionality fails is set to 80 °C.

The average (arithmetic mean) and best case percentage gains in power budgets of different mapping configurations using *PAT* strategy over *SC* for different network sizes are presented in Fig. 9.8. *PAT* achieves a surplus power budget when compared to *SC*, as the active cores are optimally arranged, balancing heat distribution across the chip. In contrast, *SC* tries to map contiguously, leading to tightly packed active cores which get heated up already at lower power consumption, resulting in a lower power budget. In addition to a better power budget, the chip always operates under safe peak operating temperature 80 °C, since the budgets are computed through TSP library which manages the upper bound on power avoiding hazardous hot spots. It can be observed that the surplus budget achieved in case of *PAT* increases with increase in amount of dark silicon on the chip. With 90 % of the chip being dark, utilizing the remaining fewer number of cores that can originally be powered (active) becomes crucial. *PAT* performs better in such scenarios, given the wider choice of dark cores that can be patterned, while *SC* remains dark silicon agonistic. The gain also increases with increase in network size, once again due to increase in scope of the chip area that can be patterned. Moreover, this is in line with the power budget gain obtained for randomly distributed tasks compared to worst case budget generated by TSP library, as in Fig. 9.3.

In view of future applications like big data, artificial intelligence, etc., requiring many-core platforms of larger network sizes, and the issue of dark silicon predicted to grow worse, patterning becomes quintessential to extract higher performance as expected from a many-core system. The potential of patterning could be further better realized with higher scaled up networks and subsequent increase in dark silicon. Nevertheless, the gain in power budget can still be realized at contemporary many-core platforms of relatively smaller network sizes. We simulated 12×12 network for 90 % dark area to observe significant gain in power budget in case of *PAT* compared to *SC* and TSP_{wc}. The average and best case (BC) gain in power budget and throughput achieved by *PAT* over *SC* and TSP_{wc} are shown in Fig. 9.10. Some of the low power and safety critical applications follow a strict and conservative approach on setting a single upper bound on power budgets. We account for such applications by comparing our proposed approach against power budget that could result from a worst case mapping configuration for a given number of active cores, generated by Pagani et al. [5]. The surplus gained when compared to worst case power budgets for different network sizes and darker chip areas is shown in Table 9.1. Understandably, PAT offers better power budget against worst case values. Although the comparison is against worst case budgets, it still establishes the impact of patterning on mitigating dark silicon to a large extent.

Since a surplus in power budget is gained through *PAT*, it can be utilized to power up more number of cores without violating any safe upper bound on power consumption, which reflects in throughput. The proposed first node selection method chooses diverse regions for different applications which has no impact on individual application's latency. However, few applications using *PAT* might have a lower individual latency at times, due to the compromise on contiguity among tasks of a patterned application. Despite the occasional latency, the overall throughput of the system using PAT still remains higher compared to that of *SC*, as the gain achieved in terms of power budget would (over) compensate for the latency. Gain in throughput for *PAT* compared to *SC* for different mesh sizes and dark regions is presented in Fig. 9.9. Throughput gain depends largely on surplus budget gained, which in turn can be used to activate more cores. Thus, throughput achieved using PAT strategy follows a similar trend to that of surplus power budget achieved (Fig. 9.10).

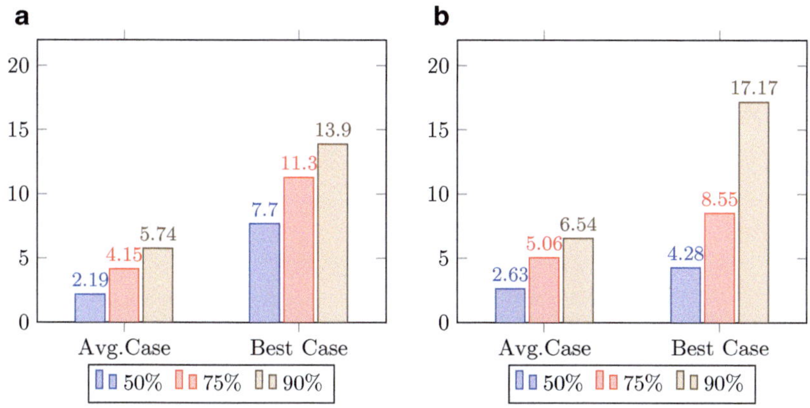

Fig. 9.8 Average and best case percentage gain in power budget of PAT over SC for different amounts of dark silicon. (**a**) Network size—16 × 16. (**b**) Network size—20 × 20

Table 9.1 Surplus power budget of PAT over TSP_{wc}

	90 % dark		75 % dark		50 % dark	
Network size	Avg.	Best	Avg.	Best	Avg.	Best
16 × 16	32.33	34.92	22.02	24.14	11.73	13.20
20 × 20	38.70	40.83	22.40	27.4	12.50	13.33

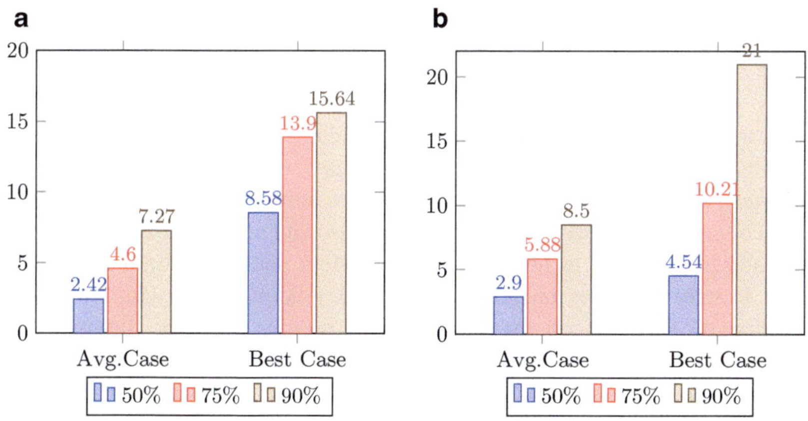

Fig. 9.9 Throughput gain of PAT over SC in percentage. (**a**) Network size—16 × 16. (**b**) Network size—20 × 20

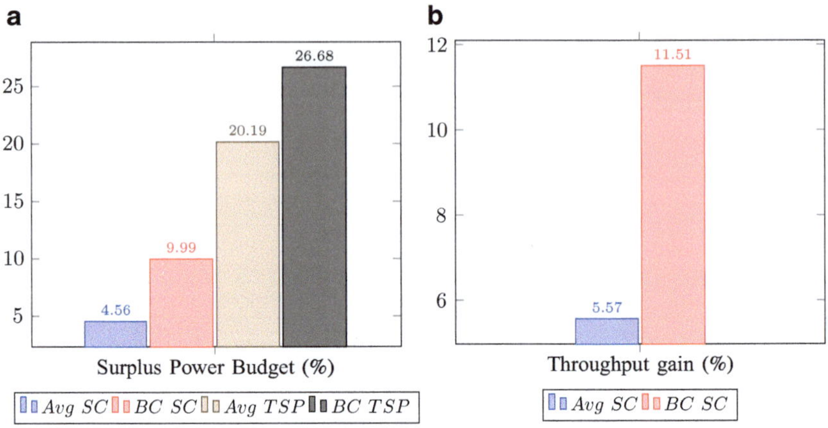

Fig. 9.10 Surplus budget and throughput gain of PAT over SC and TSP_{wc} for 12 × 12 mesh. (**a**) Surplus power budget (in %). (**b**) Gain in throughput

Summary

In this chapter, we presented a dark silicon aware run-time mapping strategy for achieving a better power budget. The proposed approach was implemented in two phases, viz., first node selection and patterning based mapping, where tasks are evenly distributed across the chip area to balance heat distribution. This lets the active cores to utilize relatively more power before they reach a maximum limit beyond which chip's functionality is harmed due to thermal violation. As a result, different applications utilize surplus power budgets better to activate more cores and thus gain in performance. Noticeably, the surplus power budget reflects in better resource utilization and throughput, while ensuring thermal safety of the chip, using a software run-time technique, with no hardware overhead. The scope of results presented on power budget gains provides only a fine line of possibilities and potential of patterning dark silicon, rather than loosing performance. An observation to be made is that gain in terms of power budget and throughput increases with increase in network sizes and amount of dark silicon on the chip, stressing the importance of dark silicon aware mappings moving into the future workload characteristics and increase in dark silicon. Platform dependent constraints such as network size, workload characteristics, ambient and peak critical temperatures would result in diverse amounts of dark silicon, at the same time offer more room for patterning to thrive. Results emphasize that patterning is a productive way to go ahead in dark silicon era, although the real potential is not thoroughly established yet.

Experimental results presented in this chapter assume the many-core platform to be homogeneous, although having a heterogenous combination of cores with different power-performance characteristics could have more potential. The surplus budget could be better used to activate even more number of cores at lower frequencies and also to run heterogeneous workloads on cores that suit them. Implementing the patterning technique and allocation of surplus budget for a heterogeneous many-core platform would be a good direction moving into the future.

References

1. M.-H. Haghbayan, A.-M. Rahmani, A. Y. Weldezion, P. Liljeberg, J. Plosila, A. Jantsch, H. Tenhunen, Dark silicon aware power management for manycore systems under dynamic workloads, in *Proceedings of IEEE International Conference on Computer Design* (2014), pp. 509–512
2. Intel Xeon Processor - Measuring Processor Power, revision 1.1. White paper, Intel Corporation, April 2011
3. Intel Corporation. Fourth Generation Mobile Processor Family Data Sheet. White paper, Intel Corporation, July 2014
4. AMD. AMD Kaveri APU A10-7800. Accessed: 2015-02-28. [Online]. Available: http://www.phoronix.com/scan.php?page=article&item=amd_a_45watt&num=1
5. S. Pagani, H. Khdr, W. Munawar, J.-J. Chen, M. Shafique, M. Li, J. Henkel, Tsp: thermal safe power: efficient power budgeting for many-core systems in dark silicon, in *Proceedings of ACM International Conference on Hardware/Software Codesign and System Synthesis* (2014), p. 10
6. E.L. de Souza Carvalho, N.L. V. Calazans, F.G. Moraes, Dynamic task mapping for MPSoCs. IEEE Des. Test Comput. **27**(5), 26–35 (2010)
7. M. Fattah, M. Daneshtalab, P. Liljeberg, J. Plosila, Smart hill climbing for agile dynamic mapping in many-core systems, in *Proceedings of IEEE/ACM Design Automation Conference* (2013)
8. C.-L. Chou, R. Marculescu, Contention-aware application mapping for network-on-chip communication architectures, in *Proceedings of IEEE International Conference on Computer Design* (2008), pp. 164–169
9. M. Fattah, P. Liljeberg, J. Plosila, H. Tenhunen, Adjustable contiguity of run-time task allocation in networked many-core systems, in *Proceedings of of IEEE Asia and South Pacific Design Automation Conference* (2014), pp. 349–354
10. C.-L. Chou, Y.O. Umit, M. Radu, Energy-and performance-aware incremental mapping for networks on chip with multiple voltage levels. IEEE Trans. Comput. Aided Des. Integr. Circuits Syst. **27**(10), 1866–1879 (2008)
11. M.A. Bender, D.P. Bunde, E.D. Demaine, S.P. Fekete, V.J. Leung, H. Meijer, C.A. Phillips, Communication-aware processor allocation for supercomputers: finding point sets of small average distance. Springer Algorithmica **50**(2), 279–298 (2008)
12. E. Carvalho, N. Calazans, F. Moraes, Heuristics for dynamic task mapping in NoC-based heterogeneous MPSoCs, in *Proceedings of IEEE/IFIP International Workshop on Rapid System Prototyping* (2007), pp. 34–40
13. M. Fattah, M. Ramirez, M. Daneshtalab, P. Liljeberg, J. Plosila, CoNA: dynamic application mapping for congestion reduction in many-core systems, in *Proceedings of IEEE International Conference on Computer Design* (2012), pp. 364–370
14. M. Deveci, S. Rajamanickam, V.J. Leung, K. Pedretti, S.L. Olivier, D.P. Bunde, U.V. Catalyurek, K. Devine, Exploiting geometric partitioning in task mapping for parallel computers, in *Proceedings of IEEE Parallel and Distributed Processing Symposium* (2014), pp. 27–36
15. S. Sano, M. Sano, S. Sato, T. Miyoshi, K. Kise, Pattern-based systematic task mapping for many-core processors, in *Proceedings of IEEE International Conference on Networking and Computing* (2010), pp. 173–178

16. A.K. Coskun, T.S. Rosing, K.C. Gross, Temperature management in multiprocessor SoCs using online learning, in *Proceedings of ACM/IEEE Design Automation Conference* (2008), pp. 890–893
17. L. Thiele, L. Schor, H. Yang, I. Bacivarov, Thermal-aware system analysis and software synthesis for embedded multi-processors, in *Proceedings of ACM/IEEE Design Automation Conference* (2011), pp. 268–273
18. Y. Liu, Y. Ruan, Z. Lai, W. Jing, Energy and thermal aware mapping for mesh-based noc architectures using multi-objective ant colony algorithm, in *Proceedings of IEEE International Conference on Computer Research and Development*, vol. 3 (2011), pp. 407–411
19. A.K. Coşkun, K.C. Gross et al., Utilizing predictors for efficient thermal management in multiprocessor SoCs. IEEE Trans. Comput. Aided Des. Integr. Circuits Syst. **28**(10), 1503–1516 (2009)
20. M. Bao, A. Andrei, P. Eles, Z. Peng, Temperature-aware task mapping for energy optimization with dynamic voltage scaling, in *Proceedings of IEEE Workshop on Design and Diagnostics of Electronic Circuits and Systems* (2008), pp. 1–6
21. T. Komoda, S. Hayashi, T. Nakada, S. Miwa, H. Nakamura, Power capping of CPU-GPU heterogeneous systems through coordinating DVFS and task mapping, in *Proceedings of IEEE International Conference on Computer Design* (2013), pp. 349–356
22. M. Shafique, S. Garg, T. Mitra, S. Parameswaran, J. Henkel, Dark silicon as a challenge for hardware/software co-design: invited special session paper, in *Proceedings of ACM International Conference on Hardware/Software Codesign and System Synthesis* (2014), p. 13
23. M. Shafique, S. Garg, J. Henkel, D. Marculescu, The EDA challenges in the dark silicon era, in *Proceedings of ACM/EDAC/IEEE Design Automation Conference* (2014), pp. 1–6
24. A. Kanduri, M.-H. Haghbayan, A.-M. Rahmani, P. Liljeberg, A. Jantsch, H. Tenhunen, Dark silicon aware runtime mapping for many-core systems: a patterning approach, in *33rd IEEE International Conference on Computer Design (ICCD)* (2015), pp. 610–617
25. D.P. Dobkin, H. Edelsbrunner, M.H. Overmars, Searching for empty convex polygons. Algorithmica **5**(1–4), 561–571 (1990)
26. H.E. White, *Modern College Physics* (Van Nostrand, Princeton, 1948)
27. TGG: Task Graph Generator. http://sourceforge.net/projects/taskgraphgen/, 2010
28. S. Li, J.H. Ahn, R.D. Strong, J.B. Brockman, D.M. Tullsen, N.P. Jouppi, McPAT: an integrated power, area, and timing modeling framework for multicore and manycore architectures, in *Proceedings of IEEE/ACM International Symposium on Microarchitecture* (2009), pp. 469–480
29. F. Fazzino, M. Palesi, D. Patti, Noxim: network-on-chip simulator. http://sourceforge.net/projects/noxim, 2008
30. L. Wang, K. Skadron, Dark vs. dim silicon and near-threshold computing extended results. University of Virginia Department of Computer Science Technical Report TR-2013-01, 2012

Chapter 10
Online Software-Based Self-Testing in the Dark Silicon Era

Mohammad-Hashem Haghbayan, Amir M. Rahmani, Antonio Miele, Pasi Liljeberg, and Hannu Tenhunen

Introduction

The aggressive technology scaling in the fabricated chips has brought to the integration of several cores within the same chip. At the same time, the drawback of such a transistor shrinking has been an increase in the susceptibility of the digital circuits to internal defects, device variability, and malfunction in execution units [1, 2]. As a matter of fact, aging and wear-out mechanisms, including time dependent dielectric breakdown (TDDB), negative bias temperature instability (NBTI), and electromigration (EM), are among the most increasingly adverse factors that can lead to timing errors and components' breakdowns causing system malfunctioning and, eventually, its overall failure. In addition, downscaling of CMOS technologies to the deep submicron levels has exacerbated the trend of high failure rate. This phenomenon has led to increased power densities and consequently operating temperatures in a device, being the main cause of the aging phenomenon. Thus, there is an increasing quest for reliability in modern computing systems.

In such a scenario, in order to handle such reliability quest, and in particular to detect and manage the occurrence of permanent failures in operational components, concurrent error detection and online testing may represent viable solutions. However, concurrent error detection is generally implemented by means

M.-H. Haghbayan (✉) • A.M. Rahmani • P. Lijeberg
University of Turku, Turku, Finland
e-mail: mohhag@utu.fi; amirah@utu.fi; pasi.liljeberg@utu.fi

A. Miele
Politecnico di Milano, Dipartimento di Elettronica, Informazione e Bioingegneria, Milan, Italy
e-mail: antonio.miele@polimi.it

H. Tenhunen
KTH Royal Institute of Technology, Stockholm, Sweden
e-mail: hannu@kth.se

A.M. Rahmani et al. (eds.), *The Dark Side of Silicon*,
DOI 10.1007/978-3-319-31596-6_10

of redundancy-based techniques [3], such as duplication with comparison (DWC) or triple modular redundancy (TMR), which present a high cost due to area occupation. For this reason they are generally considered only in the design of systems specifically targeted for critical applications, such as the aerospace appliances where the cost is secondary concern. Similarly, within the online testing field, Built-in-Self-Testing (BIST, [4, 5]) circuitries are not commonly integrated in devices targeted to the consumer market, even if they present a more limited impact on the chip area with respect to the above discussed techniques. Another strategy for online testing is Software-Based Self-Test (SBST, [2, 6]), which consists of periodic execution of specific testing routines devoted to the functional solicitation of the circuitry for the detection of permanent failures in various execution units. Since such strategy does not require any additional circuitry (or, in some situations, a reduced one), it represents the most promising solution for consumer electronic devices. Indeed, an example of its large-scale employment is in the automotive on-board computing systems [7].

Many-core systems fall under this umbrella of the digital devices requiring SBST [8]. In fact, they commonly do not feature any integrated hardware for online testing and, moreover, they are subject to a considerable stress caused by the intensive data-processing workload. However, this scenario presents two relevant issues:

1. The workload to be executed consists of applications requiring strict performance levels. This leads to the necessity of a *transparent test scheduling* because it is not possible to interrupt the nominal activities.
2. The system is characterized by a physical limit in the power budget which imposes that all the cores cannot be active at the same time at a full frequency. At the opposite a relevant portion of them have to be put dark for power limits (such phenomenon is dubbed as *dark silicon* [9, 10]). This issue implies the necessity to consider also power consumption during the test scheduling, thus necessitating a *power-aware testing approach*.

Hence, the employment of SBST in many-core systems presents at the same time new opportunities and challenges. We claim that in the scenario of the dark silicon era there is a quest for a *power-aware test scheduling approach* to detect at runtime permanent faults occurring in many-core architectures while not degrading the overall system performance.

An interesting and challenging aspect of modern many-core systems for test scheduling is the high dynamicity and heterogeneity of the executed workload. This makes the amount of dark area on the chip (i.e., total chip utilization) highly variable. Furthermore, due to the emergence of *dim silicon* [11] as a way to minimize dark areas and increase the number of active cores, the system might reach up to 100 % utilization of its cores (if the majority of running application are not performance-demanding) by making use of power management features such as Dynamic Voltage and Frequency Scaling (DVFS) [12]. This makes the behavior of such systems to be highly related to the characteristics of the workload where at different moments of time it is possible to have considerable dark areas with small resource utilization, due to the fact that some other group of cores are set on a high voltage–frequency level thus reserving the majority of the overall power budget, or small dark areas with large resource utilization by globally setting a very low voltage–frequency level. Therefore, if suitable scenarios are intelligently identified (when there is enough room in power budget), such temporary dark areas can be favorable targets for online testing in order to improve the system reliability [13, 14].

Nevertheless, DVFS knobs introduce also other issues in the testing activity, as shown in the literature [15], systems should be tested at multiple voltage–frequency settings, since faults are manifested in different ways in different configurations. Therefore, the test scheduling needs to take into account the fact that SBST routines should be executed on the various cores in different voltage–frequency levels.

Given these motivations, this chapter presents an approach for a transparent power-aware online test scheduling in many-core systems in the dark silicon era. The proposed approach benefits from the large amount of cores and the available power budget to dynamically schedule SBST routines on the idle cores that have experienced a high stress in the recent past. In particular, the approach exploits a criticality metric, computed on the basis of a measure of the utilization of the cores, to select the units to be tested. Then, a test mapping and scheduling approach selects among the candidates the actual cores to be tested on the basis of two conditions: such cores must be idle (i.e., not currently involved in the execution of an application) and there must be some available power not currently used for the execution of the running applications. Further, the test scheduling approach selects also the optimal possible voltage–frequency settings to execute the SBST routine by considering the system's power budget.

The rest of the chapter is organized as follows. Section "Related Work" reviews the related work discussing the limitations which motivate the work. In section "Adopted Many-Core Architecture" the background on many-core systems is discussed, presenting also the considered architecture and application models. Then section "Suitable Scenarios for Online Testing" describes suitable scenarios for online testing by means of a running example, showing power consumption and online testing issues. The proposed dark silicon aware online testing approach is presented in section "Dark Silicon Aware Online Testing Framework", while section "Test Scheduling for Different Voltage–Frequency Settings" proposes an enhancement to handle testing at different voltage–frequency levels. Section "Experimental Evaluation of the Approach" discusses the experimental results presenting some statistics and a comparison against a state-of-the-art approach to demonstrate the effectiveness of the proposed approach. Finally, section "Conclusions" draws the conclusions.

Related Work

SBST has been known as a useful mechanism in recent studies on online testing as it can be applied easily without any need for extra hardware resource [2, 6, 16]. Furthermore, it has been used widely for online testing in multi-/many-core system testing recently [8, 17]. The main challenge in online testing in multi-/many-core systems is to minimize the overhead of testing mechanism on the overall system performance [18]. Several works have been presented in the literature studying the impact of online error detection on the performance of multi- and many-core systems [2, 6, 8, 18–20]. In [21], an SBST scheduling algorithm is proposed for testing cores at runtime while the system is working. In [22] the authors proposed online testing algorithm for many-core systems to achieve high fault coverage for both routers and PEs. In [23, 24], a structural level process is presented to develop test software. In [4, 25–27] deterministic, random, and hybrid method were used for generating software tests. However, none of the state-of-the-art approach considers current available power budget while applying test process. In fact they are not power-aware.

Power-aware testing should not be confused with power-constrained testing. In power-constrained testing, the goal is to minimize the offline Test Application Time (TAT) by parallelization of testing the cores, e.g., using test access mechanisms, while avoiding peak power violation. Many studies have been done to achieve minimal TAT [5, 28, 29]. In fact, as the power consumption of the single core during test process is generally greater than that in normal operation mode [30], in power-constrained testing the focus is on how to test the cores without damaging it due to thermal violation. However, in power-aware testing the target is to test the core(s) when the other cores are working in their normal mode. That is why a power feedback from the system is needed to know when we have enough power budget in runtime to test the cores.

As the fault model and testing strategy for multi-core and many-core systems with advanced dynamic power management (DPM) features change, recent studies are focused on proposing new techniques for testing such systems. These studies can be classified in to groups: (1) the techniques that consider the effect of such power management capabilities to new error manifestation and (2) strategies that get benefit such power management features to control the test power while TAT minimization process [31]. Most of these strategies have been proposed for offline testing purposes. Even though we can find a limited number of online testing methods in these two categories, they do not yet consider any power feedback from the system at runtime and instead use a predefined dedicated power budget for testing. An example of power-aware optimization of the SBST has been presented in [32], where the authors propose an optimal approach to test the L1 cache in microprocessors considering power profile. However, this work is not either fully power-aware as the authors use a predefined power model of the microprocessors for different applications, which lacks an online power feedback from the system.

Using SBST in online testing can be done either in a intrusive way and in a non-intrusive one [2]. In intrusive online testing, test process is done during a fixed period in which the normal system operation is interrupted and the cores, or a subset of them, are reconfigured to *test mode* at runtime, and then run the test program. It can be concluded that, as in intrusive testing the normal operation of the system is interrupted, testing process might have negative effect on the performance of the system. On the other hand, in non-intrusive testing, each core executes the test program individually whenever the core is in idle state. As mentioned before, the power consumption needed for the test purpose is considerably higher than the power consumption of the system in the normal operation mode. As the power budget of the system is limited specially in the dark silicon era, it is not possible to perform a fully parallel intrusive testing as the power consumption can easily exceed the available budget and endanger the chip reliability. Furthermore, as the system is concurrently running multiple independent applications with different requirements, interrupting all or some of these might lead to deadline miss for some applications. Due to these facts, our focus will be on non-intrusive testing while honoring power budget.

Based on the above discussion, it can be concluded that *online testing is gradually reshaping to power-aware online testing in the dark silicon era, especially for many-core systems*. The main ground for this statement is that due to thermal and power constraints, the fraction of transistors that can operate at full frequency is decreasing with each technology generation. This highlights the fact that power budget is an extremely crucial resource in those technologies where the dark silicon phenomenon is more challenging to address (e.g., 22 or 16 nm). In such technologies, a many-core system demands an efficient power-aware online testing method capable of minimizing the usage of power for the online testing purpose. In other words, the online testing method should have the lowest negative impact on the system performance by efficiently using the power budget.

Adopted Many-Core Architecture

Figure 10.1 shows an overview of the considered architectural platform and the above software stack, composed of a runtime management layer and a set of running applications.

The target platform is the classical many-core architecture, such as the Intel Single-chip Cloud Computer (SCC) [33], or the Kalray MPPA manycore [34]. The architecture is composed of a set of homogeneous processing cores, each one provided with a private memory for instructions and data, and connected to the system's communication infrastructure. The communication infrastructure consists of an $M \times N$ 2D-mesh Network-on-Chip (NoC), using a message-passing protocol based on an X–Y deterministic wormhole routing schema.

The considered many-core architecture is generally adopted for the acceleration of intensive data-processing applications, such as image, video, or audio processing. Each application is generally implemented with a pipelined dataflow paradigm [33]. Thus, the application can be modeled by means of a task graph, where nodes represent the various computation tasks, each one characterized by a specific execution time, and the direct edges represent data dependencies (in terms of data to be transmitted) between a source task to a target one.

In order to execute an application, each task is assigned, or *mapped*, to a specific core that will execute it. In other words, we may also say that a core is *allocated* for

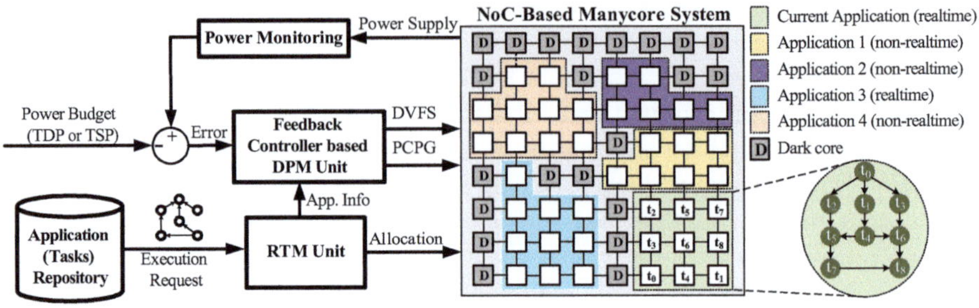

Fig. 10.1 A Network-on-Chip (NoC)-based many-core system with mesh topology supporting DPM and RTM

the execution of the task. Moreover, the execution model does not support multi-tasking, therefore at most one task can be mapped on a single core in a specific instant of time. As a motivation of this choice, Intel in 2011 [33] stated that, given the abundance of execution units in a many-core architecture, a one-to-one mapping may ease the execution management. Then, the mapped application is executed in a pipelined fashion: each core can perform a run of the hosted task each time it receives all required input messages and at the end of the execution it sends out the output messages. The NoC is in charge of routing and dispatching the messages from the senders to the receivers. Actual transmission latencies will depend on the infrastructure operating frequency, the message size, and the distance between the source and the target.

The right part of Fig. 10.1 shows an example of the described system. The architecture is an 8×8 grid of cores, on which five different applications are mapped. In the detail in the right-bottom part of the figure, the task graph of a Gaussian Elimination application (retrieved from [35]) is shown.

The left part of Fig. 10.1 shows the runtime management layer. This layer is a software module running on a controlling hardware unit, that may be a dedicated core in the NoC or an external host machine. The runtime management layer is composed of two main modules, called Runtime Mapping Unit and Dynamic Power Management Unit, that are discussed in the next.

The considered systems are generally employed in scenarios characterized by a highly variable workload. Indeed, applications arrive with an unknown trend depending on the requests of the various users. Moreover they may have different characteristics in terms of structure of the task graph and different requirements, for instance, on the minimum required throughput or on the amount of the processed data. As an intuitive example in Fig. 10.1, applications are annotated with realtime/non-realtime requirements. In order to deal with this variable scenario, the runtime management layer contains a unit devoted to the runtime mapping (RTM). This unit receives the request of applications' execution arriving from the users, and maps them on the available cores by using a specific runtime strategy (e.g., [36]) to satisfy the specified performance requirements. It may also happen that in a certain instant of time there are no enough resources to run the newly incoming application; in that case, the application will be delayed until it is not possible to satisfy its requirements.

On the other side, physical limits in circuit cooling, packaging, and power delivery in modern chips cause the many-core systems to have non-negligible power issues, expressed in terms of a limited *power budget*. According to such power budget, only a part of the available cores can be used at the same time while the other ones have to be switched off, thus causing the *dark silicon* phenomenon. For instance, Fig. 10.1 shows in gray the set of cores that are switched off (i.e., dark). Moreover, running applications may cause different power consumptions, depending on the number of allocated cores and the voltage/frequency levels at which cores work. This heterogeneity and the mentioned dynamicity in the workload will cause the amount of dark area to relevantly vary during the execution. In fact, in some situations, the allocated cores have to work at a high voltage–frequency level to provide the necessary performance in order to fulfill the required application throughput. Such cores will use a large part of the power budget, thus causing other units to be temporary set as dark. In other situations, it may happen that the set of applications to be executed do not demand high operating frequencies, and therefore it is possible to use the total chip utilization at a low voltage–frequency level, leaving no dark area on the chip. Such discussion motivates the necessity of a DPM within the runtime management layer.

Figure 10.1 shows the DPM unit within the runtime management layer. Such a unit is connected to the RTM unit to take coordinated decisions about the application mapping and power management. In particular, the aim is to achieve applications' performance requirements while respecting the power limit. The available budget is defined either at design time, by using the Thermal Design Power (TDP [9]), or dynamically managed at runtime with another feedback loop, by means of the Thermal Safe Power (TSP [37, 38]). Then, as in [39–41], the DPM unit is implemented as a feedback loop that monitors the consumed power by means of on-chip power sensors and acts on per-core DVFS and power gating knobs.

In conclusion, for each arrived application, RTM and DPM units work together in order to decide if there is enough power budget available for the execution of such application, to define a mapping and a DVFS setting to achieve the required performance while not violating the power budget. If such conditions are not satisfied, the application is delayed until some other application leaves the system and releases enough resources and power.

Suitable Scenarios for Online Testing

Nowadays, many-core system are generally employed for high performance computing in different fields spanning from data centers to high-end embedded and mobile appliances. All these scenarios are subject to highly varying workloads: different types of applications arrive with an unknown trend and are characterized by different performance requirements, variable amount of data to be elaborated, different request of processing resources, and so on. Therefore, in each instant of time, the running workload will cause a different working configuration in the many-core system, in terms of the set of currently running applications, their actual mapping on the cores' grid, and related power consumption. Figure 10.2 depicts a taxonomy of the main working situations. For each situation, Fig. 10.2 reports the allocated cores to different applications and idle cores, and, if any, the size of the new application requested to be mapped. Moreover, each subfigure presents also the related power consumption graph reporting the actual power consumption (with a solid line) and the given power budget (with a dashed line). In each of these situations, we have analyzed the possibility to perform a non-intrusive online testing on a selected candidate core. These scenarios are commented in the following paragraphs.

Scenario (a). At time t_1, three applications with strict performance requirements are running on the system. Due to the performance requirements, the active cores are set to a high frequency-voltage level, thus leading the overall power consumption to be too close to the available budget. Therefore the other cores are forced to be dark. In this case, although there are idle cores that can be tested without affecting the nominal activities of the system, there is not enough available power budget to be dedicated to online testing.

Scenario (b). At time t_2 eight applications with performance requirements exactly fit on the available cores. In such a scenario, the low power requirement caused by each of the applications allows to use 100 % of the available resources as the dim area. Consequently, even though there is available amount of power budget for the testing activity, in order to execute the SBST routine it would be necessary to intrusively interrupt one of the running applications. However, this violates our goal of transparency.

Scenario (c). At time t_3, the system is almost unloaded, since a few applications are running and the power consumption is quite low. This is the best scenario, since SBST routines can be executed by using the remaining power budget and on the idle cores, i.e., in a transparent way since the system performance is not degraded.

Scenario (d). At time t_4, the RTM unit receives a request to execute a new application having eight tasks. However, the RTM strategy [36] may decide that it is not convenient to immediately execute the applications. The reason is that, even though there is room in the power budget, the current system status is characterized a high dispersion of the idle cores. This would imply an inefficient choice in terms of performance and energy consumption due to the communication costs. Therefore, since the RTM unit delays the application execution until a contiguous region composed of at least eight cores will be available, the system can employ the available power budget to run the test process on the idle cores. Dispersed cores in such scenarios are the best candidates for being non-intrusively tested without any degradation of the system performance.

Scenario (e). At time t_5, the RTM unit receives a request to execute an application with nine tasks. In this scenario, even if the available power budget is sufficient for the execution of the application, there are not enough cores available in the system to map the arrived application. Once again, the available power budget can be exploited for non-intrusive online testing of the available cores.

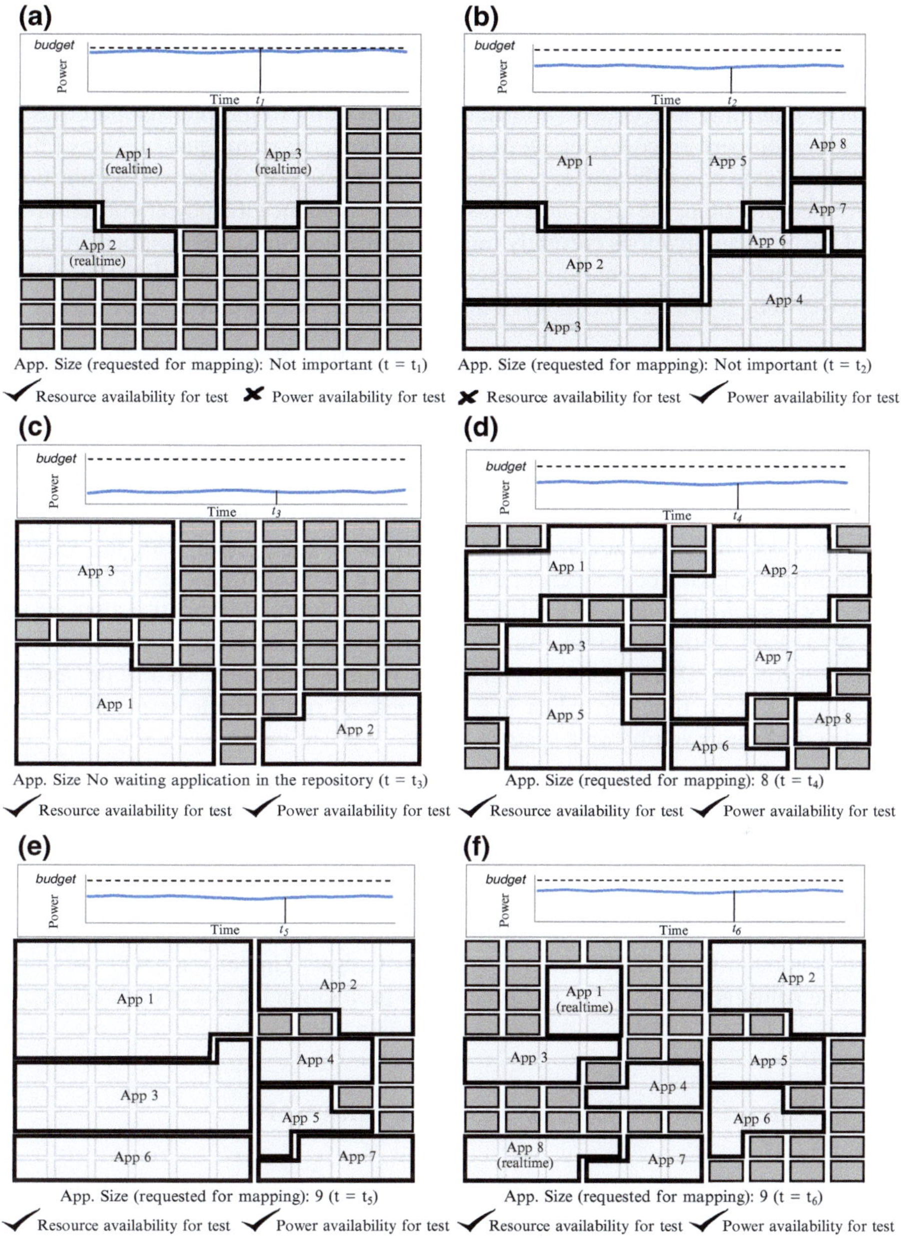

Fig. 10.2 Frequent scenarios regarding resource and power availability for test

Scenario (f). At time t_6, the RTM unit receives the request to execute a new application having nine tasks. When considering the current system status, the application can be potentially executed in that moment due to the availability of more than nine idle cores. Unfortunately, according to the pre-mapping power estimation performed by the DPM unit (with specific techniques, such as [39]), the available power budget is not large enough to support the arrived application. At the same time, the DPM unit is also not able to reduce the power consumption of the other applications currently running on the system due to their performance requirements. This scenario represents another situation in which the available power budget can be used to test a number of unallocated cores.

There are also other issues regarding online testing in the considered scenario: when conditions on the availability of resources and power are satisfied (as in scenarios from (c) to (f)), it is also necessary to choose the candidate cores to be tested. However, the concurrent test of all the idle cores generally overcomes the available power budget. Moreover, as discussed in section "Introduction", testing activities, and in particular SBST, have to be executed at several voltage–frequency levels to ensure the correct behavior of the system with the various settings [15, 42]; as a consequence, it is necessary to take into account that each of these configurations will have a different power consumption/execution time trade-off. As a result, in the scenarios (d), (e), and (f), it is also necessary to consider such aspect to run the SBST routines with a low voltage–frequency setting on several cores at the same time, or, when it is required, to run a single test with a high voltage–frequency setting on a specific core.

The accurate analysis of the presented scenarios clearly shows the promising opportunity to perform non-intrusive online testing in many-core systems. Actually, the highly variable and evolving status of the many-core system due to the dynamic workload presents periods with high resource and power utilizations and period with a low utilization. Therefore, an opportunistic online test scheduling method can take advantage of the second kind of situations in order to test the dark cores as long as there is enough room in the remaining power budget. This chapter will present a possible solution to this online test scheduling problem in the era of dark silicon.

Dark Silicon Aware Online Testing Framework

The proposed framework for dark silicon aware online testing is presented in Fig. 10.3. It is an extension of the classical runtime management framework discussed in section "Adopted Many-Core Architecture", with some additional components devoted to the execution of the test-related activities.

The goal of the proposed approach is to transparently run SBST routines during the system activities without affecting the execution of the nominal workload. Thus, the aim is to guarantee that processing cores are not affected by permanent failures and, at the same time, to maintain the required level of performance for the running

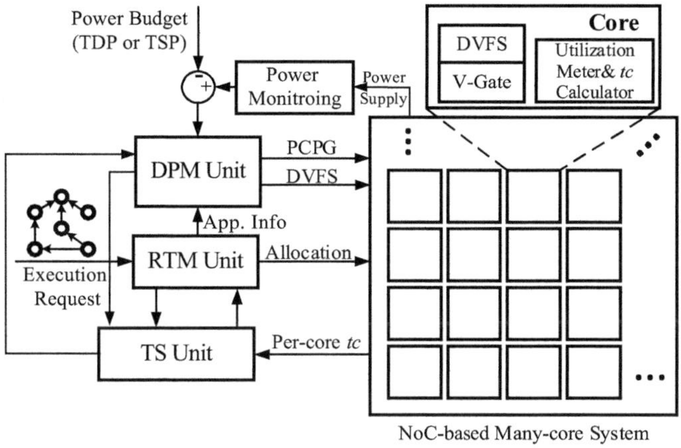

Fig. 10.3 The system architecture including the online testing framework

workload. The basic idea is to test each core with a rate proportional to the stress it has been affected due to its utilization. If a core is frequently used for execution of applications, it is highly stressed and therefore needs frequent tests. On the other side, if the core has been rarely allocated, it does not require urgent testing in the near future. The benefit of this approach is to guarantee the necessary test frequency without performing cores' over-testing that would have a negative effect on the execution of the nominal workload in terms of larger power consumption and unnecessary resources occupation, or cores' under-testing that would reduce the reliability of the system.

The proposed testing approach introduces a new component to the system, Test Scheduling Unit (TSU), that is devoted to select the cores that need to be tested according to the experienced stress and the scheduling of the testing task on those cores. The experienced stress is estimated by means of a criticality metric. It is computed according to a specific hardware component integrated within each core counting the number of executed instructions. The TSU works in a tightly coupled way with RTM and DPM units to define a proper test scheduling. In particular, the RTM unit has been slightly modified in order to take into account the fact that if a core is candidate for the test procedure, it should not be considered for mapping purposes. In the following subsections, various activities of TSU are discussed in detail together with the internal modifications to RTM unit necessary to handle the test information received by TSU.

Monitoring Cores' Stress

The first activity of the TSU is to select the cores to be tested. Such activity is performed by monitoring the stress experienced by each core.

As many modern multi-core architectures are not provided with aging sensors, in order to measure the experienced stress, some past testing approaches [8, 17] have exploited the available per-core hardware counters of the executed instructions. An example of architecture provided with such counters is the Intel SCC platform [33]. For instance, the approach proposed in [17] schedules a test on a core every time the instruction count, also dubbed as utilization metric, reaches a specified threshold, i.e., 10M, 100M, or 1B instructions. Moreover, in [8], a similar more fine-grained approach envisions the availability of counters for each execution unit in order to reduce the execution time.

Therefore, we assume that each core with coordinates (i,j) is equipped with an instruction counter, called Utilization Meter (UM), whose value α_{ij} is incremented every time an instruction is executed and is reset when the core is tested. Based on the α_{ij}, the UM computes a *test criticality* parameter tc_{ij} indicating the urgency of a core to be tested due to the experienced stress. More precisely, the parameter is computed according to the following equation:

$$tc_{i,j} = \frac{\alpha_{ij}}{\delta} - 1 \tag{10.1}$$

where δ is a threshold stating the number of executed instructions that triggers the test procedure. According to this definition, tc_{ij} assumes a value in the range $[-1; +\infty)$. As long as tc_{ij} is lower than 0.0, α_{ij} value is still below the specified threshold δ and therefore the core does not need to be tested. Then, whenever tc_{ij} exceeds 0.0, it means that the corresponding core needs to be tested at the earliest convenient moment. The UMs send tc_{ij} matrix to the TSU at fixed time intervals, by using an interrupt-based mechanism to minimize redundant communications. Then, TSU collects all candidate cores requiring to be tested so that they can be analyzed in the subsequent test-aware mapping and test scheduling phases.

Finally, when TSU starts the execution of the SBST routine on a core, it also resets the corresponding α_{ij} counter; consequently, the tc_{ij} value becomes -1.

Testing-Aware Mapping

The mapping of the nominal workload and the testing execution are two conflicting activities since both require processing resources and consume power. A classical approach of interrupting nominal execution to execute test procedures as soon as the triggering condition is violated, as in [8], causes a considerable performance degradation, especially if there is a high requests' rate. In fact, execution of test procedures uses a share of the power budget, and, moreover, the mapping-agnostic selection of cores to be tested would increase fragmentation in cores' occupation [43, 44]. On the other hand, prioritizing the mapping of the nominal workload would negatively affect the reliability of the system by delaying test procedures. Indeed, test execution should be dynamically adapted to transparently "intersect"

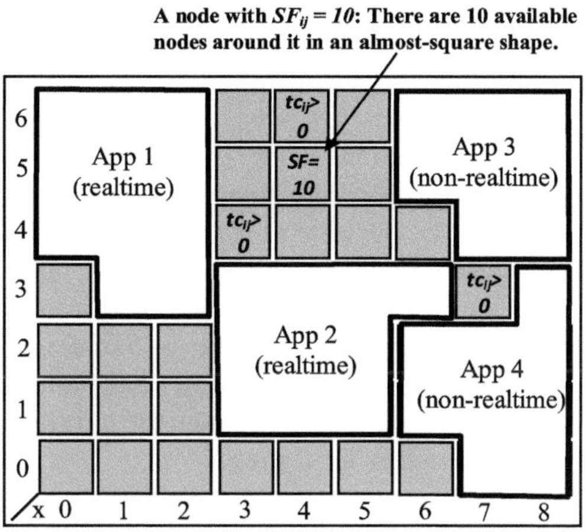

Fig. 10.4 Example of SF calculation of a node for a given system configuration

with nominal applications' execution. In this way the goal of the approach is achieved: one should not cause any performance degradation in the workload execution while satisfying reliability issues.

In the considered system, the RTM unit uses a strategy which maps the tasks of the same application on a contiguous set of cores [43, 44]. In this way, power consumed by communication and latencies are considerably reduced. In the RTM unit, such a region of neighboring cores is identified through a metric called Squared Factor, SF_{ij}, introduced in [43]. In particular, SF_{ij} metric relates to the number of almost-contiguous available nodes around a given node.

In this scenario, TSU needs to prevent the RTM unit from allocating cores with $tc_{ij} > 0.0$. Unallocated cores can be later scheduled for testing in an appropriate time, when there is enough available room in the power budget. However, if TSU directly disables the cores having $tc_{ij} > 0.0$, it may cause a dispersion of the planned contiguous allocations. For instance, let us assume that an application with ten tasks has to be executed on the system depicted in Fig. 10.4. As the SF_{ij} of the node (4,5) is 10, it will be selected to map the application onto its surrounding nodes. However, if two cores of this region have $tc_{ij} > 0.0$, the RTM unit has to allocate some available nodes from south-west region of the system which leads to a high dispersion.

To avoid such performance crippling dispersions, the RTM unit has been enhanced by means of a $newSF_{ij}$ value, which is the number of cores with $tc_{ij} > 0.0$ from the original SF_{ij} value. As a result, the $newSF_{ij}$ value shows the number of available cores around a given core that are not candidates for testing. For instance, the new SF value of the core in Fig. 10.4 will be $newSF_{ij} = 10 - 2 = 8$. Thus, the core will not be selected as the first node for mapping an application with 10 tasks, but with 8 tasks instead. It is worth mentioning that apart from the disabling of the cores with $tc_{ij} > 0.0$, this approach for computing the $newSF_{ij}$ will not necessarily prevent such cores from executing a task. Instead, the metric will only discourage using that area, possibly privileging other areas with a lower number of cores to be tested.

Test Scheduling

TSU implements test scheduling algorithm that determines the cores to be tested among the candidate ones. The core selection strategy in the scheduling algorithm is based on the following considerations. Due to limits on the available power budget, it may not be possible to test all the candidate cores at the same time. Therefore, it is necessary to define a ranking strategy to assign a priority to the testing activities. In an intuitive way, a possible ranking strategy may be: the higher the tc_{ij} is, the more critical it is to schedule an SBST routine on that core. However, by means of a systematical analysis of several possible working scenarios, we noted that if there are several cores with similar tc values, the ones with vacant vicinity should be ranked higher for testing.

The latter consideration is based on the idea that cleaning up regions of idle cores facilitates future application mappings. In fact, isolated cores with or without $tc_{ij} > 0.0$ are, nevertheless, not suitable for being allocated. More concretely, if we test a core with busy neighbors instead of the one with idle cores around, this will lead to a high dispersion of applications and hence degrading the system performance. At the opposite, if we clean up regions with a large amount of contiguous cores, we will increase the possibility to map applications and achieving higher performance. Finally, testing applications may be power-hungry. We should avoid placing them in close proximity to each other or to other running applications. Otherwise, testing several adjacent cores together can cause high power densities, and consequently high local temperatures, as shown in the example in section "Experimental Evaluation of the Approach".

Based on the above considerations, a new metric has been defined to rank candidate cores by considering at the same time test criticality and the number of idle cores in the proximity. The metric is defined as

$$tr_{ij} = tc_{ij} + \frac{\sqrt{SF_{ij}}}{total\ number\ of\ cores} \tag{10.2}$$

where SF_{ij} value is normalized to the total number of cores in the system. As a metric, SF_{ij} estimates the number of vacant cores around a given core; i.e., the larger the SF_{ij} value of a core is, the more idle cores are around it. Moreover, we use a square root of SF_{ij} value to limit its impact to the cases where tc_{ij} values are too close to each other. For instance, in case of equal tc_{ij} values in Fig. 10.4, the cores $(3, 4)$ and $(4, 6)$ will be ranked higher than the core $(7, 3)$.

Algorithm 1 shows the pseudo-code for the selection of cores to be tested. A peculiarity of the algorithm is an additional control of negative impact of testing on system performance. This is implemented by limiting the maximum number of cores that can be simultaneously tested by means of a given threshold, $\tau_{\#Test}$. The motivation is that we have to cope with a highly evolving scenario; while there might be enough power at the moment to test even more cores, this can change in

Algorithm 1: Selecting cores for test scheduling

In predefined intervals:

```
if there is available resource and power for test then {// One of the suitable scenarios shown in Figure 10.2}
    Sort available cores based on their tr_ij values;
    while there is enough power budget for test and # of (cores under test) < τ_#Test do
        Schedule the first core in the ranked list for testing;
    end while
end if
```

the near future. Other applications might enter the system, or the power demand and behavior of running applications might change.

In general, execution time of the SBST routine is short compared to applications' execution time. However, regardless of the application types, the overhead of the SBST routine is almost independent of the applications' execution time. The test criticality value of a core (tc_{ij}) depends on the number of instructions executed over time. In case of short applications, tc_{ij} becomes greater than 0 only *after* execution of several applications. While, in case of long applications, the allocated core might need to go under test after the application execution. In this case, the overhead would be again negligible compared to execution time of the application.

Finally, it is worth of noting that the tc_{ij} value increases significantly in case of executing very long applications. Such situation would be managed by means of task migration. However, we leave such an aspect as a future work.

Test Scheduling for Different Voltage–Frequency Settings

Based on the recent studies, some specific faults manifest themselves in a particular voltage–frequency (VF) settings [45]. These studies have concluded that multi-/many-core systems equipped with DVFS feature should be tested at multiple voltage levels to ensure that cores can operate reliably at different conditions. Testing at multiple voltage levels is more challenging compared to single voltage level testing as in each voltage level a separate SBST routine execution is needed and the maximum possible operating frequency is limited [46]. Applying the trivial and straightforward test scheduling and repetitively running a test process for every voltage level, drastically increase the overall TAT. Thus, this approach has a direct impact on the overall system performance. At low voltage levels, test process becomes slower as the frequency is lower than that resulting in a longer TAT. In this section, an efficient technique is proposed to test cores at different voltage levels with the aim of providing a uniform testing probability for all the levels while minimizing the performance overhead.

To apply online testing on cores running at different voltage levels, it is essential to use a test scheduling policy with the minimum negative impact on system performance. To this end, allocated core(s) need to be detected and enough power budget need to be available for the test purpose so that the upper power consumption bound will not be violated. However, as test power consumption at different voltage levels considerably varies, the suitable frequency level in each voltage level should be properly determined at runtime.

In multi-/many-core systems equipped with DVFS feature, usually for each voltage level, an upper bound for the maximum frequency that can operate at that voltage level is defined. For example, in Intel SCC platform, 7 voltage levels for each island are defined where each voltage level has a maximum possible frequency, thus forming more than 15 VF levels per island which can be changed at runtime. In each particular voltage level, different operating frequencies used for testing result in different test power/energy. As the system is tested at runtime with functional methods, and a test at a certain voltage level can be performed at different frequencies (i.e., equal or lower than the maximum frequency at the respective voltage [46]), we define a VF set as the set of different available frequencies (i.e., VF levels) that can be selected for testing at a given voltage level. At low VF levels, power consumption is lower at the cost of longer TAT, compared to high VF levels where higher power consumption is needed to achieve a shorter TAT. This raises a question whether it is more efficient to use a low VF level and save the power to have parallel testing of multiple cores or to use a higher VF level and reduce the TAT for individual cores. Our solution to address this issue is inspired by the traditional 2D rectangular packing model used in power-constrained testing. Figure 10.5 shows an example of using 2D rectangular model when three cores are tested over time at different VF levels. Each rectangle depicts a test process as a triplet (C_i, V_j, F_k) where C_i is the core to be tested, V_j is the voltage of test, and F_k is the frequency of test. The width and length of the rectangle correspond to the test power consumption and TAT, respectively, where the total summation of test power at each moment of time should not exceed the maximum available power budget for test. It can be observed that when power budget is limited and an optimal test scheduling algorithm is used, the total areas of the all test rectangles determine the overall test time. This area is the TAT-test power product which can be called as energy consumption for test. We use the energy consumption for test as a metric to choose the proper VF level for test when there is an option to select one VF level among the available VF levels in a particular VF set.

In Fig. 10.6, we have compared the normalized energy with different frequency levels when the voltage is fixed. As can be seen, by increasing the frequency up to the maximum possible frequency, the energy consumption exponentially decreases. That is because of the fact that for a constant voltage, the static power remains constant, and by decreasing the frequency, the penalty of unchanged high static power superimposes the overall power and energy accordingly. From these two observations, we propose a general rule for our test scheduling algorithm that for a given voltage level, the test frequency should be increased as much as possible while monitoring and honoring the total power budget.

Algorithm 2 shows in more detail the proposed test scheduling strategy for testing the cores at different VF levels. Algorithm 2 is the extension of Algorithm 1 to

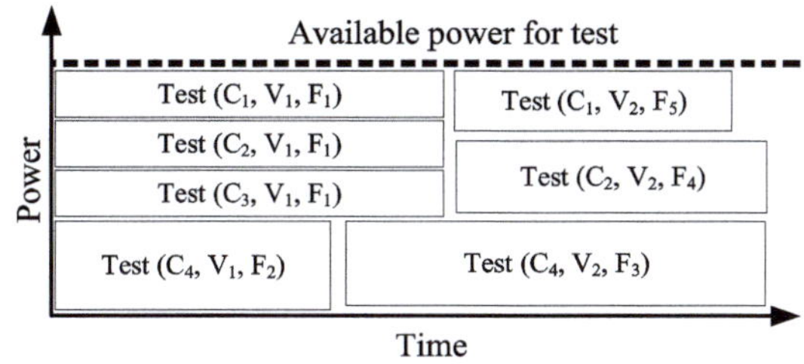

Fig. 10.5 An example of rectangular packing model for power-aware testing

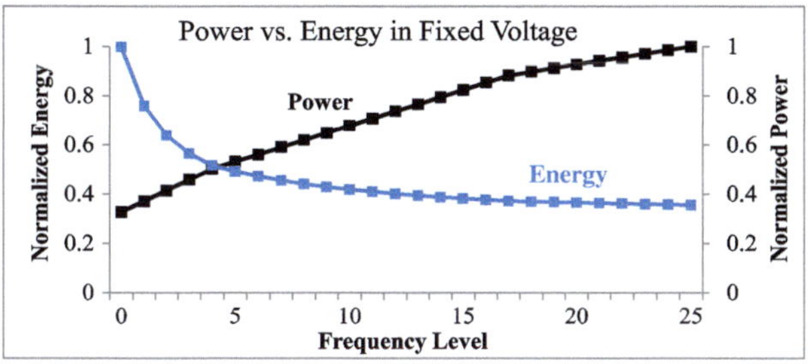

Fig. 10.6 Power versus energy in a fixed voltage level

consider VF levels in test scheduling, thus offering the system manager the option to choose two different test scheduling policies with contrasting reliability–complexity trade-offs. The input of the test scheduler is the instantaneous power consumption of the system (i.e., P_c) which is provided by the chip power sensor and the output of the test scheduler is the core(s) targeted for being tested at specified VF level(s) (i.e., set of (C_i, V_j, F_k) where V_k and F_k are the voltage and frequency of the core C_i during the test process).

First the amount of available power budget (i.e., *availablePower*) is calculated which is the available portion of power budget that can be used for test purpose (line 1 in Algorithm 2). If it is less than or equal to zero, it means there is no available power for the test purpose. If *availablePower* is greater than zero (i.e., there exists available power for the test purpose) and the number of cores under test is smaller than the maximum threshold (i.e., $\tau_{\#Test}$), the algorithm will find the first core in the list of cores sorted based on their tr_{ij} values as the target for test (lines 3–4). If such a core exists in the system, for each VF set (i.e., *VFset*) at which the core has not been tested yet, the algorithm will check if the available power can be used to test the core at that VF set or not (lines 5–12). Based on a pessimistically pre-calculated test power for each VF level, the function *minPower* returns the minimum required test power (i.e., CUT_{power}) and the corresponding voltage and frequency (i.e., (V_j, F_k)) to test the core at one of the VF levels at that specific VF set (i.e., $VFset_j$). If the test power is less than the available power, then the core and voltage–frequency for test will be added to the set of target cores for test (i.e., CUT_{set}) and *availablepower* will be updated accordingly (lines 7–11). Whenever a core is selected for test, the t_r value for other cores will be updated based on the consideration of the selected core

Algorithm 2: Selecting cores for test scheduling with VF selection algorithm

Inputs: P_c: Instantaneous power measurement from the sensor;
P_{max}: The maximum power budget (i.e., TDP or TSP);
Output: CUT'_{set}: The target core(s) to be tested at specified VF level(s) (i.e., set of (C_i, V_j, F_k));

Variables: $availablePower$: Available power for test;
CUT_{set}: Temporary variable for the target core(s) and their VF level(s) for test;
CUT_{power}: Core under test power consumption at a given VF level;
Constant: $\tau_{\#Test}$: Maximum number of core(s) under test;

Body:
1: $availablePower \leftarrow P_c$ - P_{max};
2: **while** $availablePower > 0$ and # of (cores under test) $< \tau_{\#Test}$ **do**
3: Sort available cores based on their tr_{ij} values;
4: $C_i \leftarrow$ The first core in the ranked list for testing;
5: **if** C_i is not tested in $VFset_j$ **then**
6: $(CUT_{power}, V_j, F_k) \leftarrow minPower(VFset_j)$;
7: **if** $CUT_{power} < availablePower$ **then**
8: $CUT_{set} \leftarrow (C_i, V_j, F_k)$;
9: Update tr for all cores;
10: $availablePower \leftarrow availablePower$ - CUT_{power};
11: **end if**
12: **end if**
13: **end while**
14: **while** $availablePower > 0$ and there is unselected core(s) in CUT_{set} **do**
15: select core (C_i, V_j, F_k) from CUT_{set};
16: $(CUT_{power}, V_j, F'_k) \leftarrow maxVF((C_i, V_j, F_k), availablePower)$;
17: $CUT'_{set} \leftarrow (C_i, V_j, F'_k)$;
18: update $(availablePower)$;
19: **end while**

as an occupied node. This causes the next core for test to be selected in other vacant areas.

Searching for faster test process continues as long as $\tau_{\#Test}$ threshold is reached or $availablePower$ is less than zero. $availablePower$ is the amount of power budget that can be used for test purpose. As the power for testing the cores in different VF levels can be measured in design time and it is determined in runtime, the maximum VF level at which cores can be tested without violating $availablePower$ is calculated in the algorithm through a trial-and-error process (lines 14–19). The highest possible VF level is calculated by function *maxVF*. If such a level exists, then the new voltage–frequency for testing will be added to the updated set of target cores for test (i.e., CUT'_{set}) and $availablepower$ will be updated. This process continues until either $availablePower$ is larger than zero or all the cores in CUT_{set} are selected. To determine the appropriate VF levels for test purpose we make use of the ideas applied for the traditional power-constrained testing in multi-clock domain SoCs [47]. In such works, the problem is to achieve the best TAT while for testing the cores in an SoC, while each core can be run on different VF levels. The only

Table 10.1 The system settings for different experiment setups

	Technology node (nm)	System type	Area (mm^2)	NoC size
First experimental setup	16	Medium	138	12 × 12
Second experimental setup	22	Large	232	11 × 11
Third experimental setup	32	Large	254	8 × 8

difference from such works with our problem solving attempt is that the maximum power for test for those power-constrained testing is fixed since the test process is done offline, while in our online test scheduling $availablePower$ changes during the time. However, if the test time is short enough (which is reasonable assumption as discussed in section "Test Scheduling"), we can assume that the power budget for test does not change and two problems are the same. More details regarding the efficiency of this method can be found in [47]. It is worth noting that the proposed algorithm for test scheduling is targeted for platforms featuring per-core DVFS. The extension to also consider per-cluster DVFS is left as a future work.

Experimental Evaluation of the Approach

To experimentally evaluate the proposed approach, we implemented a system-level simulation platform for the described many-core architecture together with accompanying runtime management layer and testing procedures in SystemC on the basis of Noxim NoC simulator [48]. The basic core has been characterized by using the Niagara2-like in-order core specifications obtained from McPAT [49]. Physical scaling parameters, power model, voltage–frequency scaling model, and TDP calculation were extracted from the Lumos [11], a framework to analytically quantify power/performance characteristics of devices in near-threshold operation. The physical scaling parameters have been calibrated via circuit simulations with a modified Predictive Technology Model [50]. Then, we integrated HotSpot 5.0 [51] for modeling the thermal behavior of the device. To demonstrate the efficiency of our dark silicon aware online testing approach on many-core systems, we defined three instances of the architecture by considering different technology nodes and different grid sizes as described in Table 10.1. Finally we defined a variable workload consisting of both synthetic task graphs with 4–35 tasks using TGG [35], and real applications, such as MPEG-4, UAV, and VOPD, from [52].

The proposed runtime management layer has been defined by using the RTM algorithm presented in [36] and the dark silicon aware power management (DSAPM) technique presented in [39]. In this power management strategy, a PID (Proportional Integral Derivative) controller is used for DPM that considers a fixed power budget (i.e., TDP). As an alternative, we have also integrated the

TSP calculation tool from [37] to evaluate the dynamic power budget based on the number of active cores at each moment of time.

To prepare the SBST program, we first generate deterministic test patterns from the netlist of HDL implementation of Niagara2-like cores using the technique proposed in [53]. In particular, *NetlistGen.exe* is used for generating netlists of the synthesized cores and fault simulation has been performed with PLI library in HDL environment [54]. Then, we develop test macros based on the generated deterministic test patterns. The overall coverage for the cores' datapath and controller is 79 and 63 %, respectively. The duration of the SBST routine is 9000 cycles for each core. Dynamic and static power consumption of the test process has been measured by using the adopted models [11]. Finally, we set $\tau_{\#Test}$ to 4.

In a first experiment we analyzed the throughput penalty in terms of executed instruction per unit of time of the proposed test scheduling approach when the δ is set to 10M, 100M, and 1B instructions. Moreover, we defined power budgets by using both TDP and TSP methods. The results for the two different power limits are shown in Figs. 10.7 and 10.8, respectively. As can be seen from the bar charts, the proposed online testing method has a negligible throughput penalty for both TSP- and TDP-based approaches. In all the cases except for 10M, the overhead is less than 1.5 %; when δ is set to 10M, the execution frequency of the test procedure introduces an overhead up to 6 % for the architecture designed with the 32 nm technology. An interesting aspect is that, for both TDP and TSP experiments, the minimum throughput penalty is observed for the architecture designed with the 16 nm technology (first experimental setup), that is the newest node technology, where power limitation is more challenging and the system size, i.e., the total number of cores, is larger than in the other experiments. This shows that the proposed approach will have even more opportunities in the future technologies to find suitable scenarios for online testing. It can be noticed that the penalty while using TSP as the maximum power limit is very similar to the penalty of using TDP. Finally, we can also note that the throughput penalty obtained by the proposed method is considerably lower than in the existing online testing methods reported in [17, 18]. The reason of such improvement is that our method adapts on the working status of the system. In particular, it takes advantage of non-intrusive testing of the cores that are temporarily located in the dark area by exploiting available power budget.

In the subsequent experimental sessions, we delved more into details of the performance overhead analysis of the proposed approach by comparing it against the most relevant state-of-the-art methods [55]. Notice that, this earlier method dedicates a fixed amount of power budget to the test process. Thus, we re-run the same experiment with δ set to 10M (that is the worst case scenario) for 250 s and we plotted the system throughput over time as shown in Figs. 10.9 and 10.10 when using TDP and TSP, respectively. Each of these figures compares the throughput of the proposed approach against classical DSAPM strategy without testing option and the DSAPM strategy coupled with the technique presented in [55]. It can be seen that the proposed online testing approach achieves a better performance over time compared to the DSAPM approach with dedicated power

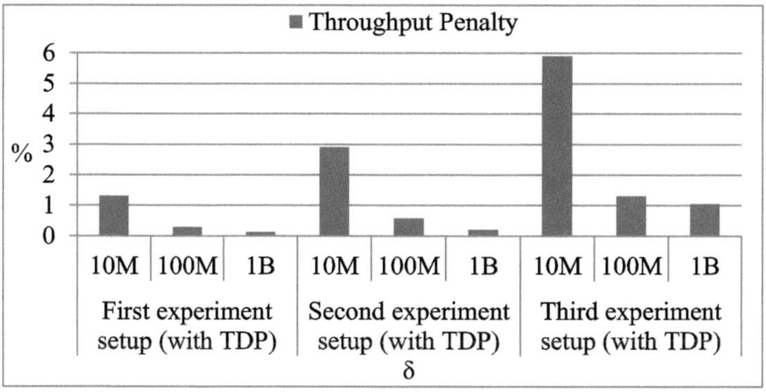

Fig. 10.7 Throughput penalty for different experiment setups and δ values while using TDP

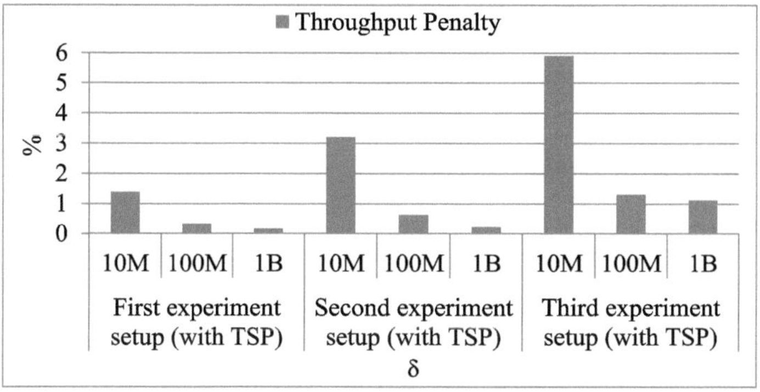

Fig. 10.8 Throughput penalty for different experiment setups and δ values while using TSP

for test procedures. In particular, the throughput penalties for DSAPM coupled with the technique presented in [55] for 16, 22, and 32 nm technologies are 23, 20, and 16 %, respectively for the TDP-based approach, and 25, 23, and 22 % for the TSP-based approach which are larger compared to the corresponding results obtained by the proposed approach, reported in the previously discussed Figs. 10.7 and 10.8. Furthermore, it can be noted that applications complete and leave the system with almost the same trend for both the DSAPM with online testing and DSAPM without online testing. This confirms the capability of the proposed approach to perform transparent scheduling by using the available power budget and resources at runtime for testing the cores with a negligible penalty on the system performance.

Within the same experimental setup we also evaluated the power consumption of the system over time. Figure 10.11 shows the power consumption of the system when running a group of random applications while using DSAPM with and without the proposed online testing approach. As it can be observed from the power curves, the total power consumption does not violate the available power (defined with TDP) for both approaches. At the same time, when the power budget is changed,

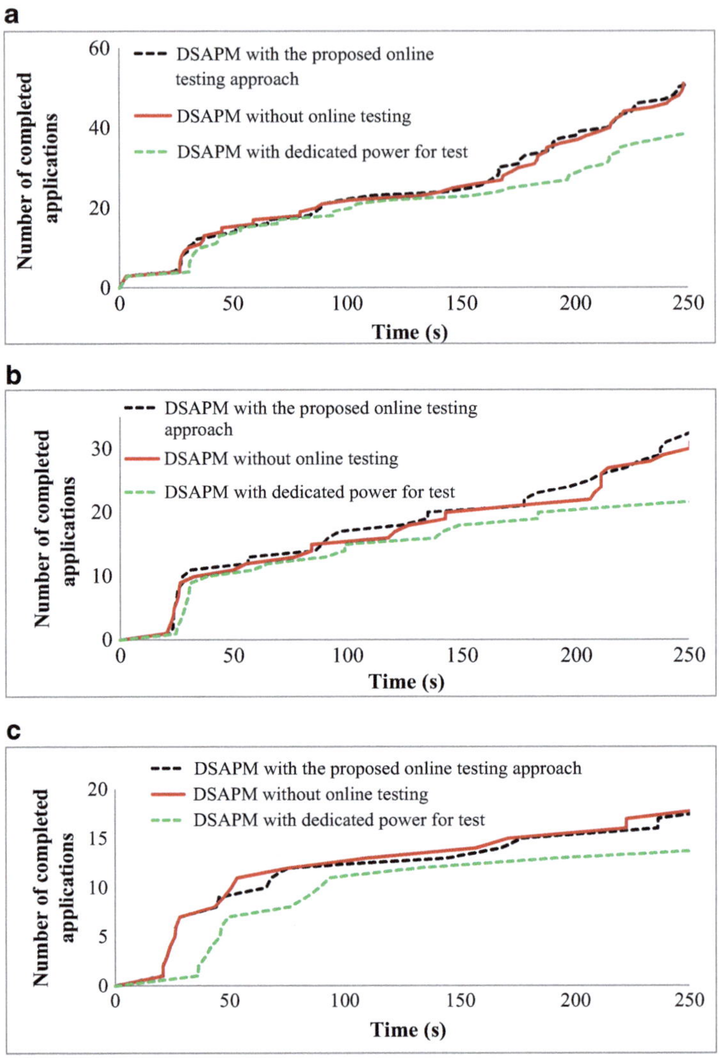

Fig. 10.9 The number of completed applications vs. time (TDP-based approach). (**a**) For 16 nm technology (first experiment setup). (**b**) For 22 nm technology (second experiment setup). (**c**) For 32 nm technology (third experiment setup)

the approach is able to adapt to a new condition. This shows that even though a dedicated power budget is *not* allocated to the test purpose, the DPM unit efficiently honors the TDP bound even when the TDP is changed at runtime. The power curves show that small throughput penalties are experienced in scenarios when the system is frequently busy and the total chip power consumption is most of the time close to the upper bound. In a last series of graphs (Fig. 10.12), it is shown

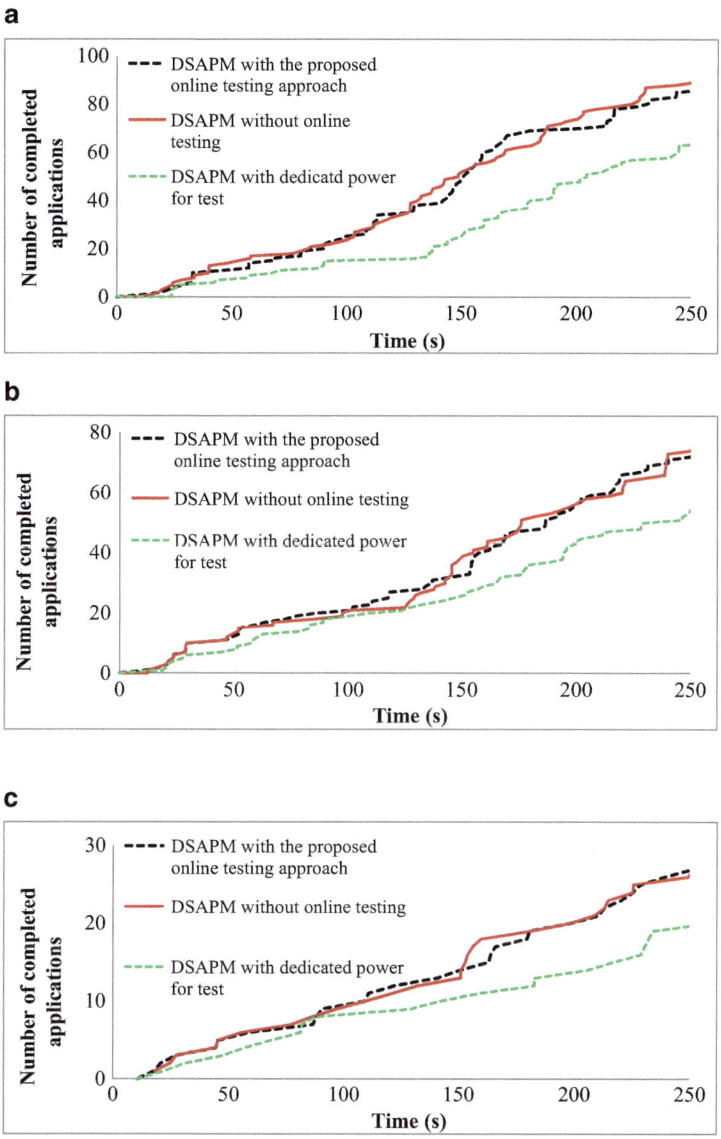

Fig. 10.10 The number of completed applications vs. time (TSP-based approach). (**a**) For 16 nm technology (first experiment setup). (**b**) For 22 nm technology (second experiment setup). (**c**) For 32 nm technology (third experiment setup)

how the actual power consumption dedicated for testing changes over time. It is worth noting that a bar chart is used since test power is not continuous but it is dedicated in specific periods. The maximum value never exceeds 3 W on the available 50 W for 16 and 22 nm technologies, and 4.5 W on the available 70 W

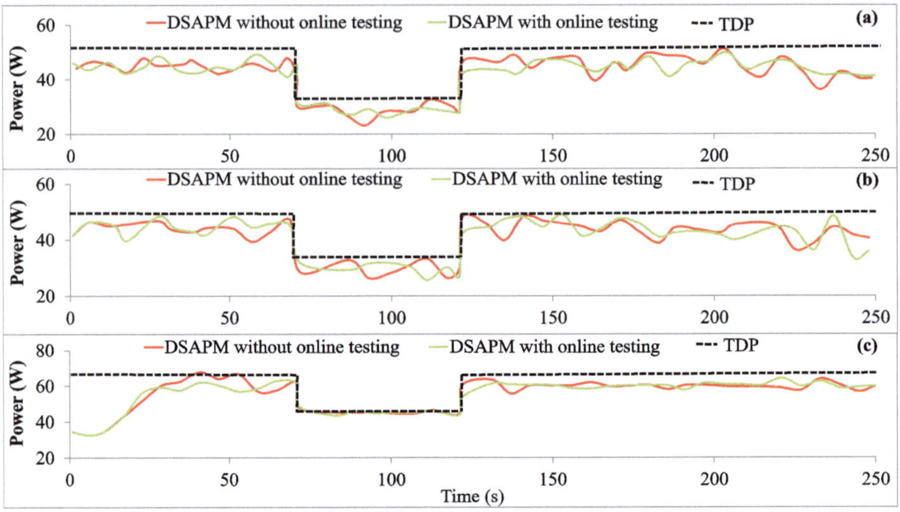

Fig. 10.11 The power consumption of the system with and without online testing approach for different experiment setups, (**a**) 16 nm, (**b**) 22 nm, and (**c**) 32 nm

for 32 nm technology. Moreover, in average such test power is around 2 % of the overall power consumption. This demonstrates that the approach is able to limit the instantaneous test power by distributing SBST routine execution over time.

We also analyzed efficiency of the test scheduling approach in avoiding temperature hotspots. In particular we analyzed the effect of using the square factor (*SF*) in Eq. (10.2). For this, several thermal snapshots are monitored during system runtime and compared against a modified version in which such parameter is not considered in Eq. (10.2), dubbed as non-thermal-aware scheduling. Figure 10.13 shows the temperature profile of the system while running non-thermal-aware and thermal-aware test scheduling at a given instant of time (with $\tau_{\#Test} = 4$). As can be seen, the non-thermal-aware scheduling selects four neighboring cores which causes high temperatures in a restricted area of the chip. At the opposite the thermal-aware strategy selects cores which are far from each other to avoid thermal hotspots.

Finally, we analyzed the effectiveness of the testing procedure at different voltage/frequency (VF) settings. We characterized the simulation platform with 6 VF sets, i.e., voltage levels, for a total of 29 VF levels. Table 10.2 reports these different VF sets, by specifying for each of them the related voltage and available frequencies in each set. The target VF level to be assigned to the core under test is chosen among all the options in each VF set. The results of the experiments, performed with the same setup discussed above, are reported in Fig. 10.14, for the three considered technologies, respectively. In particular, each pie chart reports a share of each VF set used for testing activities from the total number of tested cores at the end of the simulations. As can be noticed, VF sets are selected in almost similar way hence demonstrating the fairness of the proposed DVFS-aware test scheduling algorithm.

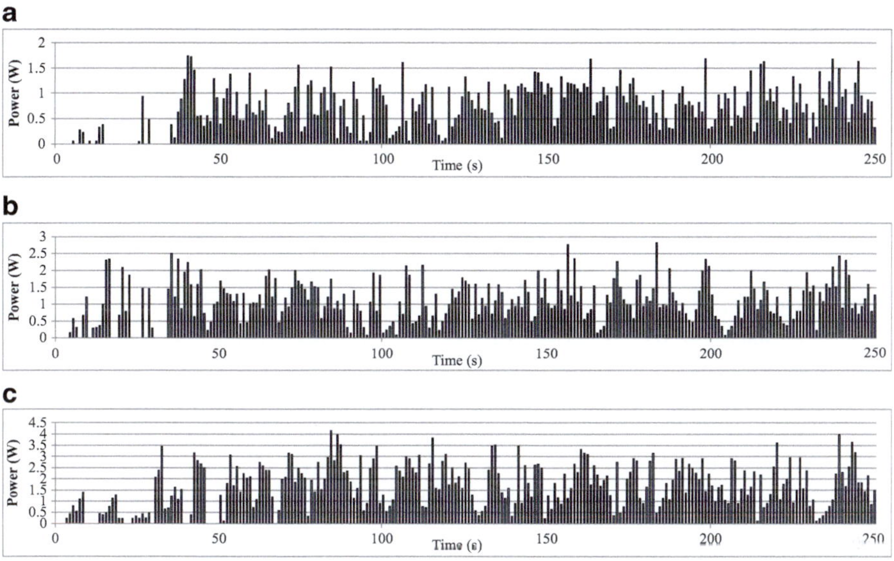

Fig. 10.12 Test power consumption of the system in 16 nm, 22 nm, and 32 nm technology, respectively. (**a**) For 16 nm technology (first experiment setup). (**b**) For 22 nm technology (second experiment setup). (**c**) For 32 nm technology (third experiment setup)

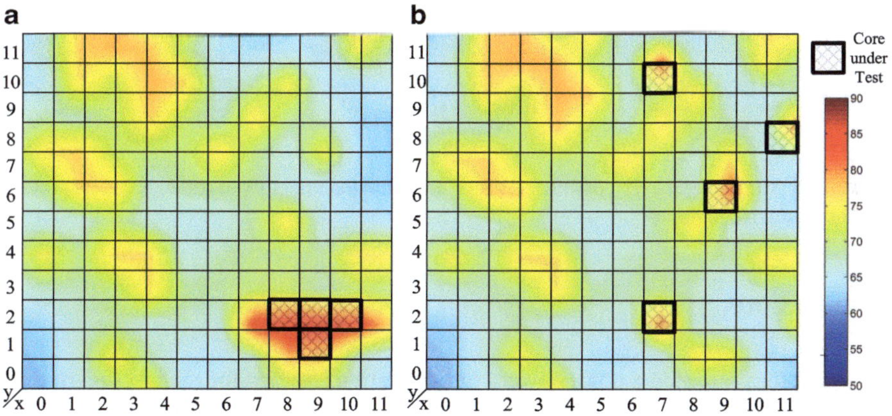

Fig. 10.13 Heat maps while running (**a**) non-thermal-aware and (**b**) thermal-aware test scheduling

As the sets with higher VF levels consume more power, their shares are a little bit lower than the sets with lower VF levels. As a conclusion the sets with lower VF levels have a better chance to use the available power than the other sets.

Table 10.2 Voltage–frequency sets for test

Technology	**16 nm**					
VF set for test	**Set 1**	**Set 2**	**Set 3**	**Set 4**	**Set 5**	**Set 6**
VF level	1–5	6–10	11–15	16–20	21–25	26–29
Voltage (V)	0.47	0.51	0.56	0.59	0.63	0.68
Frequency (GHz)	0.4–0.64	0.4–1	0.4–1.54	0.4–2	0.4–2.6	0.4–3.1
Technology	**22 nm**					
VF set for test	**Set 1**	**Set 2**	**Set 3**	**Set 4**	**Set 5**	**Set 6**
VF level	1–5	6–10	11–15	16–20	21–25	26–29
Voltage (V)	0.49	0.54	0.6	0.65	0.7	0.74
Frequency (GHz)	0.4–0.67	0.4–1.1	0.4–1.6	0.4–2.1	0.4–2.8	0.4–3.2
Technology	**32 nm**					
VF set for test	**Set 1**	**Set 2**	**Set 3**	**Set 4**	**Set 5**	**Set 6**
VF level	1–5	6–10	11–15	16–20	21–25	26–29
Voltage (V)	0.52	0.58	0.63	0.69	0.75	0.8
Frequency (GHz)	0.4–0.68	0.4–1.13	0.4–1.6	0.4–2.2	0.4–2.8	0.4–3.2

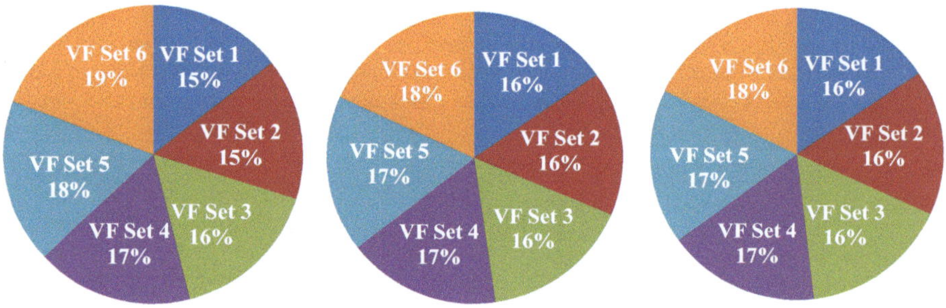

Fig. 10.14 The share of every particular VF set from the total number of tested cores (%) for 16, 22, and 32 nm technologies

Conclusions

This chapter presented a power-aware online testing strategy for many-core systems in the dark silicon era. The strategy consists of a non-intrusive online test scheduling algorithm using SBST techniques to test idle cores in the system while respecting the system's power budget. Moreover, a criticality metric is used to identify and rank cores that need testing. The goal of the approach is to guarantee prompt detection of the occurred permanent faults, while minimizing the overhead and satisfying the limited available power budget. The presented experimental results show that the proposed power-aware online testing approach can: (1) efficiently utilize temporarily unused cores and available power budget for the testing purposes, within less than 1 % penalty on system throughput and by dedicating only 2 % of the actual consumed power, (2) adapt to the current stress of the cores by using the utilization metric, and (L3) cover and balance all voltage–frequency levels during various test procedures.

References

1. JEDEC Solid State Tech. Association, Failure mechanisms and models for semiconductor devices, 2010, JEP122G
2. M. Kaliorakis, M. Psarakis, N. Foutris, D. Gizopoulos, Accelerated online error detection in many-core microprocessor architectures, in *Proceedings of VLSI Test Symposium (VTS)* (2014), pp. 1–6
3. D.P. Siewiorek, R.S. Swarz, *The Theory and Practice of Reliable System Design* (Digital Press, Bedford, 1982)
4. P. Parvathala, K. Maneparambil, W. Lindsay, FRITS - a microprocessor functional BIST method, in *Proceedings of International Test Conference (ITC)* (2002), pp. 590–598
5. M.H. Haghbayan, S. Safari, Z. Navabi, Power constraint testing for multi-clock domain SoCs using concurrent hybrid BIST, in *Proceedings of International Symposium on Design and Diagnostics of Electronic Circuits Systems (DDECS)* (2012), pp. 42–45
6. N. Foutris, M. Psarakis, D. Gizopoulos, A. Apostolakis, X. Vera, A. Gonzalez, MT-SBST: self-test optimization in multithreaded multicore architectures, in *Proceedings of International Test Conference (ITC)* (2010), pp. 1–10
7. P. Bernardi, M. Grosso, E. Sanchez, O. Ballan, Fault grading of software-based self-test procedures for dependable automotive applications, in *Proceedings of Design, Automation & Test in Europe (DATE)* (2011), pp. 1–2
8. M. Skitsas, C. Nicopoulos, M. Michael, DaemonGuard: O/S-assisted selective software-based Self-Testing for multi-core systems, in *Proceedings of International Symposium on Defect and Fault Tolerance in VLSI and Nanotechnology Systems (DFT)* (2013), pp. 45–51
9. H. Esmaeilzadeh, E. Blem, R.S. Amant, K. Sankaralingam, D. Burger, Dark silicon and the end of multicore scaling. IEEE Micro. **32**(3), 122–134 (2012)
10. M. Taylor, Is dark silicon useful? harnessing the four horsemen of the coming dark silicon apocalypse, in *Proceedings of Design Automation Conference (DAC)* (2012), pp. 1131–1136
11. L. Wang et al., Dark vs. dim silicon and near-threshold computing extended results, University of Virginia Department of Computer Science Technical Report TR-2013-01, 2012
12. A.-M. Rahmani, M.-H. Haghbayan, A. Kanduri, A. Weldezion, P. Liljeberg, J. Plosila, A. Jantsch, H. Tenhunen, Dynamic Power Management for many-core platforms in the dark silicon era: a multi-objective control approach, in *Proceedings of International Symposium on Low Power Electronics and Design (ISLPED)* (2015), pp. 1–6
13. M. Shafique, S. Garg, J. Henkel, D. Marculescu, The EDA challenges in the dark silicon era: temperature, reliability, and variability perspectives, in *Proceedings of Design Automation Conf. (DAC)* (2014), pp. 185:1–185:6
14. M. Haghbayan, A. Rahmani, P. Liljeberg, J. Plosila, H. Tenhunen, Online testing of many-core systems in the dark silicon era, in *Proceedings of International Symposium on Design and Diagnostics of Electronic Circuits Systems (DDECS)* (2014), pp. 141–146
15. X. Kavousianos, K. Chakrabarty, Testing for SoCs with advanced static and dynamic power-management capabilities, in *Proceedings of Conference on Design, Automation & Test in Europe (DATE)* (2013), pp. 737–742
16. D. Gizopoulos, A. Paschalis, Y. Zorian, *Embedded Processor-Based Self-Test* (Kluwer Academic Publishers, Boston, 2004)
17. K. Constantinides, O. Mutlu, T. Austin, V. Bertacco, Software-Based Online Detection of Hardware Defects Mechanisms, Architectural Support, and Evaluation, in *Proceedings of International Symposium on Microarchitecture (MICRO)* (2007), pp. 97–108

18. S. Nomura, M. Sinclair, C.-H. Ho, V. Govindaraju, M. de Kruijf, K. Sankaralingam, Sampling + DMR: Practical and low-overhead permanent fault detection, in *Proceedings of International Symposium on Computer Architecture (ISCA)* (2011), pp. 201–212
19. A. Apostolakis, D. Gizopoulos, M. Psarakis, A. Paschalis, Software-based self-testing of symmetric shared-memory multiprocessors. IEEE Trans. Comput. **58**(12), 1682–1694 (2009)
20. M. Kaliorakis, N. Foutris, D. Gizopoulos, M. Psarakis, A. Paschalis, Online error detection in multiprocessor chips: a test scheduling study, in *Proceedings of International On-Line Testing Symposium (IOLTS)* (2013), pp. 169–172
21. B. Khodabandeloo, S. Hoseini, S. Taheri, M.H. Haghbayan, M.R. Babaei, Z. Navabi, Online test macro scheduling and assignment in MPSoC design, in *Proceedings of Asian Test Symposium (ATS)* (2011), pp. 148–153
22. M. Kakoee, V. Bertacco, L. Benini, A distributed and topology-agnostic approach for on-line NoC testing, in *Proceedings of International Symposium on Networks on Chip (NOCS)* (2011), pp. 113–120
23. N. Kranitis, A. Paschalis, D. Gizopoulos, G. Xenoulis. Software-based self-testing of embedded processors. IEEE Trans. Comput. **54**(4), 461–475 (2005)
24. L. Chen, S. Dey, Software-based self-testing methodology for processor cores. IEEE Trans. Comput. Aided Des. Integr. Circuits Syst. **20**(3), 369–380 (2001)
25. P. Kabiri, Z. Navabi, Effective RT-level software-based self-testing of embedded processor cores, in *Proceedings of International Symposium on Design and Diagnostics of Electronic Circuits Systems (DDECS)* (2012), pp. 209–212
26. M. Psarakis, D. Gizopoulos, M. Hatzimihail, A. Paschalis, A. Raghunathan, S. Ravi, Systematic software-based self-test for pipelined processors, in *Proceedings of Design Automation Conference (DAC)* (2006), pp. 393–398.
27. T. Lu, C. Chen, K. Lee, Effective hybrid test program development for software-based self-testing of pipeline processor cores. IEEE Trans. Very Large Scale Integr. VLSI Syst. **19**(3), 516–520 (2011)
28. Y. Xia, M. Chrzanowska-Jeske, B. Wang, M. Jeske, Using a distributed rectangle bin-packing approach for core-based soc test scheduling with power constraints, in *Proceedings of International Conference on Computer Aided Design (ICCAD)* (2003), pp. 100–105
29. H. Yu, S. Reddy, C. Wu-Tung, P. Reuter, N. Mukherjee, T. Chien-Chung, O. Samman, Y. Zaidan, Optimal core wrapper width selection and SOC test scheduling based on 3-D bin packing algorithm, in *Proceedings of International Test Conference (ITC)* (2002), pp. 74–82
30. R. Chou, K. Saluja, V. Agrawal, Scheduling tests for VLSI systems under power constraints. IEEE Trans. Very Large Scale Integr. VLSI Syst. **5**(2), 175–185 (1997)
31. P. Venkataramani, V. Agrawal, ATE test time reduction using asynchronous clock period, in *Proceedings of International Test Conference* (2013), pp. 1–10
32. G. Theodorou, N. Kranitis, A. Paschalis, D. Gizopoulos, Power-aware optimization of software-based self-test for L1 caches in microprocessors, in *Proceedings of International On-Line Testing Symposium (IOLTS)* (2014), pp. 154–159
33. J. Howard et al., A 48-Core IA-32 message-passing processor with DVFS in 45nm CMOS, in *Proceedings of International Solid-State Circuits Conference (ISSCC)* (2010), pp. 108–109
34. Kalray, Kalray MPPA Manycore (2014), http://www.kalrayinc.com/
35. TGG: Task Graph Generator. http://sourceforge.net/projects/taskgraphgen/. Last update: 2013-04-11, 2013
36. M. Fattah, P. Liljeberg, J. Plosila, H. Tenhunen, Adjustable contiguity of run-time task allocation in networked many-core systems, in *Proceedings Asia and South Pacific Design Automation Conference (ASP-DAC)* (2014), pp. 349–354
37. S. Pagani, H. Khdr, W. Munawar, J.-J. Chen, M. Shafique, M. Li, J. Henkel, TSP: thermal safe power: efficient power budgeting for many-core systems in dark silicon, in *Proceedings of International Conference on Hardware/Software Codesign and System Synthesis (CODES)* (2014), pp. 10:1–10:10
38. M. Shafique, S. Garg, T. Mitra, S. Parameswaran, J. Henkel, Dark silicon as a challenge for hardware/software co-design. in *Proceedings of International Conference on Hardware/Software Codesign and System Synthesis (CODES)* (2014), pp. 13:1–13:10

39. M.-H. Haghbayan, A.-M. Rahmani, A. Weldezion, P. Liljeberg, J. Plosila, A. Jantsch, H. Tenhunen, Dark silicon aware power management for manycore systems under dynamic workloads, in *Proceedings of International Conference on Computer Design (ICCD)* (2014), pp. 509–512
40. K. Ma, X. Wang, PGCapping: exploiting power gating for power capping and core lifetime balancing in CMPs, in *Proceedings of International Conference on Parallel Architectures and Compilation Techniques (PACT)* (2012), pp. 13–22
41. Z. Chen, D. Marculescu, Distributed reinforcement learning for power limited many-core system performance optimization, in *Proceedings of Design, Automation & Test in Europe (DATE)* (2015), pp. 1521–1526
42. F. Vartziotis, X. Kavousianos, K. Chakrabarty, R. Parekhji, A. Jain, Multi-site test optimization for multi-Vdd SoCs using space- and time- division multiplexing, in *Proceedings of Conference on Design, Automation & Test in Europe (DATE)* (2014), pp. 1–6
43. M. Fattah, M. Daneshtalab, P. Liljeberg, J. Plosila, Smart hill climbing for agile dynamic mapping in many-core systems, in *Proceedings of Design Automation Conference (DAC)* (2013), pp. 1–6
44. C.-L. Chou, U. Ogras, R. Marculescu, Energy- and performance-aware incremental mapping for networks on chip with multiple voltage levels. IEEE Trans. Comput. Aided Des. Integr. Circuits Syst. **27**(10), 1866–1879 (2008)
45. N. Ali, M. Zwolinski, B. Al-Hashimi, P. Harrod, Dynamic voltage scaling aware delay fault testing, in *Proceedings of European Test Symposium (ETS)* (2006), pp. 15–20
46. X. Kavousianos, K. Chakrabarty, A. Jain, R. Parekhji, Test schedule optimization for multicore SoCs: handling dynamic voltage scaling and multiple voltage Islands. IEEE Trans. Comput. Aided Des. Integr. Circuits Syst. **31**(11), 1754–1766 (2012)
47. T. Yoneda, K. Masuda, H. Fujiwara, Power-constrained test scheduling for multi-clock domain SOCs, in *Proceedings of Design, Automation & Test in Europe (DATE)* (2006), pp. 1–6
48. F. Fazzino, M. Palesi, D. Patti, Noxim: network-on-chip simulator. http://sourceforge.net/projects/noxim, 2008
49. S. Li, J. H. Ahn, R. Strong, J. Brockman, D. Tullsen, N. Jouppi, McPAT: an integrated power, area, and timing modeling framework for multicore and manycore architectures, in *Proceedings of International Symposium on Microarchitecture (MICRO)* (2009), pp. 469–480
50. B. Calhoun, S. Khanna, R. Mann, J. Wang, Sub-threshold circuit design with shrinking CMOS devices, in *Proceedings of International Symposium on Circuits and Systems (ISCAS)* (2009), pp. 2541–2544
51. K. Skadron, M. Stan, K. Sankaranarayanan, W. Huang, S. Velusamy, D. Tarjan, Temperature-aware microarchitecture: modeling and implementation, in *ACM Transactions on Architecture Code Optimization* (2004), pp. 94–125
52. M. Fattah, A.-M. Rahmani, T. Xu, A. Kanduri, P. Liljeberg, J. Plosila, H. Tenhunen, Mixed-Criticality Run-Time Task Mapping for NoC-Based Many-Core Systems, in *Proceedings of International Conference on Parallel, Distributed and Network-Based Processing (PDP)* (2014), pp. 458–465
53. M. Haghbayan, S. Karamati, F. Javaheri, Z. Navabi, Test pattern selection and compaction for sequential circuits in an HDL environment, in *Asian Test Symposium (ATS)* (2010), pp. 53–56
54. Z. Navabi, *Digital System Test and Testable Design: Using HDL Models and Architectures* (Springer, Berlin, 2010)
55. M.-H. Haghbayan, A.-M. Rahmani, P. Liljeberg, J. Plosila, H. Tenhunen, Energy-efficient concurrent testing approach for many-core systems in the dark silicon age, in *Proceedings of International Symposium on Defect and Fault Tolerance in VLSI and Nanotechnology Systems (DFT)* (2014), pp. 270–275

Part III
Design and Management: Communication Perspective

Chapter 11
Adroit Use of Dark Silicon for Power, Performance and Reliability Optimisation of NoCs

Haseeb Bokhari, Muhammad Shafique, Jörg Henkel, and Sri Parameswaran

Introduction and Motivation

Network-on-Chips (NoCs) provide a scalable communication fabric for connecting a large number of resources within a chip. NoC can contribute significantly to the total chip power, e.g., up to 18 % and 33 % in Intel SCC [1] and RAW [2] architectures, respectively. To reduce the power consumption of NoCs, various power saving techniques for NoC at system-, architecture- and circuit-level have emerged, among which Dynamic Voltage and Frequency Scaling (DVFS) is prominent [3, 4]. System-level DVFS managers apply reduced voltage and frequency (VF) levels in an NoC to save power. According to several recent studies [5, 6], the energy saving potential of DVFS is diminishing for two reasons:

- With shrinking node size, the difference between nominal voltage and the threshold voltage is decreasing. As the operating voltage V approaches threshold voltage, the transistor delay increases exponentially, resulting in huge performance penalties [5]. This limits the factor by which V can be scaled down. Transistors with lower threshold voltages can be used in the circuit to increase the gap between the nominal and threshold voltages, which results in an increased leakage power.
- DVFS does not influence the intrinsic physical properties of the circuit such as gate sizes, capacitances, etc., which are fixed at design time.

H. Bokhari (✉) • S. Parameswaran
UNSW, Sydney, NSW, Australia
e-mail: hbokhari@cse.unsw.edu.au; sridevan@cse.unsw.edu.au

M. Shafique • J. Henkel
KIT, Karlsruhe, Germany
e-mail: muhammad.shafique@kit.edu; henkel@kit.edu

A.M. Rahmani et al. (eds.), *The Dark Side of Silicon*,
DOI 10.1007/978-3-319-31596-6_11

With the help of the following NoC synthesis case study, we show how the problem of diminishing returns of DVFS can be alleviated.

Motivational Example

Multi-Vt Optimisation

Modern fabrication foundries characterise cell libraries for various gate threshold voltage (Vt) values such as normal Vt (NVt), Low Vt (LVt) and High Vt (HVt). LVt cells can switch at faster speeds than HVt cells. However, LVt cells can be up to 5× leakier than their HVt counterparts. Modern CAD tools exploit the power-delay characteristics of multi-Vt cell libraries and slacks in path delays to synthesise power efficient circuits [7]. A typical CAD tool optimises the circuit by using LVt cells on critical paths to meet the target latency, while inserting HVt cells on the non-critical paths to reduce leakage power. For example, in Fig. 11.1, a square and its annotation represent the type of cell and its delay. With multi-Vt circuit optimisation (Fig. 11.1), a mix of LVt, NVt and HVt cells is used to meet the target latency rather than just the NVt cells. Thus, the circuit on the right will have different gate sizes, capacitances and leakage power than the circuit on the left.

Multi-Vt Optimisation of NoC Routers

We exploited the multi-Vt circuit optimisation available in CAD tools to synthesise architecturally identical NoC routers for a set of target VF levels: [1 GHz, 0.9 V], [750 MHz, 0.81 V], [500 MHz, 0.81 V] and [250 MHz, 0.72 V]. Figure 11.2 reports the network power for operation at [500 MHz, 0.81 V] and [250 MHz, 0.72 V]

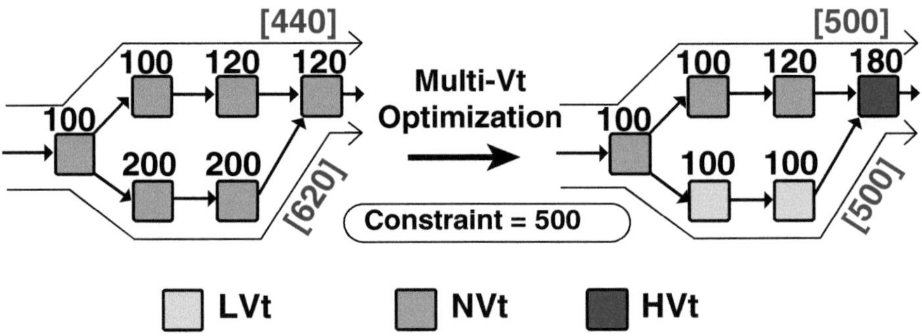

Fig. 11.1 An example of multi-Vt circuit optimisation

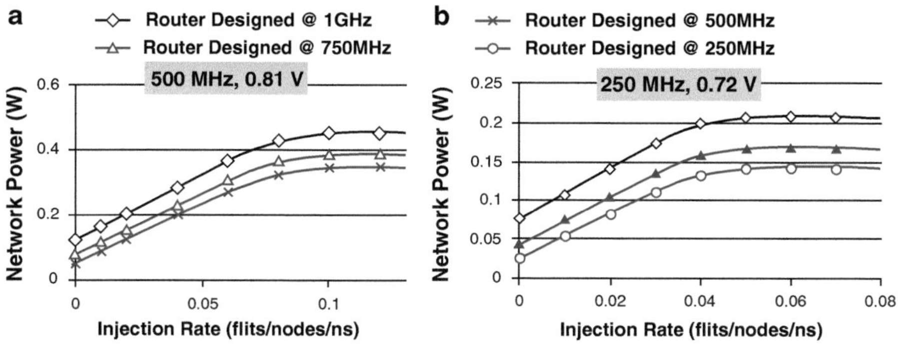

Fig. 11.2 Network-on-Chip (NoC) power for transpose traffic for (**a**) (*left*) [500 MHz, 0.81 V] VF level, (**b**) (*right*) [250 MHz, 0.72 V] VF level

(details of our experiments are presented in section "Experimental Setup and Results"). We can observe that for operation at [500 MHz, 0.81 V], the NoC designed particularly for [500 MHz, 0.81 V] VF level is on average 35 % and 16 % more power efficient than applying DVFS on an NoC designed for [1 GHz, 0.9 V] and [750 MHz, 0.81 V], respectively. Similarly, for operation at [250 MHz, 0.72 V], the NoC designed particularly for [250 MHz, 0.72 V] VF level is on average 42 % and 22.7 % more power efficient than DVFS applied to NoC designed for [1 GHz, 0.9 V] and [750 MHz, 0.81 V], respectively. This shows that, unlike traditional NoC with a single layer of routers, it may be beneficial in terms of power to have multiple layers of routers in an NoC such that each layer is optimised for a particular VF level.

Multi-Vt Optimisation, Multiple Routers and Dark Silicon

The provision of multiple layers of routers in an NoC may cost a significant amount of silicon; however, we can leverage *dark silicon* [8, 9] to alleviate this overhead. Several researchers have proposed to leverage dark silicon to provide extra application-specific accelerators [10], extra low-power processors [11], etc., for better energy efficiency. In fact, the authors of [12] state that *"in eight years, we will be faced with designs that are 93.75 % dark!"*, and *"we will spend exponentially increasing amounts of silicon area to buy energy efficiency"*. Therefore, even with the addition of extra accelerators [10] and processors [11], *we believe that there will still be dark silicon available which could be exploited for the energy efficient design of other system components such as NoC.*

DarkNoC Architecture

In this section, we focus on the architectural details of our darkNoC architecture and study the interplay of application characteristics, architecture and dark silicon budget. We discuss the fine-grain details of an NoC architecture, where architecturally homogenous routers which have been optimised specifically for particular VF ranges are seamlessly integrated at the architecture-level to complement system-level DVFS managers, and exchange dark silicon for energy efficiency in NoC.

We consider NoCs with predefined topologies as shown in Fig. 11.3a. The NoC consists of n routers, which are divided into regions. Each region has m routers, and separate voltage and frequency (VF) rails with their own VF regulators. The value of m can vary from 1 to n, depending on the granularity of the regions. For communication between VF regions, bi-synchronous links [13] must be used. However, in this work we assume that VF regions can only communicate if they are operating at the same VF level. A designer can opt to use the darkNoC architecture in any of the VF regions based upon their requirements. The following paragraphs explain the darkNoC architecture from the perspective of a single VF region.

darkNoC Layers

A region contains l network layers, where a network layer consists of m routers as shown in Fig. 11.3b. At a given time, only one of the network layers is *illuminated* (activated), while the rest of the network layers remain *dark* (deactivated). When a network layer is illuminated, all of its routers are active, and thus, at any given time, only m routers are active. Figure 11.4 shows an example of transitioning from network layer 0 to 3 in a region with four network layers.

Each network layer is optimised at design time to operate in a certain VF range. That is, multi-Vt circuit optimisation of CAD tools is used to optimise all the routers of a network layer for a particular VF range. All the layers in a region are managed by a hardware-based darkNoC Layer Manager (*dLM*) as shown in Fig. 11.3b. The function of the *dLM* is to switch between network layers when directed by the system-level DVFS manager.

darkNoC Routers Stack

Each node in Fig. 11.3b represents a stack of l routers to realise l network layers of the region. For example, for the four network layers in Fig. 11.4, each router stack has four routers and all the routers marked 2 belong to the network layer 2. As shown in Fig. 11.3b, all the routers in a stack share the same set of link wires with the neighbouring stacks of routers, *which reduces verification costs.*

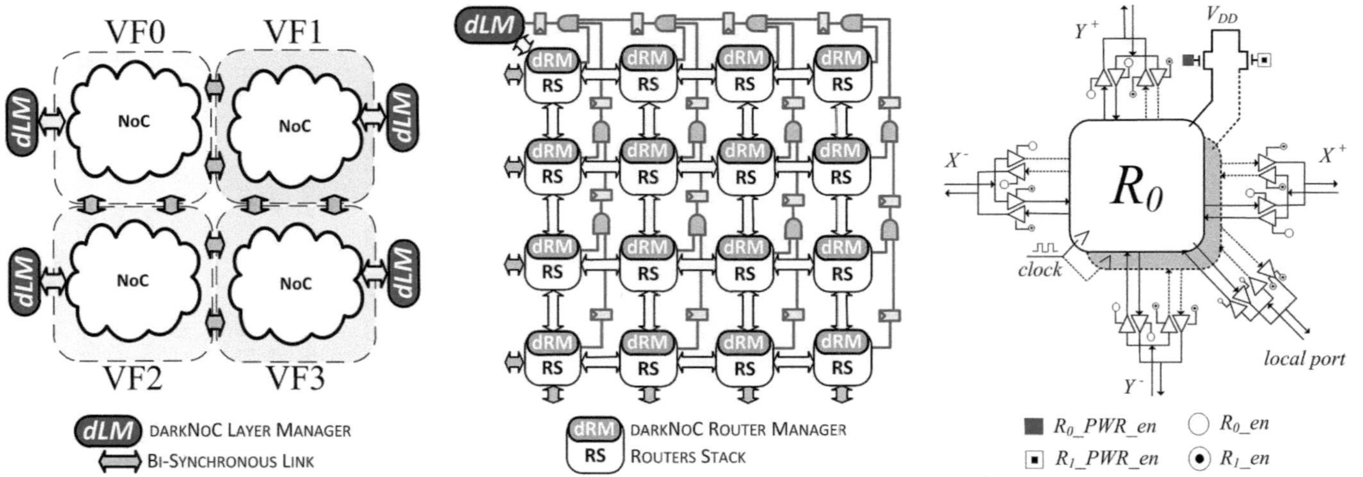

Fig. 11.3 darkNoC architecture: (**a**) (*left*) NoC divided into VF regions, (**b**) (*middle*) a single VF region and (**c**) (*right*) a stack of (two) VF optimised routers

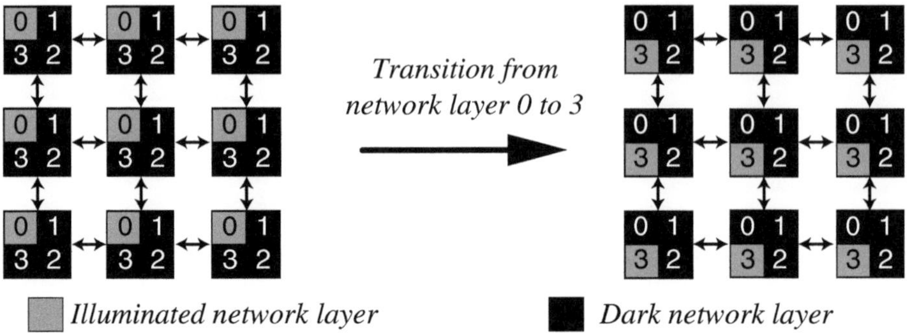

Fig. 11.4 An example darkNoC architecture with four network layers

Each router in a stack has separate controls for power gating and enabling input/output ports. For example, a stack of two routers is shown in Fig. 11.3c, where R_0*_PWR_en* and R_0*_en* controls are used to power-on or off and enable or disable input/output ports of router 0. Thus, a router in a stack is illuminated in two steps: powering it on and enabling its input/output ports. The local port of a router stack is connected to a tile (which may consist of a processor, memory, etc.) through a network interface (NI). The NI has the capability to stop injection of packets from its tile, which we exploit during the switching of network layers. To manage a router stack, we use a hardware-based darkNoC Router Manager (*dRM*) as shown in Fig. 11.3b. The function of *dRM* is to switch between routers when directed by the *dLM*.

darkNoC Layer Switch-Over Mechanism

The switch-over between network layers is an important design requirement for our darkNoC architecture. The main challenges are: (a) the lossless data communication property of packet-switched buffered NoC should be preserved, (b) the switch-over mechanism should be transparent to software, and (c) the switch-over mechanism should be efficient in terms of time and energy overhead. In our solution, the *dLM* and the darkNoC Router Managers (*dRMs*) autonomously coordinate with each other to realise a switch-over mechanism with the aforementioned requirements.

At a high level, the *dLM* ensures correct switch-over between two network layers in the region. The *dLM* uses the injection channel of the NI of the router stack it is attached to, and sends different types of control flits to *dRMs* through the NoC channels. Further, the *dLM* gathers the buffer occupancy information from the routers stacks using a single bit AND-gate network as shown in Fig. 11.3b. Similar AND-gate networks have been used in NoC for congestion detection [14]. The *dRM* of a router stack is attached to the ejection port of the NI, and receives control flits from the *dLM*. Based upon the control flits from the *dLM*, a *dRM* can: (a) power-on or off a router, (b) enable or disable input/output ports of a router or (c) enable or

Algorithm 1: dLM Function

```
1  nodes = m; // number of routers in a region
2  WaitForSwitchOverCommand();
3  for i=nodes to 0 do
4    └ SendDisableInjControlFlitTodRM(i);
5  WaitForNetworkFlush();
6  for i=nodes to 0 do
7    └ SendSwitchRouterControlFlitTodRM(i);
8  WaitForNetworkFlush();
9  for i=nodes to 0 do
10   └ SendEnableInjControlFlitTodRM(i);
```

Algorithm 2: dRM Function

```
   // Phase 1
1  WaitForDisableInjControlFlitFromdLM();
2  SwitchOnTargetRouter();
3  DisableInjectionFromNI();
4  if RouterLocatedAtRegionBorder then
5    └ SetStopFlagForNeighbourRouters();
6  WaitForRouterFlush();
7  SetEmptyFlag();
   // Phase 2
8  WaitForSwitchRouterControlFlitFromdLM();
9  DisableActiveRouterPorts();
10 EnableTargetRouterPorts();
11 PowerOffActiveRouter();
12 SetEmptyFlag();
   // Phase 3
13 WaitForEnableInjControlFlitFromdLM();
14 EnableInjectionFromNI();
15 if RouterLocatedAtRegionBorder then
16   └ ResetStopFlagForNeighbourRouters();
```

disable injection of packets into the NI. Further, a *dRM* generates a single bit flag by analysing the buffer occupancies of its router to indicate the presence or absence of in-transit packets, which is propagated through the AND-gate network to the *dLM*.

Details of dLM and dRM Algorithms 1 and 2 report the functionalities of *dLM* and *dRM*, respectively, during switch-over of a network layer. Figure 11.5 illustrates the switch-over mechanism for a router stack (other stacks similarly switch-over their routers in parallel). On detecting a new switch-over command (from a system-level DVFS manager), the *dLM* starts sending control flits to all the *dRMs* to initiate the switch-over mechanism (Algorithm 1, lines 2–4). A *dRM* enters into its first phase on receiving the control flits from the *dLM* (Algorithm 2, lines 2–5; Fig. 11.5 @t_1). The *dRM*: (1) powers-on the target router, and (2) stops NI from injecting any

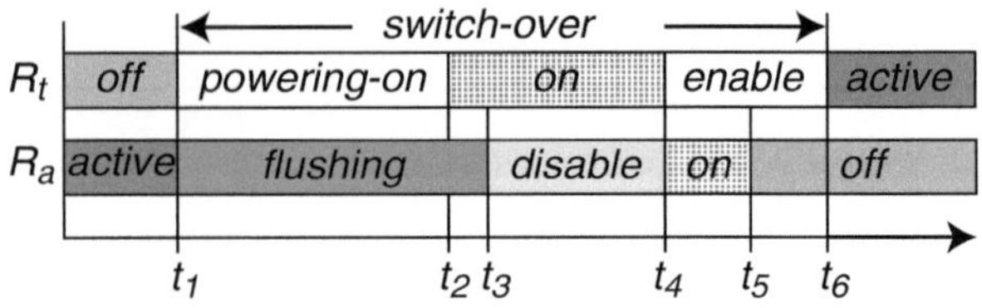

Fig. 11.5 Timing diagram for switch-over from router R_a to R_t

new packets. If the *dRM* is controlling one of the border routers, then it also sets a flag to stop neighbouring routers of other regions from injecting any new packets. The reason behind stopping the injection of packets is to ensure that all the in-transit packets reach their destination before the actual switch-over starts. Once the router buffers are flushed and no packets are injected by the neighbouring routers, the *dRM* sets the empty flag, which is propagated through the AND-gate network. It is important to note that the powering-on of the target router and the flushing of the currently active router (Fig. 11.5, $t_1 - t_3$) are simultaneous to speed up the switch-over mechanism.

Once the empty flags from all the *dRMs* have been propagated to the *dLM* through the AND-gate network, the *dLM* concludes the flushing of the currently active network layer. Then it starts another round of control flits to all the *dRMs* to start the actual switch-over of routers (Algorithm 1, lines 5–7). On receiving the control flit, a *dRM* enters into its second phase (Algorithm 2, lines 8–12), where it: (1) disables the input/output ports of the active router (Fig. 11.5, @t_3), (2) enables the input/output ports of the target router (Fig. 11.5, @t_4) and (3) powers-off the (previously) active router (Fig. 11.5, @t_5). Note that the order of (1) and (2) is essential to avoid short-circuit conditions. At the end, the *dRM* sets the empty flag, which is propagated through the AND-gate network.

Once the *dLM* detects (from the AND-gate network) that the network is empty again, it starts the last round of control flits to all the *dRMs* (Algorithm 1, lines 8–10). This control flit forces a *dRM* to enter its third phase (Algorithm 2, lines 13–16), where the injection of packets into NI is enabled (Fig. 11.5, @t_6). Further, if the *dRM* is controlling one of the border routers, then it enables injection of packets from the neighbouring routers of other regions. With this, the network layer switch-over mechanism is complete.

Time and Energy Overhead Now we present an analysis of the time and energy overhead of switch-over mechanism in darkNoC architecture. For this analysis, we assume that the transition of VF level is done after the switch-over from an active network layer to a target network layer. We do not include the time and energy overhead of VF scaling because this overhead is present in the traditional single layer NoC as well. The time overhead in clock cycles of switch-over can be estimated as

$$T_{so} = in - transit - packets(m, load) + no - load(m)$$
$$+ control - flits(m, load)$$

The *in-transit-packets* term represents the time to flush the active network layer which depends on the number of routers *m* and the network load. The *no-load* term captures the time spent in switch-over of all the routers which only depends on *m*, while the term *control-flits* represents the time spent by *dLM* in sending control flits to *dRMs* and is dependent on both *m* and the network load. Since the whole switch-over mechanism occurs at the frequency of the active network layer, it can be used to convert the switch-over clock cycles to actual time.

The energy overhead of switch-over mechanism can be estimated as

$$E_{so} = power - on - energy(L_t, m) + control - flits(L_t, L_a, m)$$
$$+leakage - power(L_t, L_a, m) \times T_{so}$$

where L_a and L_t refer to the active and target network layers, respectively. The first term *power-on-energy* is the energy overhead of powering-on a network layer, which depends on the type of the target network layer and the number of routers *m* in it. During a typical switch-over, the control flits from the *dLM* use both the active and target network layers, and their energy consumption is captured in the *control-flits* term. The last term includes the leakage energy of the active and target layers, since both the layers consume leakage power for almost the whole duration of the switch-over mechanism.

We experimented with 4×4 and 8×8 mesh sizes, transpose random (TR) and uniform random (UR) traffic patterns and varying traffic injection rates (details are in section "Experimental Setup and Results"). The results for time and energy overhead are presented in Figs. 11.6 and 11.7, respectively. Three trends are evident. The time and energy overhead: (1) increases with an increase in mesh size due to a higher number of control flits and hops, and longer delays in the AND-gate network, (2) increases with an increase in the traffic injection rates due to longer flushing times of in-transit packets and delays of control flits and (3) heavily depends on the network load as trends are different for TR and UR traffic patterns. It is important to note that the time and energy overhead of our switch-over mechanism is not significant compared to the total execution time of typical applications and energy consumption of the NoC (see section "Experimental Setup and Results").

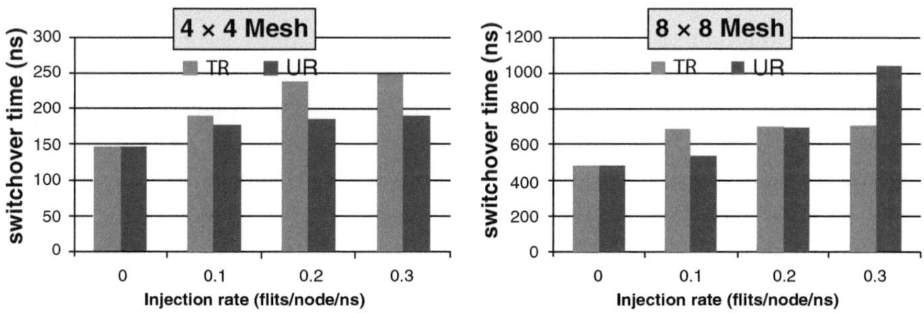

Fig. 11.6 Time overhead of switch-over mechanism for NoC running @ 500 MHz

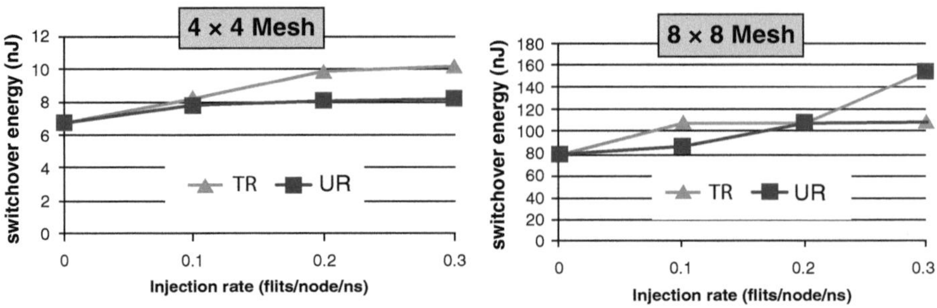

Fig. 11.7 Energy overhead of switch-over from [500 MHz, 0.81 V] to [750 MHz, 0.81 V] layer

Table 11.1 NoC architectural details

Topology	4×4 mesh
Router architecture	5 port router (4 neighbours + 1 local), 3 stage pipeline [LT+BW, RC+SA, ST]
Input buffer	8 flit deep (no virtual channel)
Routing	Dimension order XY
Link flow control	On/Off
Link width	64 bit data, 8 bit control
Switch arbiter	Matrix arbiter

Table 11.2 Synthesis results for a single router

Frequency	1 GHz	750 MHz	500 MHz	250 MHz
Voltage	0.9 V	0.81 V	0.81 V	0.72 V
Area [μm^2]	49,200	46,313	45,750	45,225
HVt cells [%]	44.7	60.5	90.7	92.8
NVt cells [%]	12.5	13.6	4.3	3.9
LVt cells [%]	42.8	25.9	5.0	3.3

Experimental Setup and Results

NoC Synthesis

We use a packet-switched wormhole NoC architecture for experiments. The details of the NoC architecture are reported in Table 11.1. The proposed router is implemented in Verilog RTL, and synthesised using Synopsys Design Compiler for commercial TSMC 45 nm libraries characterised for HVt, NVt and LVt cells. We enabled leakage and dynamic power optimisation in Synopsys Design Compiler, which automatically enabled multi-Vt optimisation. We explored the following four VF levels: [1 GHz, 0.9 V], [750 MHz, 0.81 V], [500 MHz, 0.81 V] and [250 MHz, 0.72 V], and the synthesis results are reported in the last four columns of Table 11.2.

We can observe that the target VF level significantly affects the type and number of cells used during optimisation. For example, the router optimised for

Table 11.3 MPSoC architectural details

Topology	4 × 4 mesh
Processors	Tensilica in-order LX4 @ 1 GHz
L1 I/D caches	4 KB, 2-way set associative, 16-byte line size
DRAM	2 memory controllers, 40 ns latency
NoC VF levels	[1 GHz, 0.9 V], [750 MHz, 0.81 V], [500 MHz, 0.81 V], [250 MHz, 0.72 V]

Table 11.4 Details of application mixes with four copies of each application

AM-1	mpeg2enc, sha, mpeg2dec, g721enc
AM-2	h264enc, mpeg2dec, sha, jpeg_dec
AM-3	jpeg_enc, sha, g721enc, g721dec
AM-4	g721enc, mpeg2enc, g721dec, mpeg2dec

[1 GHz, 0.9 V] contains 42.8 % LVt cells, which decreases to 5.0 % and 3.3 % for optimisation at [500 MHz, 0.81 V] and [250 MHz, 0.72 V], respectively. Moreover, the cells are sized according to the target VF level, which has resulted in different silicon areas for different target VF levels (row 2). These results corroborate our motivational example that it is worthwhile to explore multi-Vt optimisation in NoCs for energy efficiency.

Experimental Setup

MPSoC and Applications We used a 16-node MPSoC laid out as a 4×4 2D-mesh, whose architectural details are in Table 11.3. Four VF levels are assumed for the NoC, whose architectural details are in Table 11.1. Note that the VF level of processors is not changed in our experiments. We used eight applications from the MediaBench suite and created diverse multi-programmed application mixes (AM) whose details are in Table 11.4. For an application mix, four copies of each application in that mix were used. We also used synthetic traffic injection models consisting of Uniform Random (UR), Bit Complement (BC) and Transpose (TR).

Simulation Setup We used two-step system simulation methodology where the memory access trace of each application executing on a processor is collected from the Xtensa instruction set simulator. These memory access traces are then simulated through a closed-loop cycle-accurate NoC and DRAM simulator. Our NoC simulator also modelled different VF levels accurately for the NoC. We also modelled the *dLM*, *dRM* and AND-gate network in the NoC simulator. For application mixes, each processor was executed for at least two million instructions. For synthetic traffic injection models, the NoC was warmed up for 100,000 clock cycles, followed by collection of statistics for 80,000 clock cycles. The injection rates were measured in $flits/node/ns$.

NoC Power Estimation We used Synopsys PrimeTime tool to analyse the netlist generated by Synopsys Design Compiler to compute leakage and dynamic power

Table 11.5 NoC configurations used in experiments

Name	Configuration	Relative chip area
baselineNoC	Traditional NoC with [1 GHz, 0.9 V] layer	1 ×
darkNoC1	[1 GHz, 0.9 V] + [750 MHz, 0.81 V]	1.13×
darkNoC2	[1 GHz, 0.9 V] + [500 MHz, 0.81 V]	1.13×
darkNoC3	[1 GHz, 0.9 V] + [750 MHz, 0.81 V] + [500 MHz, 0.81 V]	1.26×

of an NoC. We first apply this analysis to compute power values at the nominal VF level of the NoC. Then, scaled VF conditions are applied to compute power values at VF levels lower than the nominal VF level. These computed power values are used in the NoC simulator to compute the total NoC energy consumption. This methodology is also used for all the network layers in a darkNoC.

NoC Configurations We used different NoC configurations in our experiments, which are listed in Table 11.5. The **baselineNoC** represents the traditional single layer NoC designed for the highest VF level, [1 GHz, 0.9 V]. *For fairness of comparison, we synthesised baselineNoC with multi-Vt optimisation.* The **darkNoC1** and **darkNoC2** configurations have 2 VF-optimised network layers each, while **darkNoC3** configuration is a superset of darkNoC1 and darkNoC2. In Table 11.2, the difference in silicon area and distribution of multi-Vt cells between [500 MHz, 0.81 V] and [250 MHz, 0.72 V] optimised routers (columns 4 and 5) is small. This means that two network layers optimised for [500 MHz, 0.81 V] and [250 MHz, 0.72 V] will have nearly identical properties, and thus similar energy efficiencies. Therefore, a designer can use synthesis results for early elimination of network layers in a darkNoC. That is why we did not include [250 MHz, 0.72 V] optimised network layer in our experiments.

The last column of Table 11.5 reports the relative chip area of NoC configurations, measured as a *sum of the area of the processors, caches, network layers, and the dLM and dRMs*. A designer can use darkNoC1 or darkNoC2 if they have dark silicon budget of a single network layer. Likewise, they can use darkNoC3 if the dark silicon budget permits the inclusion of two network layers.

System-Level DVFS Manager We used the state-of-the-art memory latency based NoC DVFS technique introduced by Chen et al. [15] for the baselineNoC. We used a control interval of 50K clock cycles for the DVFS. For the darkNoC configurations, we modified their technique as follows. If the DVFS manager decides to switch to the *i-th* VF level and there is a network layer optimised for the *i-th* VF level, then the DVFS manager requests the *dLM* to switch-over to that particular network layer. On the other hand, if there is no network layer optimised for *i-th* VF level, then the DVFS manager requests the *dLM* to switch-over to the network layer optimised for the closest yet higher VF level, and will scale the VF of the selected network layer. For example, in darkNoC1, if the system-level DVFS manager decides to operate NoC at [250 MHz, 0.72 V], then the [750 MHz, 0.81 V] network layer will be scaled down to operate at [250 MHz, 0.72 V] rather than the [1 GHz, 0.9 V] network layer.

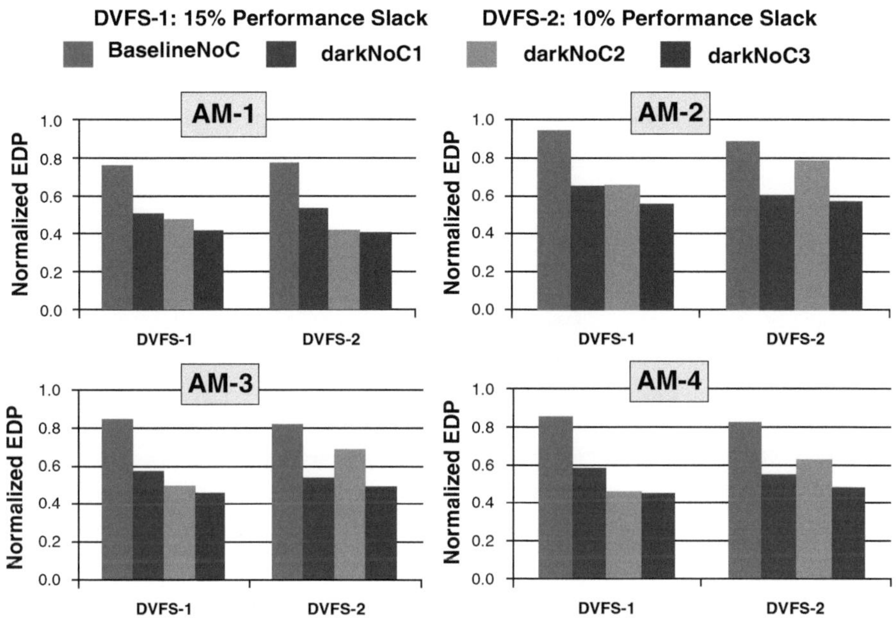

Fig. 11.8 NoC energy-delay product (EDP) normalised w.r.t. traditional NoC operating at the highest VF level

We created two flavours of DVFS managers based upon application requirements: **DVFS-1 with a target performance loss of 15 % and DVFS-2 with a target performance loss of 10 %.** Note that our NoC simulator *included the time and energy overhead of network layer switch-over* for the darkNoC configurations.

Results and Discussion

Overall Trend in NoC Energy-Delay Product (EDP) Savings Figure 11.8 shows the savings in NoC EDP for the four application mixes of Table 11.4, four NoC configurations of Table 11.5 and two DVFS managers. Overall, darkNoC configurations show an improvement in EDP over baselineNoC.

For AM-2, the DVFS-1 reduced EDP by only 5 % in baselineNoC. On the other hand, EDP is reduced by 34 % and 44 % by DVFS-1 in darkNoC1 and darkNoC3, respectively.

Likewise, for AM-3, DVFS-1 in baselineNoC saved 15 % EDP, whereas DVFS-1 in darkNoC1 and darkNoC3 reduced EDP by 45 % and 52 %, respectively. In summary, darkNoC provides up to 56 % EDP savings (AM-1, darkNoC3, DVFS-2) over baselineNoC. These results show that DVFS alone can be ineffective in reducing NoC energy due to increasing leakage power to dynamic power ratio. Thus, our darkNoC architecture complements system-level DVFS managers.

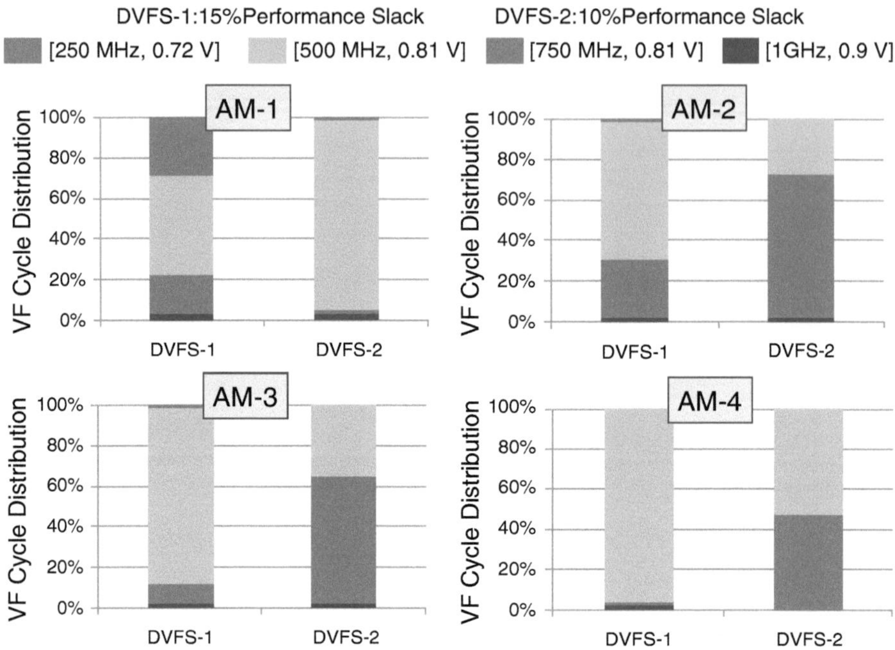

Fig. 11.9 Distribution of NoC VF cycles

Interplay of Applications, DVFS Manager and darkNoC Configurations EDP savings vary across DVFS managers and darkNoC configurations. For example, in AM-2, AM-3 and AM-4, use of DVFS-2 with darkNoC1 is far more beneficial than DVFS-2 with darkNoC2. Likewise, for AM-3 and AM-4, DVFS-1 with darkNoC2 results in more EDP savings than DVFS-1 with darkNoC1.

To further explain these results, in Fig. 11.9, we report the distribution of NoC VF cycles, which is the percentage of total execution cycles the NoC operates at a particular VF level. For AM-2, DVFS-2 operates the NoC at [750 MHz, 0.81 V] and [500 MHz, 0.81 V] levels for 71.4 % and 27.4 % of the time, respectively. Thus, darkNoC1 results in better EDP savings because it has a [750 MHz, 0.81 V] optimised network layer. In the case of AM-4, DVFS-1 with darkNoC3 provided only 1 % EDP improvement over darkNoC2, at the expense of an additional network layer. This is because the NoC is operated at [500 MHz, 0.81 V] level for 95 % of the time, and thus a darkNoC with [500 MHz, 0.81 V] optimised network layer is more than sufficient.

Insights From these results, we can conclude that savings in EDP depends on:

- darkNoC configuration
- System-level DVFS manager
- Applications

There is a complex interaction between these factors. For example, the energy efficiency of a darkNoC configuration (the number and type of network layers) heavily depends upon the dark silicon budget and distribution of VF cycles, which in turn depends upon the target performance loss set in the system-level DVFS manager and the executing applications. Taking it one level further, the decisions taken by the system-level DVFS manager depend on the number and type of network layers in the darkNoC configuration, which results in a cyclic dependency between these factors. Thus, the problems (highlighted in the Introduction) of: (1) determining the number and type of network layers, and (2) enhancement of the system-level DVFS manager to exploit multiple network layers are interdependent. Consequently, these problems result in a multidimensional design space exploration problem.

SuperNet Architecture

The cost of designing an SoC has increased dramatically with shrinking node sizes as the laws of physics are challenged [16–18]. The cost trend of chip design is shown in Fig. 11.10. A significant amount of financial and human resources are required to design and verify chips designed in smaller nodes. Therefore, according to the laws of economy, the only way to achieve profitability and beat the initial design cost is to increase the number of units sold. To increase the sales volume, a chip has to target a wider set of applications. This means that the SoC has to fulfil the design requirements for different usage scenarios. For example, designing a multimedia SoC for a battery-powered smart phone requires a special emphasis on low energy operation. On the other hand, using the same SoC for a home entertainment system requires special attention to high performance. A similar SoC used for engine management or a braking system in a car has to operate reliably throughout the car's lifetime. This poses a serious challenge to designers who must account for various operational scenarios in a single chip.

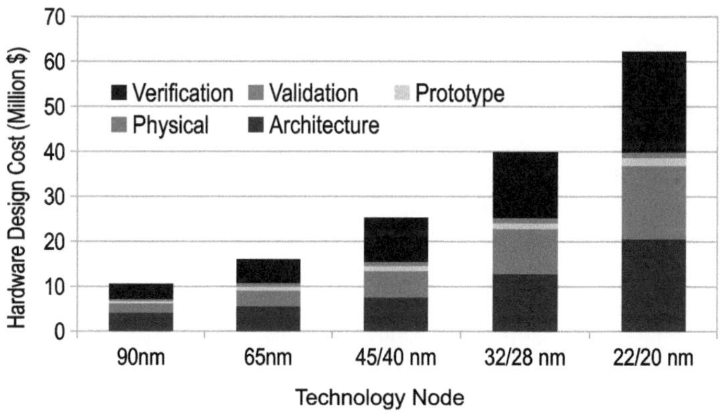

Fig. 11.10 Hardware design cost trend (source: Semico research [18])

It is important to reconsider the on-chip interconnect in the context of such complex multipurpose SoC designs. Packet-switched NoCs are seen as a scalable interconnect to support the growing communication demand of SoC components. Like other on-chip components, it is essential to study the effects of downscaling trends on NoCs. Research studies show that leakage power is the major share of total NoC power, and this is expected to increase for smaller nodes [19]. On the other hand, deep transistor integration adversely affects the reliability of digital circuits. A soft error in NoC can have adverse effects, such as partial disconnection of SoC component, erroneous execution of application or system-level deadlock [20]. To maintain high performance, NoCs are designed either for worst-case or average-case scenarios. However, a design approach for worst-case delay can lead to wasted energy for less demanding applications [21].

The above mentioned design problem becomes more complex when NoCs must be designed for multipurpose SoCs. Traditionally NoCs are designed with a specific goal of either low-power operation, high performance [21] or high reliability [22]. For example, an NoC for a high performance chip can be designed with a wider channel width and higher clock frequency to fulfil the requirement of high bandwidth and low latency[21]. Such a high performance NoC will be impractical for low-power battery-powered devices. Similarly, low-power NoC can be designed with narrow channels and lowest possible voltage guard-bands. However this NoC will neither be good enough for application with high bandwidth required to support high performance applications nor operate reliably under harsh operational conditions. Therefore, to our best of our knowledge, there is no single design proposal that covers all three possible operational requirements discussed earlier.

A very naive solution to the above stated design problem is to integrate an NoC for each possible design goal and use those NoCs depending on the usage scenario. Such designs are not a suitable choice for two reasons. The first reason is that having multiple mutually exclusive NoCs will carry a higher silicon cost. Despite the fact that a large amount of *dark silicon* is available in future chips due to power budget limitations, NoCs still need to compete for this *dark silicon* [23] area with other on-chip components such as application accelerators, etc. The second reason is that every on-chip component has to be extensively verified before fabrication [16] and therefore naively adding more NoCs can potentially increase the chip design cost. In this section we show that in place of such inefficient NoC architecture, we can design a system with the minimal number of NoCs that can be used to achieve the same goal through the adroit reuse of one set of NoCs for different operational modes.

Our solution to the above stated problem is the *SuperNet* NoC architecture. The architecture consists of parallel architecturally homogenous, but voltage–frequency (VF) optimised heterogeneous NoCs. For energy efficient mode, we leverage the two NoCs for DVFS by switching on only one NoC depending on application requirements or static configuration. In performance mode, we classify the network packets under the category of critical and non-critical packets, and steer the packets accordingly into either low VF or high VF NoC. In the reliability mode, the two

NoCs work in dual lock step mode where one NoC carries the actual data while the other NoC carries the error correction codes. This scheme improves the reliability by mitigating the effects of *soft errors* in the control and data path.

Architecture

Our target architecture is a general-purpose tiled multi-core chip as shown in Fig. 11.11, which closely matches commercial SoC chips such as Intel Single Chip Cloud Computer (SCC) [1] architecture. The processing core is a simple in-order core with private L1 Data and Instruction caches. Cache load and store requests are injected into the NoC by cores. Memory controllers are placed at the border routers as shown in Fig. 11.11 to serve cache load and store requests. We target an $N \times N$ mesh topology based NoC architecture as it is commonly used by commercial and academic multi-core chips. Each core runs a different application, therefore at a given time up to N^2 applications are executed in parallel.

Multiple VF Optimised NoCs

SuperNet architecture consists of two parallel NoCs that are optimised for different VF levels using *multi-vt optimisation* [19]. We assume that two voltage supplies $V1$ and $V2$ are available, where $V1 > V2$. These voltages are associated with two frequencies f and $f/2$. At each node we integrate two parallel routers as shown in Fig. 11.11. Although both routers are architecturally homogeneous, they are optimised for a VF level using *multi-vt optimisation*. One of the routers is designed for VF level $[V1,f]$ whereas the other router is designed for $[V2,f/2]$. Using these routers, two parallel NoCs are realised as follows:

- **NoC A:** This NoC consists of routers designed for $[V1,f]$. *NoC A* can be operated at either $[V1,f]$ or $[V2,f/2]$ depending on the operating mode.
- **NoC B:** This NoC consists of routers designed for $[V2,f/2]$. *NoC B* can only be operated at $[V2,f/2]$.

Description of NoC Components

Fabric Manager Fabric Manager (*FM*) is responsible for managing all the NoC components in the MPSoC. The *FM* uses the NoC channels for communication with Local Managers and reads the response through a 3-wire AND network as shown in Fig. 11.12a. The *FM* can be implemented as a specialised hardware or it can be a part of existing chip management firmware. *FM* autonomously coordinates with Local Managers in the NoC to implement different operational modes. This eases the burden of writing software routines to manage resources.

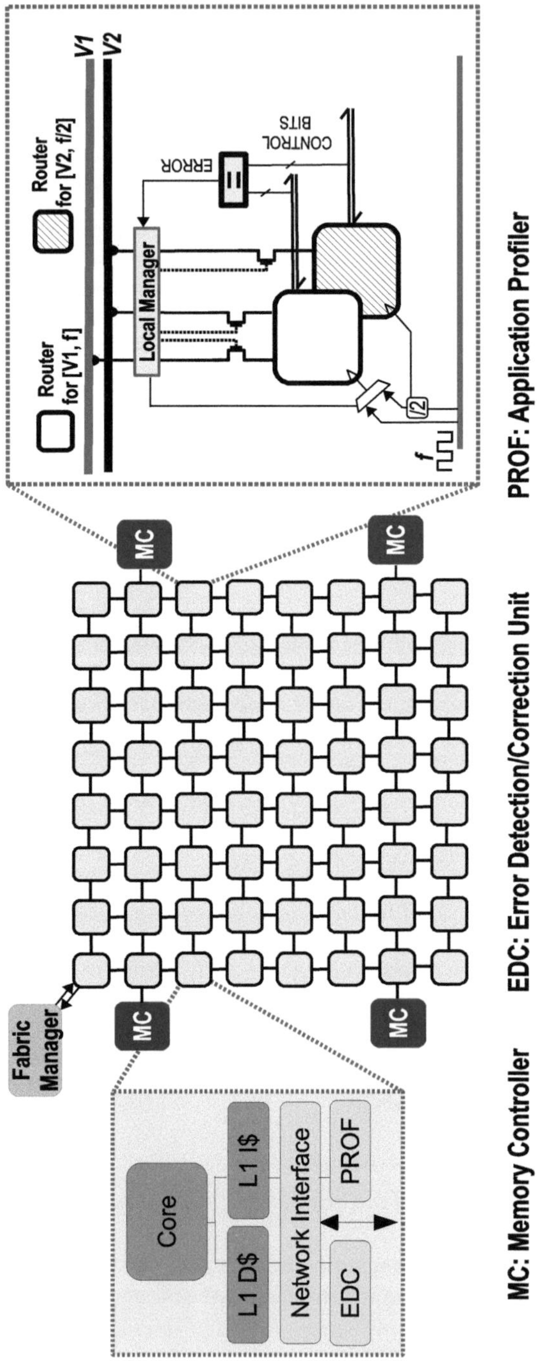

Fig. 11.11 Overview of MPSoC with *SuperNet* interconnect

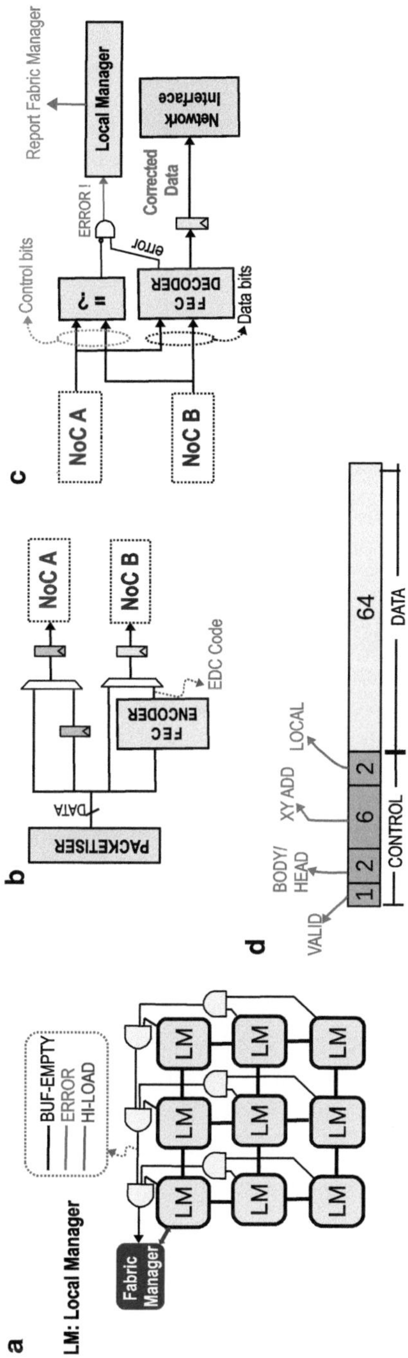

Fig. 11.12 Details of *SuperNet* components. (**a**) Communication between LM and FM. (**b**) SECDED encoding. (**c**) SECDED decoding. (**d**) Physical channel (flit)

Local Manager Local Managers (*LMs*) are responsible to manage the routers at mesh node and to forward information from the local mesh node to *FM*. *LM* drives the power gating enable and disable signals for each router. *LM* also enables and disables the traffic injection from the local core's port. *LM* monitors the buffer empty condition, application profiler status and error status from the lock step comparator (more on this in sections "Energy Efficient Mode", "Performance Mode" and "Reliability Mode"). This information is communicated back to the *FM* through a simple 3-wire AND network as shown in Fig. 11.12a. *LMs* receive their commands from *FM* through the NoC for setting up the operational mode.

Application Profiler Unit Each core contains an Application Profiler Unit. This profiler unit (marked PROF in Fig. 11.11) keeps track of vital information on application behaviour. Each profiler unit is augmented with two saturating counters, an instruction counter and an L1 cache miss counter. The profiler unit resets the counters after every preset control interval and saves the value from the previous interval. The counter information is used by *FM* to implement either the Energy Efficient Mode or the Performance Mode.

Error Detection and Correction Unit Every Network Interface (NI) is augmented with an Error Detection and Correction Unit (marked EDC in Fig. 11.11) for use in the Reliability Mode. The EDC unit is responsible for encoding the outgoing data from NI with a forward error correction (FEC) code (Fig. 11.12b), and detecting and correcting any errors in the incoming data (Fig. 11.12c). We use single error correction and double error detection (SECDED) codes for each 8-bit data, which results in a 5-bit codeword. In our architecture each flit contains 64 bits of data as shown in Fig. 11.12d. Therefore for every 64-bit flit, we generate a 40-bit codeword. Similarly on the receiver side, a decoder checks and tries to correct errors in the flit data. Therefore, through EDC unit, we can correct up to eight single bit errors (as long as they occur in separate data bytes) and detect up to 16 bit errors (maximum of two errors in each byte). To save power, these units can be power gated by *LM* in case the Reliability Mode is not activated.

Equivalence Checker Equivalence Checker (blocks with "=" sign in Figs. 11.11 and 11.12a) is used for implementing the Reliability Mode. The checker circuit asserts that the *control bits* (Fig. 11.12d) of the two NoC channels are equal. These checkers are placed at two locations: (1) each of the five outgoing router channels, (2) incoming channel of each EDC unit. In case the control parts of the channels are not equal, the checker unit raises an error signal which is read by the *LM*.

Mode Selection

When the chip is switched on, *FM* by default initialises the *SuperNet* in the normal mode (i.e., only *NoC A* is powered on with [*V*1,*f*]). Depending on various factors such as power budget and reliability requirement, the chip user can decide to activate a certain *SuperNet* mode. Therefore at runtime the user can execute the chip's firmware or OS system call to write the mode change command to the *FM*'s setting register. Once this happens, the *FM* autonomously implements the desired operational mode.

Energy Efficient Mode

We base our Energy Efficient Mode on the fact that not all applications need a low latency NoC. One important metric (to measure the dependence of applications on NoC) is cache misses per kilo instructions (MPKI) [21]. Applications with higher MPKI values spend much time waiting for the cache load miss to be served from the memory controller through the NoC. The graph for MPKIs for two applications *h264enc* and *mpeg2enc* is shown in Fig. 11.13. In our experiments we found that the MPKI value for *h264enc* is 40× the MPKI of *mpeg2enc*. Therefore, for an MPSoC with *h264enc* being executed, NoC has to be clocked at a higher frequency. On the other hand, for low MPKI applications such as *mpeg2enc*, NoC can be clocked at a lower frequency to save energy. Furthermore, the MPKI values can also fluctuate, depending on the application behaviour. Therefore, depending on an application's current MPKI, the NoC can be clocked at different frequencies.

VF Selection Scheme *SuperNet* implements a coarse-grain DVFS method with two VF levels. When $[V1,f]$ is used, *NoC A* is operational and *NoC B* is power gated. Whereas, when $[V2,f/2]$ is used, *NoC B* is used and *NoC A* is switched off. Bokhari et al. [19] showed that using NoC that is optimised for a specific VF level is more energy efficient than using a scaled VF on an NoC that has been designed for a higher VF level. MPKIs are calculated for each application using instruction and cache miss counters in the profiling unit. If the MPKI of the application falls below a pre-set *threshold* value, the profiling unit de-asserts the HI-LOAD signal, which is sensed by *LM* and broadcasted to *FM*. If all applications in the system

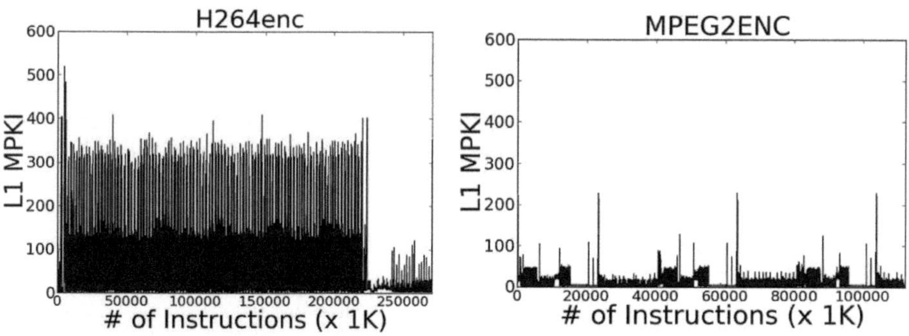

Fig. 11.13 Misses per kilo instructions (MPKI) for *h264enc* and *mpeg2enc*

de-assert the signal, *FM* decides to lower the NoC frequency to $f/2$. The value of threshold can be chosen to mark a tradeoff between performance and power. On the other hand, if any application in the system asserts HI-LOAD signal (Fig. 11.12a), the *FM* switches back to higher frequency (f). Another possible mode that can be used by *FM* is to keep the NoC running at low frequency irrespective of application characteristics (i.e., static VF selection) in case of low-power requirements, or if the performance mode has been executed for too long and the chip is too hot.

Mode Setup To start the Energy Efficient Mode, the *FM* sends control flits to each *LM* to stop any further transaction in the NoC and informs the *LM* about initiation of the mode along with the *threshold value* to be used. *FM* then waits for all existing transactions to finish by monitoring the global *BUFF-EMPTY* signal (Fig. 11.12a). At the start of the mode, *LM*s switch on the router for *NoC A*.

VF Switching Procedure Whenever the *FM* decides to switch between VF levels, the currently active NoC has to be switched off after ensuring that there are no in-flight packets in the system. *FM* starts the procedure by sending a *Stop NI* command to all *LMs* to stop any further injection of packets. The global *BUFF-EMPTY* signal (Fig. 11.12a) is raised when there are no more packets in the NoC. Then *FM* sends the next batch of control flits to *LM* to activate the required NoC and power gate the existing NoC. *FM* sends the final batch of control flits to *LMs* to resume normal operation. This concludes the VF switch-over process.

Performance Mode

The Performance Mode for *SuperNet* is based on the fact that not all packets in the NoC are critical to the performance of the application. We have classified the possible packet types in Table 11.6. The *LoadRequest* packet is generated by the core in case a memory load instruction causes a cache miss. The memory controller replies to the request by sending the required cache line through *LoadRequestReply* packet. In case a cache line is evicted, the core generates a *StoreRequest* packet for the memory controller. In response to the *StoreRequest*, the memory controller sends the *StoreRequestAck* packet as an acknowledgement. We assume an MPSoC with an in-order core and therefore the core stalls in case of a cache load miss. Thus, it is critical to deliver the *LoadRequest* and *LoadRequestReply* as quickly as possible. The process of sending the evicted cache line to the memory controller happens

Table 11.6 Packet classification

Type	Length	Critical?
Load request	1 flit	Yes
Load request reply	(cache line size)/8 + 1 flits	Yes
Store request	(cache line size)/8 + 1 flits	No
Store request ack	1 flit	No

in parallel to the execution, and therefore the *StoreRequest* and *StoreRequestAck* packets are not critical for application performance. When only one NoC is used, all packets compete for shared channel resources. Therefore due to contention, packets delivery can be delayed. This delay can possibly affect the system performance if delay for critical packets increases. The situation becomes even worse when applications have high MPKIs.

Packet Steering Based on Criticality We leverage the two NoCs available in *SuperNet* to implement the performance mode. *FM* switch on both *NoC A*(with $[V1,f]$) and *NoC B* at the same time. At runtime, the cores inject *LoadRequest* packets in the *NoC A* and *StoreRequest* in the *NoC B*. The memory controller injects the *LoadRequestReply* packets in the *NoC A* and *StoreRequestAck* packet in the *NoC B*. The intuition behind the steering scheme is to reduce the contention between critical and non-critical packets by physical separation. However, this scheme is implemented at the cost of increased NoC power compared to normal operation because two NoCs are used instead of one NoC. Moreover the performance mode is only useful for applications where network load is high due to a high application MPKI. Therefore, *FM* can enable or disable the performance mode based on the power budget and application characteristics.

Mode Setup To start the Performance Mode, the *FM* sends control flits to each *LM* to stop any further transaction in the NoC and informs the *LM* about initiation of the mode. *FM* then waits for all the existing transactions to finish by monitoring the global *BUFF-EMPTY* signal (Fig. 11.12a). In parallel, *LM* also switches on *NoC A* and *NoC B* routers in case one of them was not already switched on. *FM* sends control flits to each *LM* in response to which NI is enabled and configured to steer traffic into a specific NoC based on traffic classification.

Reliability Mode

The Reliability Mode for *SuperNet* enables the detection and correction of soft errors in the NoC. The scheme can provide protection against soft errors in both the data path and control path. In this mode both NoCs are used at the same time. Reliability Mode uses an *NoC A* and *NoC B* in parallel by operating them at $[V2, f/2]$.

Data Path Protection The router's data path is protected through the use of strong SECDED codes in an FEC mode. The EDC unit is used to generate a (13,8) BCH code (8-bit data generating 5-bit code). In *SuperNet* we only encode the data at the time of packet injection and only try to check and correct the data at the receiver node. A more error resilient approach is to perform error encoding and decoding at each router in the path of the packet. However this scheme would increase the energy cost due to additional hardware at each router. Moreover, per-router encoding and decoding will increase the router's pipeline stages, which will eventually result in an increase in network latency and hence performance penalty. The actual packets are injected in *NoC A* and the error correction codes are injected in *NoC B* simultaneously as shown in Fig. 11.12b.

Control Path Protection In *SuperNet*, we employ the idea of dual modular redundancy (DMR) and lock step processing to protect the router control path against the soft errors. Both NoCs are architecturally homogenous and are operated at the same frequency. Furthermore, during the setup of Reliability Mode, switch allocators for all the routers are reset to a known state. Therefore, routers from different NoCs at a given mesh node are expected to generate the same output. Each NoC channel consists of control bits and data bits as shown in Fig. 11.12d. We perform equivalence check on the control bit output on outgoing router channels in a given direction as shown in Figs. 11.11 and 11.12c. Any soft error in the control circuit in either of the routers will result in in-correct data written on a port or no output at all. For example, an error in routing logic will route the flit to the wrong port. If the equivalence check fails, an error is raised for *LM*.

Mode Setup To start reliability mode, the *FM* sends control flits to each *LM* to stop any further transaction in the NoC. The *FM* then waits for all the existing transactions to finish by monitoring the global *BUFF-EMPTY* signal (Fig. 11.12a). Then, *FM* sends control flits to *LM* to switch on the routers for *NoC A* (at VF $[V2, f/2]$) and *NoC B*. The last set of control flits are sent to *LM* to reset the router switch allocators to a known state and continue normal operation.

Runtime Fault Recovery At runtime, whenever a non-correctable error occurs in the data path or lock step router execution fails, an error is raised by the *LM* for *FM* as shown in Fig. 11.12c. In response to the error, *FM* sends control flits to each *LM* to stop any new transaction from the NI. Once the network is free of any traffic, *FM* sends control flits to *LM*s to reset the switch arbiter to a known state. *FM* sends the last set of control flits to *LM*s to continue normal operation with these conditions:

- If the core has sent a *LoadRequest* packet and the memory controller has not sent a reply through *LoadRequestReply* packet, the local NI again sends the *LoadRequest* (Table 11.6).
- If the core has sent a *StoreRequest* packet and the memory controller has not sent a reply through *StoreRequestAck* packet, the *StoreRequest* packet is sent again.

Performance Overhead The main performance overhead of *SuperNet*'s Reliability Mode is due to the fact that NoC is operated at a reduced frequency of $f/2$ which causes an increase in the network latency. The performance penalty of increased NoC latency will be more significant for applications that have high MPKIs. Furthermore, a performance overhead is incurred when there is a non-correctable fault (data path or control) which requires retransmission from source.

Experiments and Results

Experiment Setup

NoC Synthesis For experiments, we use a two-stage pipelined wormhole switched router with 64-bit data channel width. Architectural details of the NoC router used in our study are given in Table 11.7. In the first stage, flits travel on links and get written in the input buffer. The second stage consists of route computation, switch arbitration and switch transversal. The router description is written at the RTL (Verilog). We synthesised the RTL using Synopsys Design Compiler version *H-2013.03-SP4*. We used commercial TSMC 45 nm libraries that are characterised for LVt, NVt and HVt. We enable leakage and dynamic power optimisation which automatically invokes multi-Vt optimisation. We target two VF levels for the router design for experiments, i.e., [1 GHz, 0.9 V] and [500 MHz, 0.81 V]. We report the synthesis results for NoC in Table 11.8. We also synthesised the error detection and correction hardware, results of which are shown in Table 11.9.

MPSoC Details We target a 64-core general-purpose MPSoC organised in a 8×8 2D-mesh. Architectural details for the MPSoC are included in Table 11.10. The processors are an in-order Tensilica Xtensa core with private L1 instruction and data caches. The core stalls when there is an outstanding cache load miss. The frequency for the cores is fixed at 1GHz. For the baseline system, we assume that there are four memory controllers placed at the boundaries as shown in Fig. 11.11. To stress the

Table 11.7 NoC architectural details

Topology	8×8 mesh
Router architecture	5 port router (4 neighbours + 1 local), 2 stage pipeline [LT+BW, RC+SA+ST]
Input buffer	8-flit deep (no virtual channel)
Routing	Dimension order XY
Link flow control	Wormhole switching with On/Off flow control
Link width	64 bit data
Switch arbiter	Matrix arbiter

Table 11.8 Synthesis results for a single router using multi-Vt optimisation

Frequency	1 GHz	500 MHz
Voltage	0.9 V	0.81 V
Area [μm^2]	33,041	28,976
HVt Cells [%]	45	82
NVt Cells [%]	7	6
LVt Cells [%]	48	12

Table 11.9 Synthesis results for error detection and correction units

	ECC decode	ECC encode	Equivalence
Area [μm^2]	1814	760	46

Table 11.10 MPSoC configuration

Topology	8 × 8 mesh
Processors	Tensilica in-order LX4 @ 1 GHz
L1 I/D caches	16 KB, 2-way set associative, 16 byte line size
Memory controller	4 memory controllers with perfect L2 cache
NoC VF levels	[1 GHz, 0.9 V], [500 MHz, 0.81 V]

NoC, we assume that the memory controllers are augmented with perfect L2 caches. This technique has been used previously for testing NoC architectures [24, 25].

Benchmark Applications We use a diverse set of benchmark applications from MediaBench and SPEC2006 packages to evaluate our proposed scheme. We created multi-programmed workloads using the applications from these two suites by running a copy of the selected application on each core. Each core executes 50 million instructions on average for experimentation.

Simulator We use an in-house cycle-accurate simulator for our study. The simulation is carried out in two steps: First, applications are executed on the Xtensa instruction set simulator which generates time-stamped cache load/store miss traces. In the full system simulation, these memory traces are replayed in a closed-loop simulator which accurately models the NoC latency effect on application execution. Our simulator accurately models the low-level hardware details for NoC VF switching, operation of *LM* and *FM*, and memory controller queuing delays.

VF Switching Setup We explore designs with dual voltage coarse-grain VF selection scheme. Two possible VF levels for NoC are [1 GHz, 0.9 V] and [500 MHz, 0.81 V]. For Energy Efficient Mode, we used a control interval of 50K clock cycles for the VF selection algorithm based on analysis presented by Chen et al. [15]. For power characterisation of routers, we passed the netlist generated by synthesis tool to Synopsys PrimeTime. The netlist generated for [1 GHz, 0.9 V] router was evaluated at [1 GHz, 0.9 V] and [500 MHz, 0.81 V], while the netlist for [500 MHz, 0.81 V] router was evaluated at the design VF only. The energy values extracted from PrimeTime tool are fed into the simulator to calculate the NoC energy.

Experimental Results

Performance Mode The results for experiments with the performance mode are shown in Fig. 11.14. We report the values for *NoC Power* and *Average IPC*. All the results are normalised to a system where only [1 GHz, 0.9 V] NoC is used. The performance mode for *SuperNet* is clearly useful for applications with high MPKI. For the application *h264enc*, the average IPC is improved by 17 %. Similarly for another application *bzip2*, the average IPC is improved by 14 %. This improvement comes at a cost of increase in NoC power by 13 % and 11 % for *h264enc* and *bzip2*, respectively. A small improvement of 3 % is observed for applications *lbm* and

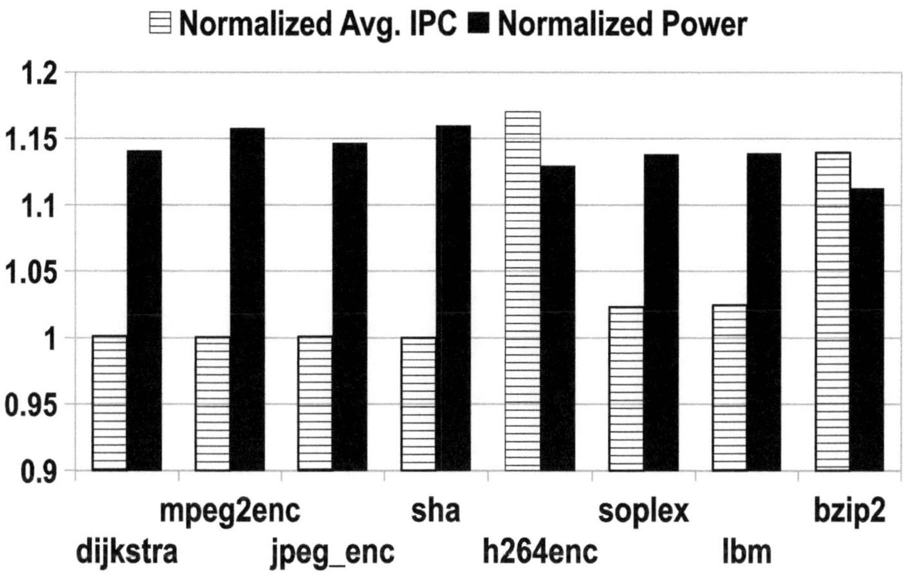

Fig. 11.14 Results for performance mode

soplex at a cost of 14 % increase in NoC power. No improvement was observed for other applications (i.e., *mpeg2enc*, *dijkstra*, *jpeg_enc* and *sha*) as the MPKIs for these applications are moderate to low.

Based on these results we advocate the judicious use of *SuperNet*'s Performance Mode for applications with very high MPKIs. Therefore *FM* can be enhanced to use the already existing application profiling information to selectively switch on the performance mode for applications with high MPKIs.

Energy Efficient Mode The results for experiments with the energy efficient mode are shown in Fig. 11.15. We report the values for the *NoC Power* and the *Average IPC* for all the cores. All results are normalised to a system where only [1 GHz, 0.9 V] NoC is used with a fixed VF level. For comparison, we also report results for architecture where a single VF-optimised NoC is used with DVFS. From Fig. 11.15, it is evident that using multiple multi-vt optimised NoCs provide better power savings than using DVFS with a single NoC. We performed a parameter sweep on the values of *threshold* for the VF selection routine. From the results in Fig. 11.15, a *threshold* value of 20 gives the best results across all benchmarks. For example, by using energy efficient mode of *SuperNet* we are able to save NoC power by 75 % for *mpeg2enc*, 65 % for *jpeg_enc* and 81.5 % for *sha* applications. The power saving comes at a cost of a small decrease in *IPC* by 0.9 %, 6 % and 0.1 % for *mpeg2enc*, *jpeg_enc* and *sha* applications, respectively. This is because NoC is sometimes running at a lower frequency, which increases the application execution time. Note that these three applications have low MPKIs. Even with applications with high MPKIs such as *dijkstra*, *lbm* and *bzip2*, the NoC power is saved by 25 %,

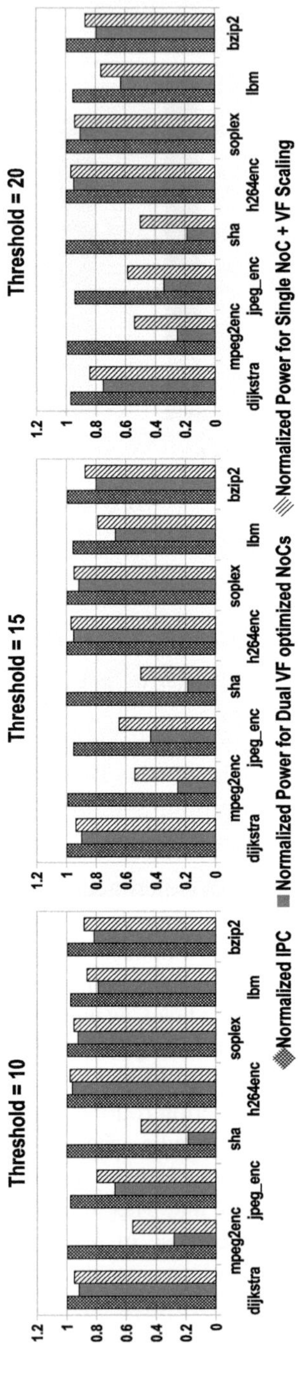

Fig. 11.15 Results for energy efficient mode with different thresholds

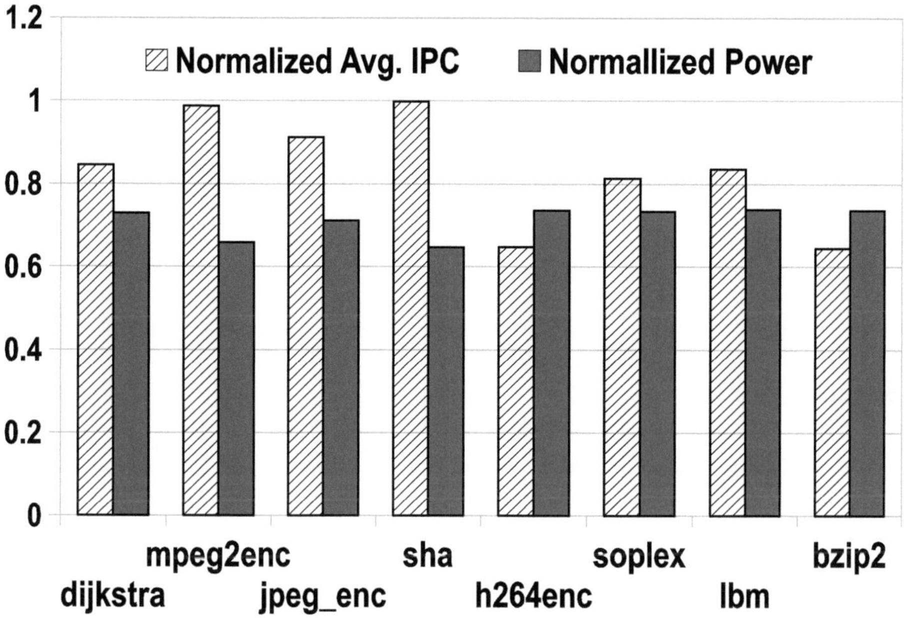

Fig. 11.16 Results for reliability mode

37 % and 21 %, respectively. This NoC power saving comes at a cost of reduction of *IPC* by 3 %, 4 % and 0.6 % for *dijkstra*, *lbm* and *bzip*2 applications, respectively. A small saving in NoC power of 5 % and 9 % is observed for applications *h*264*enc* and *soplex*, respectively.

It is very important to select an appropriate value for *threshold*. We can observe from Fig. 11.15 that choosing a low *threshold* value results in low-power savings, whereas choosing a high *threshold* value results in low-power savings and considerable degradation in *IPC*. Therefore, it is expected that designers will profile the applications of interest to choose a suitable *threshold* value.

Reliability Mode The results for experiments with the reliability mode are shown in Fig. 11.16. We report the values for *NoC Power* and *Average IPC*. All results are normalised to a system where only [1 GHz, 0.9 V] NoC is used without activating any reliability features. Running *SuperNet* in reliability mode can cause some degradation in chip performance for applications with moderate to high MPKIs. For example, for applications *mpeg*2*enc*, *jpeg_enc* and *sha*, the *IPC* is reduced by 1.3 %, 9 % and 0.3 %, respectively. On the other hand, the performance of high MPKI applications *h*264 and *bzip*2 is strongly affected by the reliability mode, as the *IPC* is dropped by 35 % for both applications. The reason for the drop in performance is the fact that the NoC is operated at a lower frequency of 500 MHz. Overall, the relative NoC power is low due to two reasons: (1) reduced VF level, and (2) longer execution times.

Discussion

In our experiments, the *SuperNet* architecture is operated exclusively in one of three possible modes. One way to reduce the performance overheard of the reliability mode would be to virtually partition the *SuperNet* into reliable and non-reliable regions [26] and map only applications with high susceptibility [27] to errors in the reliable NoC region. Similarly, regions where reliability mode is not activated can be operated in either the performance mode or the energy efficient mode depending on power budget and application characteristics.

Related Work

Multi-Vt Optimisation of NoCs Pullini et al. [28] investigated the advantages of using multi-Vt circuit optimisation in terms of performance and power for a single layer NoC architecture. In contrast, we exploit the multi-Vt circuit optimisation in the context of multiple network layers and DVFS. Various NoCs with multiple network layers have been proposed in [21, 29, 30]. However, all of these works customised the architecture of the network layers or the routers, in contrast to the work described in this chapter, where we integrate architecturally homogenous yet VF-optimised network layers. Our work is inspired by Allred et al. [31] and Chakraborty and Roy [32], where the authors exploited multi-Vt circuit optimisation for designing processors in the context of dark silicon. However, one cannot infer from their work that multi-Vt circuit optimisation will be beneficial for NoCs, and what should be the architecture of an NoC with multiple VF-optimised network layers.

Multiple NoCs There are several commercial and academic chips that leverage multiple NoCs. Tilera's iMesh architecture [30] uses five parallel NoCs for different class of traffic, i.e., cache coherence messages, memory controller access, I/O port access, and user space message passing. TRIPS [29] architecture uses multiple networks for operand and data forwarding. The reason for using multiple NoCs in TRIPS and Tilera iMesh architecture is to provide isolation between different classes of traffic. Balfour and Dally [33] showed that adding parallel NoCs can potentially reduce the network latency and improves the bandwidth. Mishra et al. [21] extended the idea by using two heterogeneous networks, where one network is used for bandwidth sensitive applications and another network is used for latency sensitive applications. In contrast, in performance mode for *SuperNet*, we divide the traffic from all the applications based on packet type instead of the application to which the traffic belongs. In addition to the performance improvement, *SuperNet* can also be used to provide reliability or energy efficiency by using the same set of NoCs.

System-Level Power Saving Techniques Over the years various system-level power management techniques have been introduced. DVFS for on-chip networks has been proposed at various granularities—such as link-level [34], router-level

[3, 4] and region-level [15, 35]. However the advantage of the DVFS scheme has been reducing due to NoC's high leakage power in smaller nodes and diminishing room for reducing supply voltages [6]. To improve the efficiency of DVFS schemes, in this chapter, we propose integrating multiple NoCs, where each NoC is optimised using multi-vt circuit technique. At runtime, along with changing VF levels, a different NoC is activated. However we also use these multiple NoCs to improve system performance and reliability. Run-time power gating has also been proposed to reduce leakage power of NoCs. Matsutani et al. [36] proposed an ultra fine-grained power-gating technique for routers. Similar techniques at other granularities have been explored in [37–39]. All of these techniques are orthogonal to our work and can be tweaked to exploit the provision of multiple network layers in an NoC.

Fault Tolerance in NoC Many researchers have studied the topic of soft error detection and correction for NoCs.[1] The soft errors are caused by random logic flips on the input/output of a logic gate. This leads to temporary glitches in an NoC router's data path and/or control path [22]. BulletProof router [40] uses a Cyclic Redundancy Check (CRC) to detect and correct the soft errors in data path, and selective triple modular redundancy (TMR) for detecting faults in the control paths of the routers. Prodromou et al. [20] proposed a scheme based on hardware invariance checking to detect faults in router control path. Park et al. [41] explored the use of hop-by-hop error detection and correction by using SECDED codes. In the case of errors which were not correctable, flits are retransmitted from downstream routers. Commercial NoC vendor, Arteris, has also introduced an error resilience solution called FlexNoC Resilience Package [26] that combines end-to-end error data correction and port checking. Murali et al. [42] analysed different error detection and correction schemes and concluded that end-to-end fault detection, correction and retransmission based on time-out is a good tradeoff between performance, energy and area overhead. None of the techniques discussed above studied the soft error resilience in the context of reusing the existing multiple NoCs. In contrast to the previous solutions, we use the additional NoC to transmit a strong byte-based SECDED code for end-to-end error detection and correction, and use the control path of the two NoCs for lock step dual modulo redundancy (DMR) protection in the control path.

Conclusion

In this chapter, we explored two different NoC architectures that exploit dark silicon to improve power, performance and reliability metrics. Our proposed architectures are based on observation that use of multi-vt optimisation during circuit synthesis can result in strikingly different energy efficiency for different VF levels.

---

[1]We are limiting our discussion to mitigating soft errors. An overview of fault tolerance schemes for NoCs is presented by Radetzki et al. [22].

First, an NoC architecture called *darkNoC* is introduced. darkNoC consists of multiple NoCs, where each NoC is optimised at design time using multi-vt optimisation for specific VF level. The main purpose of the architecture is to improve the efficiency of the DVFS scheme. We show that instead of applying DVFS to a router that has been designed to operate at a higher VF level, using a router that has been designed for lower VF level is more energy efficient. Based on this observation, we propose using dark silicon to integrate multiple NoCs, and propose a scheme to switch between these NoCs in parallel to switching between different VF levels. The switch-over mechanism is developed using a combination of a global *Region Manager (dRM)* and local *Local Managers (dLMs)* which coordinate autonomously for seamless transition between NoC planes. Through experiments, we show that the switch-over process can take about 400–1000 ns for an 8×8 mesh depending on various factors. Our experiments with real application show that the darkNoC architecture can provide up to 52 % saving in NoC EDP for certain benchmarks, whereas, a state-of-the-art DVFS scheme only saved 15 % EDP. This shows the efficacy of our proposed architecture. We also discuss the affect of application characteristics, dark silicon budget available and DVFS schemes on overall NoC EDP savings.

We then presented another dark silicon inspired NoC architecture, *SuperNet*, that optimises energy, performance and reliability of on-chip interconnect in exchange for dark silicon. We are motivated by the fact that chip design cost is rising and it is therefore becoming normal practice to design SoCs for multiple use scenarios. *SuperNet* consists of multiple parallel NoC planes that are optimised for VF levels $[V1,f]$ and $[V2,f/2]$. At a given time, SuperNet can be configured to run in energy efficient mode, reliability mode or performance mode, depending on available power and optimisation goals. In energy efficient mode, depending on the MPKI values of all the cores, the DVFS manager decides to run NoC at a low or high VF level and, depending on the VF level, the NoC designed for that VF is switched on while the other NoCs remain switched off. The performance mode is based on the fact that not all network traffic is critical for an application's performance. To avoid interference between critical and non-critical traffic, the data packets associated with cache load operations are injected into the high VF NoC, and data packets associated with cache miss operation are injected into the low VF NoC. The reliability mode defends against soft error in both data path elements and the control circuit of NoCs. Both NoCs are operated at the lower VF level. The reliability mode provides protection against soft errors in the router's data path through byte oriented SECDED codes that can correct up to 8-bit errors and detect up to 16-bit errors in a 64-bit flit, whereas the router's control path is protected through DMR lock step execution.

The research work presented in this chapter can be extended further. In the *darkNoC* architecture, the routers used in all NoCs are architecturally homogenous. However, it will be interesting to study the effect of heterogeneous router microarchitecture. Some of the router parameters can be changed, such as channel width, pipeline stages, complexity of switch allocators and buffer sizing. Heterogeneous NoC architectures can potentially provide better control over the dark silicon area and power saving tradeoff. In *SuperNet* architecture, the NoC is exclusively operated in one of the three possible modes. However, *SuperNet* is can be easily modified to operate in multiple modes at the same time by dividing the physical fabric into multiple logical networks and operating each logical network in modes chosen independently. For example, half of the chip can be used to map applications that are safety-critical whereas the other half can be used to run normal applications.

References

1. P. Salihundam, S. Jain, T. Jacob, S. Kumar, V. Erraguntla, Y. Hoskote, S.R. Vangal, G. Ruhl, N. Borkar, A 2 tb/s 6 × 4 mesh network for a single-chip cloud computer with DVFS in 45 nm cmos. J. Solid-State Circuits **46**(4), 757–766 (2011)
2. J.S. Kim, M.B. Taylor, J. Miller, D. Wentzlaff, Energy characterization of a tiled architecture processor with on-chip networks, in *Proceedings of the 2003 International Symposium on Low power Electronics and Design, ISLPED '03* (ACM, New York, 2003), pp. 424–427
3. P. Bogdan, R. Marculescu, S. Jain, R. Gavila, An optimal control approach to power management for multi-voltage and frequency islands multiprocessor platforms under highly variable workloads, in *2012 Sixth IEEE/ACM International Symposium on Networks on Chip (NoCS)* (2012), pp. 35–42
4. A. Mishra, R. Das, S. Eachempati, R. Iyer, N. Vijaykrishnan, C. Das, A case for dynamic frequency tuning in on-chip networks, in *42nd Annual IEEE/ACM International Symposium on Microarchitecture, 2009. MICRO-42* (2009), pp. 292–303
5. K. Swaminathan, E. Kultursay, V. Saripalli, V. Narayanan, M. Kandemir, S. Datta, Steep slope devices: from dark to dim silicon. IEEE Micro **99**(PrePrints) 1 (2013)
6. S. Garg, D. Marculescu, R. Marculescu, U. Ogras, Technology-driven limits on DVFS controllability of multiple voltage-frequency island designs: a system-level perspective, in *46th ACM/IEEE Design Automation Conference, 2009. DAC '09* (2009), pp. 818–821
7. T. Karnik, Y. Ye, J. Tschanz, L. Wei, S. Burns, V. Govindarajulu, V. De, S. Borkar, Total power optimization by simultaneous dual-vt allocation and device sizing in high performance microprocessors, in *2002 Proceedings of 39th Design Automation Conference* (2002), pp. 486–491
8. H. Esmaeilzadeh, E. Blem, R.St. Amant, K. Sankaralingam, D. Burger, Dark silicon and the end of multicore scaling, in *2011 38th Annual International Symposium on Computer Architecture (ISCA)* (2011), pp. 365–376
9. M. Taylor, A landscape of the new dark silicon design regime. IEEE Micro **33**, 1–1 (2013)
10. N. Goulding-Hotta, J. Sampson, G. Venkatesh, S. Garcia, J. Auricchio, P. Huang, M. Arora, S. Nath, V. Bhatt, J. Babb, S. Swanson, M. Taylor, The GreenDroid mobile application processor: an architecture for silicon's dark future. IEEE Micro **31**(2), 86–95 (2011)
11. Y. Turakhia, B. Raghunathan, S. Garg, D. Marculescu, Hades: architectural synthesis for heterogeneous dark silicon chip multi-processors, in *Proceedings of the 50th Annual Design Automation Conference, DAC '13* (ACM, New York, 2013), pp. 173:1–173:7
12. M.B. Taylor, Is dark silicon useful?: Harnessing the four horsemen of the coming dark silicon apocalypse, in *Proceedings of the 49th Annual Design Automation Conference, DAC '12* (ACM, New York, 2012), pp. 1131–1136
13. A. Strano, D. Ludovici, D. Bertozzi, A library of dual-clock FIFOs for cost-effective and flexible MPSoC design, in *2010 International Conference on Embedded Computer Systems (SAMOS)* (2010), pp. 20–27
14. S. Ma, N.E. Jerger, Z. Wang, DBAR: an efficient routing algorithm to support multiple concurrent applications in networks-on-chip, in *Proceedings of the 38th Annual International Symposium on Computer Architecture, ISCA '11* (ACM, New York, 2011), pp. 413–424

15. X. Chen, Z. Xu, H. Kim, P.V. Gratz, J. Hu, M. Kishinevsky, U. Ogras, R. Ayoub, Dynamic voltage and frequency scaling for shared resources in multicore processor designs, in *Proceedings of the 50th Annual Design Automation Conference, DAC '13* (ACM, New York, 2013), pp. 114:1–114:7
16. Synopsys, Design Solutions for 20 nm and Beyond, White Paper (2013), http://www.synopsys.com/
17. J. Warnock, Circuit design challenges at the 14nm technology node, in *Proceedings of the 48th Design Automation Conference, DAC '11* (ACM, New York, 2011), pp. 464–467
18. Semico Research - How Reducing Verification and Validation Time Can Improve Profitability (2014), http://www.semico.com/
19. H. Bokhari, H. Javaid, M. Shafique, J. Henkel, S. Parameswaran, darkNoC: designing energy-efficient network-on-chip with multi-vt cells for dark silicon, in *Proceedings of the 51st Annual Design Automation Conference on Design Automation Conference, DAC '14* (ACM, New York, 2014), pp. 161:1–161:6
20. A. Prodromou, A. Panteli, C. Nicopoulos, Y. Sazeides, NoCAlert: an on-line and real-time fault detection mechanism for network-on-chip architectures, in *2012 45th Annual IEEE/ACM International Symposium on Microarchitecture (MICRO)* (2012), pp. 60–71
21. A.K. Mishra, O. Mutlu, C.R. Das, A heterogeneous multiple network-on-chip design: an application-aware approach, in *Proceedings of the 50th Annual Design Automation Conference, DAC '13* (ACM, New York, 2013), pp. 36:1–36:10
22. M. Radetzki, C. Feng, X. Zhao, A. Jantsch, Methods for fault tolerance in networks-on-chip. ACM Comput. Surv. **46**, 8:1–8:38 (2013)
23. M. Shafique, S. Garg, T. Mitra, S. Parameswaran, J. Henkel, Dark silicon as a challenge for hardware/software co-design: invited special session paper, in *Proceedings of the 2014 International Conference on Hardware/Software Codesign and System Synthesis, CODES '14* (ACM, New York, 2014), pp. 13:1–13:10
24. T. Moscibroda, O. Mutlu, A case for bufferless routing in on-chip networks. SIGARCH Comput. Archit. News **37**, 196–207 (2009)
25. C. Fallin, C. Craik, O. Mutlu, Chipper: a low-complexity bufferless deflection router, in *2011 IEEE 17th International Symposium on High Performance Computer Architecture (HPCA)* (2011), pp. 144–155
26. Arteris Resilience Package, http://www.arteris.com/flexnoc-resilience-package
27. T. Li, M. Shafique, J.A. Ambrose, S. Rehman, J. Henkel, S. Parameswaran, Raster: runtime adaptive spatial/temporal error resiliency for embedded processors, in *Proceedings of the 50th Annual Design Automation Conference, DAC '13* (ACM, New York, 2013), pp. 62:1–62:7
28. A. Pullini, F. Angiolini, S. Murali, D. Atienza, G. De Micheli, L. Benini, Bringing NoCs to 65 nm. IEEE Micro **27**(5), 75–85 (2007)
29. K. Sankaralingam, R. Nagarajan, H. Liu, C. Kim, J. Huh, D. Burger, S. Keckler, C. Moore, Exploiting ILP, TLP, and DLP with the polymorphous trips architecture. IEEE Micro **23**(6), 46–51 (2003)
30. D. Wentzlaff, P. Griffin, H. Hoffmann, L. Bao, B. Edwards, C. Ramey, M. Mattina, C.-C. Miao, J.F. Brown III, A. Agarwal, On-chip interconnection architecture of the tile processor. IEEE Micro **27**, 15–31 (2007)
31. J. Allred, S. Roy, K. Chakraborty, Dark silicon aware multicore systems: employing design automation with architectural insight. IEEE Trans. Very Large Scale Integr. (VLSI) Syst. **22**, 1–1 (2013)
32. K. Chakraborty, S. Roy, Architecturally homogeneous power-performance heterogeneous multicore systems. IEEE Trans. Very Large Scale Integr. (VLSI) Syst. **21**(4), 670–679 (2013)
33. J. Balfour, W.J. Dally, Design tradeoffs for tiled CMP on-chip networks, in *Proceedings of the 20th Annual International Conference on Supercomputing, ICS '06* (ACM, New York, 2006), pp. 187–198
34. L. Shang, L.-S. Peh, N. Jha, Dynamic voltage scaling with links for power optimization of interconnection networks, in *Proceedings of the Ninth International Symposium on High-Performance Computer Architecture, 2003. HPCA-9 2003* (2003), pp. 91–102

35. G. Liang, A. Jantsch, Adaptive power management for the on-chip communication network, in *9th EUROMICRO Conference on Digital System Design: Architectures, Methods and Tools, DSD 2006* (2006), pp. 649–656
36. H. Matsutani, M. Koibuchi, D. Ikebuchi, K. Usami, H. Nakamura, H. Amano, Performance, area, and power evaluations of ultrafine-grained run-time power-gating routers for CMPs. IEEE Trans. Comput. Aided Des. Integr. Circuits Syst. **30**(4), 520–533 (2011)
37. G. Kim, J. Kim, S. Yoo, Flexibuffer: reducing leakage power in on-chip network routers, in *2011 48th ACM/EDAC/IEEE Design Automation Conference (DAC)* (2011), pp. 936–941
38. A. Samih, R. Wang, A. Krishna, C. Maciocco, C. Tai, Y. Solihin, Energy-efficient interconnect via router parking, in *2013 IEEE 19th International Symposium on High Performance Computer Architecture (HPCA2013)* (2013), pp. 508–519
39. R. Das, S. Narayanasamy, S.K. Satpathy, R.G. Dreslinski, Catnap: energy proportional multiple network-on-chip, in *Proceedings of the 40th Annual International Symposium on Computer Architecture*, ISCA '13 (ACM, New York, 2013), pp. 320–331
40. K. Constantinides, S. Plaza, J. Blome, B. Zhang, V. Bertacco, S. Mahlke, T. Austin, M. Orshansky, Bulletproof: a defect-tolerant CMP switch architecture, in *The Twelfth International Symposium on High-Performance Computer Architecture* (2006), pp. 5–16
41. D. Park, C. Nicopoulos, J. Kim, N. Vijaykrishnan, C. Das, Exploring fault-tolerant network-on-chip architectures, in *International Conference on Dependable Systems and Networks, 2006. DSN 2006* (2006), pp. 93–104
42. S. Murali, T. Theocharides, N. Vijaykrishnan, M.J. Irwin, L. Benini, G.D. Micheli, Analysis of error recovery schemes for networks on chips. IEEE Des. Test Comput. **22**(5), 434–442 (2005)

Chapter 12
NoC-Aware Computational Sprinting

Jia Zhan and Yuan Xie

Introduction

The continuation of technology scaling leads to a *utilization wall* challenge [24]: to maintain a constant power envelope, the fraction of a silicon chip that can be operated at full frequency is dropping exponentially with each generation of process technology. Consequently, a large portion of silicon chips will become dark or dim silicon, i.e., either idle or significantly under-clocked. However, most previous work focuses on energy-efficient core/cache design while the impact of on-chip interconnect is neglected. In fact, Network-on-chip (NoC) plays a vital role in message passing and memory access that directly influences the overall performance of many-core processors. Recent studies show that NoCs can draw a substantial percentage of a chip's power, by up to 10–36 % [9, 17, 23, 25]. Therefore, how to design the interconnection network is critical to tackle the challenges of multicore scaling in the dark silicon age.

Recently, Raghavan et al. [20] proposed *computational sprinting*, in which a chip improves its responsiveness to short burst of computations through temporarily exceeding its sustainable thermal design power (TDP) budget. All the cores will be operated at the highest frequency/voltage to provide instant throughput during sprinting, and after that the chip must return to the single-core nominal operation to cool down. While such mechanism sheds light upon how "dark" cores can be utilized for transient performance enhancement, it exposes two major design issues: First, the role of interconnect is neglected. NoCs consume a significant portion of chip power when all cores are in sprinting mode. When switching back to the nominal mode, only a single core is active. However, the network routers and links cannot be completely powered down, otherwise a gated-off node would block

J. Zhan (✉) • Y. Xie
University of California, Santa Barbara, CA, USA
e-mail: jzhan@ece.ucsb.edu; yuanxie@ece.ucsb.edu

A.M. Rahmani et al. (eds.), *The Dark Side of Silicon*,
DOI 10.1007/978-3-319-31596-6_12

packet forwarding and the access of the local but shared resources (e.g., cache and directory). As a result, the ratio of network power over chip power rises substantially and may even lead to higher NoC power than that of the single active core. Second, the mode-switching lacks flexibility and only provides two options: nominal single-core operation and maximum all-core sprinting. Depending on the workload characteristics, an intermediate number of active cores may provide the optimal performance speedup with less power dissipation. Also, potentially the sprinting process can last longer with a less number of cores turned on simultaneously. To address these two issues, we propose *fine-grained sprinting*, in which the chip can selectively sprint to any intermediate stages instead of directly activating all the cores in response to short-burst computations. The optimum number of cores to be selected depends on the application characteristics. Scalable applications may opt to a large number of cores that can support highly parallel computation, whereas other applications may mostly consist of sequential programs and would rather execute on a small number of cores. Apparently, *fine-grained sprinting* can flexibly adapt to a variety of workloads. In addition, landing on intermediate sprinting stages can save chip power and slow down the heating process by power gating the remaining inactive on-chip resources, which are capable of sustaining longer sprint duration for better system performance.

Fine-grained sprinting opens up an opportunity to better utilize the on-chip resources for power-efficiency, but it also poses challenges on designing the interconnect backbone. Inherently it incurs three major concerns: *(1) how to form the topology which connects the selected number of cores during sprinting when dark cores and active cores co-exist? (2) how to construct a thermal-aware floorplan of the on-chip resources (cores, caches, routers, etc.) for such sprinting-based multicores? and (3) what would be an appropriate NoC power-management scheme?* To answer these questions, we propose a topological sprinting mechanism with deadlock-free routing support, and a fast heuristic floorplanning algorithm to address the thermal problem during sprinting. Moreover, this sprinting scheme naturally enables power gating on network resources in the dark silicon region.

In summary, we propose **NoC-sprinting**, which provides topological/routing support for fine-grained sprinting and employs network power-gating techniques for combating dark silicon. Overall, this work makes the following contributions:

- Explores challenges and opportunities of designing NoC in the dark silicon era, from the perspectives of both performance and power.
- Investigates the pitfalls of the conventional all-core sprinting which fails to fulfill diverse workload characteristics, and proposes *fine-grained computational sprinting* for better power-efficiency.
- Proposes NoC support to enable fine-grained computational sprinting, including topology construction, routing, floorplanning, and power management.
- Conducts thermal analysis of different computational sprinting schemes and shows how *NoC-sprinting* improves with the sprint duration.

The rest of this chapter is organized as follows: Section "Challenges and Opportunities" shows some background of dark silicon and computational sprinting, and identifies major challenges in designing on-chip interconnect in the dark silicon age. Section "Our Method: NoC-Sprinting" introduces NoC-sprinting which provides NoC support for fine-grained computational sprinting to tackle those challenges. A comprehensive experimental evaluation is conducted in section "Architectural Evaluation" to validate the proposed NoC-sprinting scheme. Finally, section "Conclusion" concludes the paper.

Challenges and Opportunities

In this section, we briefly introduce the background of dark silicon, and how computational sprinting can mitigate this problem. Then we show the remaining challenges of designing on-chip network in the dark silicon age.

Dark Silicon and Computational Sprinting

Conventionally in multicore scaling, the power gain due to the increase of transistor count and speed can be offset by the scaling of supply voltage and transistor capacitance. However, in today's deep sub-micron technology, leakage power depletes the power budget. We cannot scale threshold voltage without exponentially increasing leakage. Consequently, we have to hold a constant supply voltage, and hence produce a shortfall of energy budget to power a chip at its full frequency. This gap accumulates through each generation and results in an exponential increase of inactive chip resources—dark silicon. For example, prediction from ARM [18] says only 9 % of the transistors may be able to be activated by 2020.

Instead of shrinking the chip or sacrificing transistor density, *computational sprinting* [20] embraces dark silicon by leveraging the extra transistors transiently when performance really counts. To improve the utilization of the dark silicon, it allows a chip to temporarily exceed its TDP and activate all the cores simultaneously to provide instantaneous throughput for short-burst parallel computation, after which it returns to the nominal single-core operation. Special phase change materials should be used as heat storage to support such temporary intense sprinting, leveraging the property that temperature stays constant during the melting phase of the material. Figure 12.1 demonstrates the nominal single-core operation as well as the sprint mode for a 16-core system. The temperature rises from the ambient environment when the sprint starts at t_{sprint}, and then extra thermal energy is absorbed by the melting process of the phase change material, which keeps the temperature at T_{melt}. After the material is completely melted, the temperature rises again until T_{max} where the system terminates all but one core (t_{one}) to sustain the operation. The system starts to cool after all work is done at t_{cool}. Note that numbers in the curve mark different sprint phases and will be analyzed in section "Architectural Evaluation".

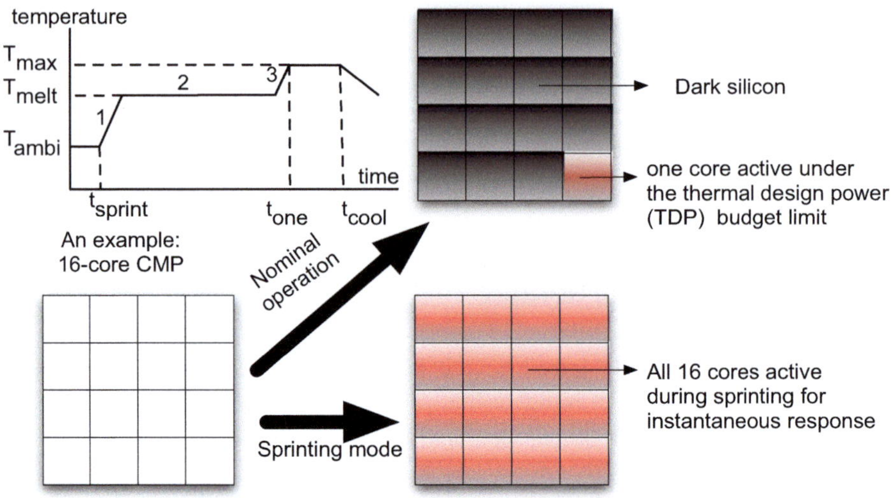

Fig. 12.1 How computational sprinting tackles the dark silicon problem: During nominal operation, only one core is active under the TDP constraint, whereas the rest cores are *dark silicon*. In sprinting mode, all the cores will be activated to provide instantaneous throughput for a short amount of time

Conventional computational sprinting still focuses on computation, whereas the role of interconnect is neglected. Here we demonstrate two key challenges that require careful consideration when designing NoC for sprinting-based multicores in the dark silicon age.

NoC Power Gating

Power gating is an efficient power-saving technique by completely shutting down the idle cores to reduce leakage. However, as more and more cores turn "dark," the network components such as routers and links also become under-utilized. As mentioned, NoC dissipates 10–36 % of total chip power [9, 17, 23, 26]; additionally, the more cores become dark, the larger the ratio of network power over the total chip power. This observation also points out the flaw of the conventional computational sprinting which turns off all but one core during nominal operation, while neglecting the impact of NoC.

To give a brief overview of network power, we simulate a classic wormhole router with a network power tool DSENT [22]. The flit width is 128 bits. Each input port of a router comprises two virtual channels (VC) and each VC is 4-flit deep. The power value is estimated with an average injection rate of 0.4 flits/cycle. Figure 12.2 shows the router power breakdown when varying the operating voltage (1, 0.9, and 0.75 V) and frequency (2, 1.5, and 1.0 GHz) under 45 nm technology. We can see

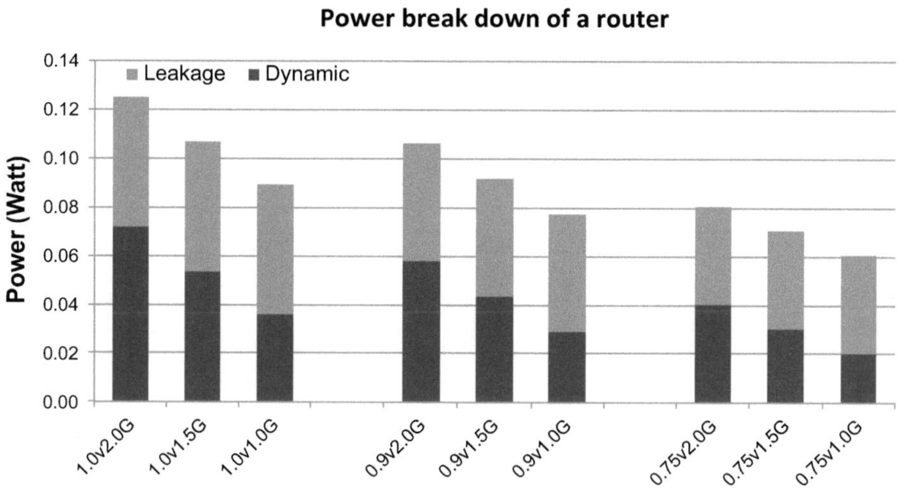

Fig. 12.2 Power breakdown (dynamic power vs leakage power) of a single router when varying the operating voltage and frequency

that leakage power contributes a significant portion to the total network power. In addition, the ratio of leakage power increases as we scale down the supply voltage and frequency, and even exceeds that of dynamic power in some cases. For example, leakage occupies around 66.4 % with a supply voltage of 0.75 V and frequency of 1 GHz. This also reflects the power crisis in the dark silicon age due to the increasing leakage.

Conventional computational sprinting proposes to operate at a single core during nominal execution and activate all the otherwise dark cores in case of intense parallel computation. To give a general picture of the power distribution during such nominal operation when only one core is active whereas the rest are gated off, Fig. 12.3 shows the chip power breakdown when scaling the total core count based on the Niagara2 [19] processor. We evaluate the power dissipation with McPAT [14] for cores, L2 caches, memory controllers (MC), NoC, and others (PCIe controllers, etc.). We assume that idle cores can be gated off (dark silicon) while *other on-chip resources stay active or idle.*

As shown in Fig. 12.3, NoC occupies a significant portion of total chip power when the multicore system is operating at nominal mode. Specifically, it accounts for 18 %, 26 %, 35 %, and 42 % of chip power, respectively, for 4-core, 8-core, 16-core, and 32-core CMP chips when they are operating at nominal mode. In contrast, the power ratio for the single active core keeps decreasing as the "dark silicon" grows. It even drops to as low as 7 % in a 32-core system in which the NoC power actually dominates. Therefore in this scenario, it is inappropriate to only measure core power when power budget is the design limitation. Moreover, while not considered in this work, any orthogonal cache power-gating techniques [5] will further raise the power ratio of the interconnect.

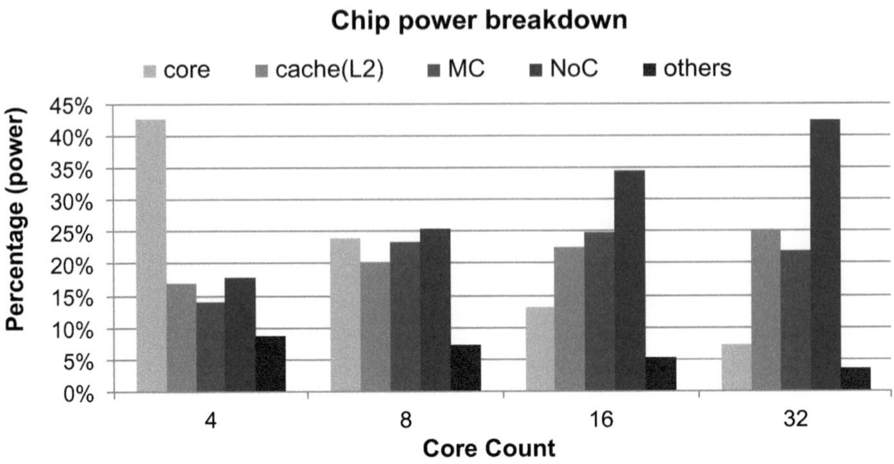

Fig. 12.3 Chip power breakdown during nominal operation in sprinting-based multicores, i.e., only one core is active while all the other cores are in dormant mode. The percentages denote the component power (core, cache, NoC, MC, and others) over the total chip power

NoC power gating is heavily dependent on the traffic. In order to benefit from power gating, an adequate idle period (namely, "break-even time") of routers should be guaranteed to make sure they are not frequently woken up and gated off. Recently researchers have proposed various schemes [4, 6, 16, 21] to mitigate the latency overhead caused by frequent router wake-up. However, these techniques do not account for the underlying core status and will result in sub-optimal power-gating decisions.

Workloads-Dependent Sprinting

A straightforward sprinting mechanism is to transiently activate all the dark cores at once. However, this scheme fails to explore the sporadic workload parallelism and thus may waste power without sufficient performance gain. Therefore, it becomes very critical to design a power-efficient sprinting scheme that can use the available resources to fulfill sporadic workload activities. Depending on the applications that are executing, the optimal number of cores required to provide maximal speedup may vary, especially for multi-threaded applications that have various scalability. Here we use PARSEC 2.1 [2] as an example. We simulate CMP systems using gem5 [3] and observe the performance speedup when varying the core count.[1] For clarity, Fig. 12.4 selects a few results that can represent different workload characteristics.

[1]The detailed evaluation methodology is described in section "Architectural Evaluation".

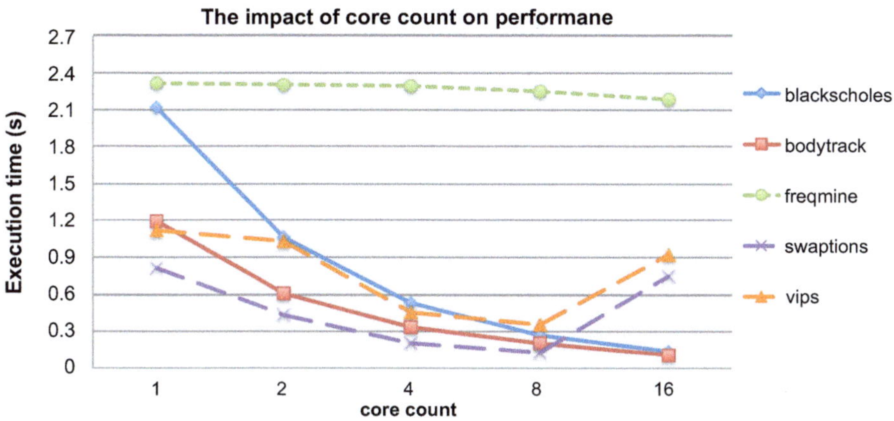

Fig. 12.4 Execution time of running multi-threaded applications provided in the PARSEC benchmark suite. We sweep the number of active cores to observe correlation between core count and system performance

As shown in Fig. 12.4, some benchmarks (e.g., *blackscholes* and *bodytrack*) achieve significant performance speedup as the number of cores increases. In contrast, for *freqmine*, the execution time is almost identical at different configurations, which implies its serial program benefits little from the extra cores. In addition, there are some applications (e.g., *vips* and *swaptions*) that achieve obvious speedup as the core count increases within a small range but then slow down gradually, and further suffer from delay penalty after exceeding a certain number. This is because adding more cores than required by the application parallelism may incur significant overheads that may offset and even hurt performance. The overheads include thread scheduling, synchronization, and long interconnect delay due to the spread of computation resources. Therefore, for sprinting-based multicores, activating all the dark cores is not a universal solution for all applications.

Our Method: NoC-Sprinting

As illustrated above, an efficient sprinting mechanism should be able to provide different levels of parallelism desired by different applications. Depending on workload characteristics, the optimal number of cores required to provide maximal performance speedup varies. This also raises challenges in designing a high performance and low power interconnect to support the sprinting process.

Fine-Grained Computational Sprinting

We first propose *fine-grained sprinting*, a flexible sprint mechanism that can activate a subset of network components to connect a certain number of cores for different workloads.

Specifically, during execution, the CMP system may experience a short burst of computation due to the arrival of new applications or the abrupt fluctuation of a running program. As such, the system will quickly react to such intense computation and determine the optimal number of cores that should be offered for instantaneous responsiveness. Then the system will activate the required number of cores while the others remain "dark." There are some existing work [7, 13] on adapting system configurations like core count/frequency to meet runtime application requirements. Our mechanism can be integrated with those approaches for runtime power saving. *However, since our focus is on how to design interconnect under such circumstances, we assume that these application parallelisms can be learnt in advance or monitored during runtime execution.*

Irregular Topological Sprinting and Deadlock-Free Routing

Under the nominal operation, only a single core (namely *master core*) remains active. There are different choices of placement for the master core. We list a few examples here, but real implementations should not be limited by these mentioned conditions. Firstly it could be placed in the center of the chip to reduce the transmission latency for thread migration. Another example is to select the core running the OS as the master core since it is always activated. The core next to the memory controller is also a good candidate if the application generates intensive memory accesses. Without loss of generality, we choose the top-left corner node as the master node which is closest to the memory controller, i.e., Node 0 as shown in Fig. 12.5a.

After the system transfers to the sprinting mode, a number of cores will be activated and keep running for a short duration. *Fine-grained computational sprinting* requires topological support from the following aspects:

- Pay-as-you-go: fine-grained activation of any number of cores.
- Short communication delay between different nodes, especially to the master node where the memory controller resides.
- Routing should be simple and deadlock-free which does not incur significant control complexity or hardware overhead.

To achieve these goals, we propose to start from the master node, and connect other nodes to the network in ascending order of their Euclidean distances to the master node. For example, the red nodes in Fig. 12.5a demonstrate the topology of an 8-core sprinting. Note that we use Euclidean distances instead of Hamming

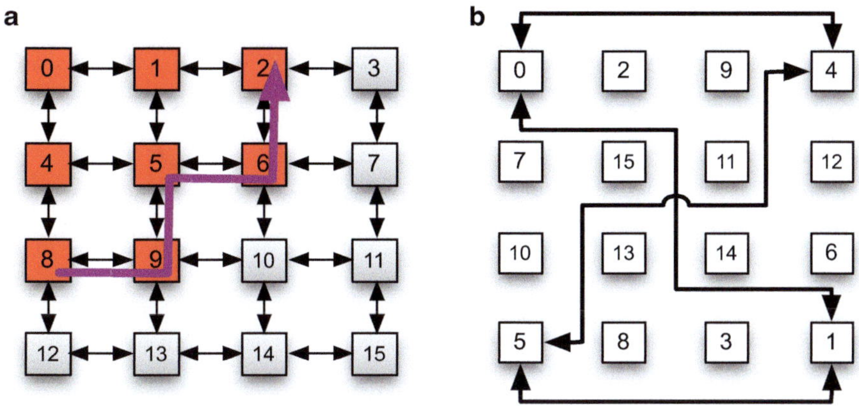

Fig. 12.5 NoC support for fine-grained computational sprinting, including topology, routing, and floorplanning. (**a**) Logical connection of a 16-node mesh network. The *red* nodes are active nodes which are connected by an irregular network topology. A Convex dimension-order routing (CDOR) algorithm is designed for deadlock-free routing. (**b**) Physical allocation for the original logical network in (**a**). The same logical connection is kept but the wires are re-organized for this new floorplan. Here, only links for 4 nodes are shown for clarity

distances here. The latter may guarantee a shortest routing distance between the newly added node to the master node, but would generate longer inter-node communication to other nodes. For example, both cases would choose node 0, 1, and 4 as 3-core sprinting. But if 4-core sprinting is triggered, the method with Hamming distance may possibly choose node 2 whereas the method with Euclidean distance would generate a better choice by accommodating node 5. Algorithm 1 generates the order of N nodes used for *topological sprinting*.

Algorithm 1: Irregular topological sprinting

Result: A linked-list L of routers to be activated
Initialize: $D[i] = 0, i = 0, 1, 2 \ldots N-1$. The coordinate for R_k is (x_k, y_k).
for $i \leftarrow 1$ **to** $N-1$ **do**
  $D[i] = \sqrt{(x_i - x_0)^2 + (y_i - y_0)^2}$;
end
Sort $R[i](i = 0, 1 \ldots N-1)$ in ascending order of $D[i](i = 0, 1 \ldots N-1)$ and put them into a linked-list L. Break ties according to the order of indexes.

The *fine-grained sprinting* process will generate an irregular network topology to connect active cores. Meanwhile, it guarantees that chosen nodes would form a convex set in the Euclidean space, i.e., the topology region contains all the line segments connecting any pair of nodes inside it. Flich et al. [8] proposed a distributed routing algorithm for irregular NoC topologies but their algorithm requires 12 extra bits per switch. Adapted from their approach, we extend the

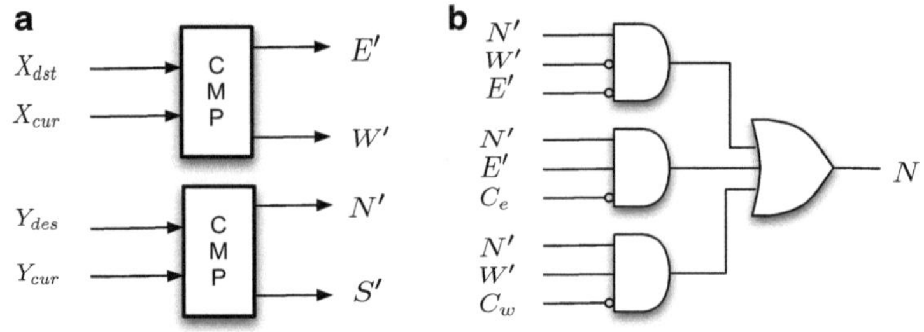

Fig. 12.6 Routing logic to support convex DOR algorithm in the network. (**a**) Two comparators per switch to determine the routing direction. (**b**) Routing logic of the North port as an example. Similar logic for other ports

Dimension-Order-Routing (specifically, X–Y routing) algorithm for such convex topologies (CDOR). Specifically, two connectivity bits (C_w and C_e) are leveraged to indicate whether a router is connected to its western or eastern neighbors. As in conventional DOR, we assume that X and Y coordinates of the final destination are stored in the packet header (X_{des} and Y_{des}), and each switch knows its X and Y coordinates (through X_{cur} and Y_{cur} registers at each switch). The origin of the coordinate system is located at the top-left corner of the 2D mesh. Messages are routed from the current router to the destination router, according to the offsets of coordinates and the two connectivity bits per router. Figure 12.5a shows a routing path from the source to its destination. The detailed routing algorithm is described in Algorithm 2. Furthermore, Fig. 12.6 depicts the routing logic design, which includes two comparators per switch and the routing circuit for computing the North port. The routing logic for other ports can be designed similarly based on Algorithm 2. We implemented CDOR on behavioral Verilog. Synthesized results using Synopsys Design Compiler (45 nm technology) show that it adds less than 2 % area overhead compared to a conventional DOR switch.

Algorithm 2: Convex dimension order routing (CDOR)

if $X_{des} > X_{cur}$ *and* $C_e = 1$ **then**
| output_port = east;
else if $X_{des} < X_{cur}$ *and* $C_w = 1$ **then**
| output_port = west;
else if $Y_{des} > Y_{cur}$ **then**
| output_port = north;
else if $Y_{des} < Y_{cur}$ **then**
| output_port = south;
else
| output_pot = local;
end

DOR (such as deterministic X–Y routing) is deadlock-free because some of the turns are eliminated such as SE, NW, NE, and SW (S, W, N, and E represent south, west, north, and east, respectively). In our CDOR, although the NE turn may happen, it is deadlock-free. For example, as shown in Fig. 12.5a, an NE turn happens at Node 5 but this also indicates that the east output port of its southern neighbor 9 is not connected. Therefore a WN turn cannot happen and thus eliminates a cycle that may cause a deadlock.

Thermal-Aware Floorplanning

The key design constraint of *fine-grained sprinting* is the TDP. The above topological sprinting process does not consider thermal behavior to simplify control and routing logic. Therefore, here we propose a design-time floorplanning algorithm that can be seamlessly integrated with the topological sprinting process while providing better thermal distribution to avoid hot spots. Thus, it sustains a longer sprint duration by slowing down the heating process.

Consider a 4-core sprinting in a 16-node mesh network as shown in Fig. 12.5a. We may opt to choose the top-left four nodes for better performance, but alternatively may prefer the four scattered corner nodes from the thermal point of view. To overcome this dilemma, *we still maintain the original logic connectivity of the mesh network in consistent with the topological sprinting process, but propose a heuristic algorithm that reallocates the physical location of each node.*

As shown in Algorithm 3, our floorplanning algorithm treats the 2D mesh network as a graph, and allocates the nodes based on the list generated from Algorithm 1. In our annotations, G represents the original logical network, S contains the set of nodes that have already been explored in G, G' represents the physical floorplan, and S' is the corresponding set of occupied nodes in G'. At each iteration, it picks up a node R_k in $G-S$, and maps it to a node in $G'-S'$ that has the maximum weighted sum of Euclidean distances to all the nodes in S' for the optimal thermal distribution. This process is described in Function *MaxWeightedDistance* as in Algorithm 4. Note that the weight of a distance is inversely proportional to the Hamming distance between R_k and the node in S. The rationale behind this scheme is that, the longer the Hamming distance in logical connectivity, the less chance these two nodes would be selected together during sprinting and accumulate heat dissipation, thus they can be placed closer in the physical floorplan.

The floorplanning algorithm heuristically allocates a node to the physical location that has the maximum weighted sum of Euclidean distances to all the other occupied physical tiles. This frees the sprinting process and routing algorithm from the thermal concern, i.e., only the logical connectivity of mesh network needs to be considered during topological sprinting. Moreover, the mapping order is based on a given list L, therefore it works with any algorithms regarding to the topological sprinting. Figure 12.5b shows the final floorplan of the physical network and only links for four-core sprinting are shown for clarity. Note that the floorplanning

Algorithm 3: Thermal-aware heuristic floorplanning algorithm

```
Result: Positions for all nodes
Initialize: Original floorplan f: {R_0, R_1 ... R_{N-1}}. Transformed floorplan f': {R'_0, R'_1
... R'_{N-1}}. The coordinate for R_k or R'_k is (x_k, y_k).
Set S = φ, S'={R'_0,R'_1 ... R'_{N-1}}. Queue Q = φ. List L from Algorithm 1
Goal: Find the mapping Pos() from f to f'.
Pos(R_0) = 0(Master Node); Put R_0 in S; Delete R'_0 from S';
Put all unexplored adjacent nodes of R_0 into Q based on List L;
while Q ≠ φ do
    R_k = Q[0]; Delete Q[0] from Q;
    Pos(R_k) = MaxWeightedDistance(S, S', R_k);
    Delete R'_{Pos(R_k)} from S'; Put R_k in S;
    Put all unexplored adjacent nodes of R_k into Q based on List L;
end
```

Algorithm 4: MaxWeightedDistance(S, S', R_k)

```
Initialize: Sum = 0; Max = 0;
for every node R'_i in S' do
    for every node R_j in S do
        w_ij = 1/(|x_k − x_j| + |y_k − y_j|);
        d_ij = sqrt((x_i − x_{Pos(R_j)})^2 + (y_k − y_{Pos(R_j)})^2);
        s_ij = w_ij * d_ij;
    end
    Sum = Σ s_ij;
    if Sum > Max then
        Max = Sum; Pos(R_k) = i;
    end
end
Return Pos(R_k);
```

algorithm will increase the wiring complexity and generate long links. A standard method of reducing delay of long wires is to insert repeaters in the wire at regular intervals. Recently, Krishna et al. [12] have validated such clockless repeated wires that allow multi-hop traversals to be completed in a single clock cycle.

Network Power Gating

With our proposed *fine-grained computational sprinting*, the network power-gating scheme becomes straightforward. *Since the topological sprinting algorithm activates a subset of routers and links to connect the active cores, we gate off the other network components as shown in the shaded nodes of Fig. 12.5a.* Moreover, the proposed CDOR algorithm routes packets within the active network and thus avoids unnecessary wakeup of intermediate routers for packet forwarding. This further increases the idle period of the dark region for longer power gating.

However, we still need to consider the Last-Level-Cache (LLC) architecture for network power gating. For private per-core LLC, centralized shared LLC, or distributed shared LLC connected with a separate network (NUCA), our power-gating mechanism works perfectly without the need for any further hardware support. However, for tile-based multicores where each tile comprises of a shared bank of LLC, there may be some packet accesses to dark nodes for cache resources. Therefore, some complimentary techniques such as bypass paths [4] can be leveraged to avoid completely isolating cache banks from the network. We accommodate this method in our design.

Architectural Evaluation

We use the gem5 [3] full system simulator to setup a sprinting-based multicore architecture with 16 ALPHA CPUs. Each CPU is the out-of-order architecture and only supports single thread. Each core has a split private L1 cache (instruction & data cache each 64 KB in size). All the 16 cores share an NUCA (Non-Uniform Cache Access) L2 cache of size 4 MB, which is distributed into 16 banks. We use Ruby [15] to model a detailed memory system, Garnet [1] to model a 4 × 4 mesh network, and DSENT [22] for network power analysis. The detailed system configurations are listed in Table 12.1.

We evaluate *NoC-sprinting* with multi-threaded workloads from PARSEC [2] by assuming that the chip can sustain computational sprinting for 1 s in the worst case, which is consistent with [20]. Later we will analyze how *NoC-sprinting* influences the sprint duration. We first start running the benchmarks in a simple mode and take checkpoints when reaching the parallel portion of the program. Then, the simulation restores from checkpoints and we record the execution time of running a total of one billion instructions for each benchmark. In addition, we construct synthetic traffic for further network analysis.

Table 12.1 System and interconnect configuration

Core count/freq.	16, 2 GHz	Topology	4 × 4 2D Mesh
L1 I & D cache	Private, 64 KB	Router pipeline	Classic five-stage
L2 cache	Shared & tiled, 4 MB	VC count	4 VCs per port
Cacheline size	64 B	Buffer depth	4 buffers per VC
Memory	1 GB DRAM	Packet length	5 flits
Cache-coherency	MESI protocol	Flit length	16 bytes

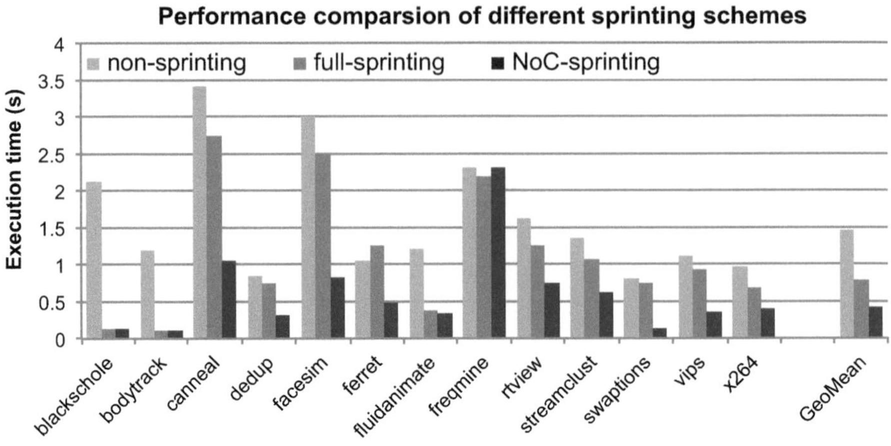

Fig. 12.7 Execution time comparison with different sprint mechanisms: Non-sprinting does not apply any computational sprinting; Full-sprinting is to always sprint at 16 cores; NoC-sprinting is our proposed technique that provides fine-grained computational sprinting with NoC support

Performance Evaluation

Here we evaluate how NoC-sprinting improves the system responsiveness. In comparison, one naive baseline design (*non-sprinting*) is to always operate with one core under TDP limit. Another extreme case (*full-sprinting*) is to activate all the 16 cores during sprinting. While the methods to predict the application parallelism [7, 13] is beyond the scope of this paper, we conduct off-line profiling on PARSEC to capture the internal parallelism of the benchmarks. Figure 12.7 shows the execution time of PARSEC workloads with different sprinting schemes.

We can see that *NoC-sprinting* cuts down the execution time substantially compared to *non-sprinting*. It achieves 3.6× speedup on average for all the applications. In comparison, *full-sprinting* fails to provide the maximal speedup in some cases with an average 1.9× speedup. It is because increasing the active core count in some programs would incur overheads that may outweigh achievable benefits after a saturating point is reached. These overheads come from OS scheduling, synchronization, and long interconnect delay due to the spread of computation resources. These results verify that NoC-sprinting can flexibly select the optimal level of sprinting to satisfy the application parallelism, and thus it provides better responsiveness for short burst of parallel computation.

Core Power Dissipation

Instead of waking up all the dark cores for quick response, *NoC-sprinting* provides better power-efficiency by allocating *just enough* power to support the maximal performance speedup. Since the triggered topology directly determines the number of cores to be powered on, here we first explore its impact on the core power dissipation. Apart from *full-sprinting*, we also compare *NoC-sprinting* with a naive *fine-grained sprinting* scheme which does not employ any power-gating techniques, i.e., the system only cares about selecting the optimal number of cores for actual execution and leaves the others idle.

As shown in Fig. 12.8, except for *blackscholes* and *bodytrack* which achieve the optimal performance speedup in *full-sprinting* and hence leave no space for power gating, *NoC-sprinting* cuts down the most power across all the other applications. Compared to *full-sprinting*, *fine-grained sprinting* saves 25.5 % power even though power gating is not applied. More promisingly, *NoC-sprinting* achieves 69.1 % core power saving on average for all applications.

Analysis of On-Chip Networks

NoC-sprinting provides customized topology, routing, floorplanning, and efficient power-gating support for fine-grained sprinting. Therefore in this subsection, we evaluate network performance and power to see how the NoC behaves during the sprinting process.

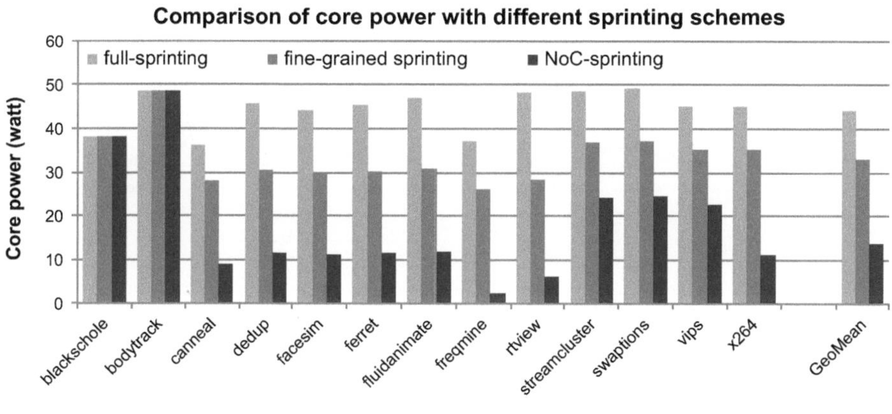

Fig. 12.8 Power consumption of cores with different sprinting schemes. Here *fine-grained sprinting* executes on the selected active cores but still leaves the idle cores consuming leakage power, whereas *NoC-sprinting* applies power gating to both idle cores and NoC

Network Latency *Full-sprinting* activates the entire network and would possibly lose some performance speedup in the long interconnect. In contrast, *NoC-sprinting* uses a subset of routers to directly connect the active cores, which avoids unnecessary network traversals in the dark nodes with the support of CDOR routing algorithm. As an example, Fig. 12.9 shows the average network latency for running PARSEC with different sprinting schemes. Apparently, *NoC-sprinting* shortens the communication latency for most applications. Overall, it cuts down the network latency by 24.5 %.

Network Power As Fig. 12.3 shows, network power becomes more and more significant as cores turn dark. Therefore, optimizing NoC power dissipation becomes an urgent issue in order to combat the power shortage in the dark silicon age.

Figure 12.10 shows the total network power consumption during the sprint phase of running PARSEC. As we can see, *NoC-sprinting* successfully cuts down the network power if an intermediate level of sprinting is selected. On average, it saves 71.9 % power compared to *full-sprinting*. This is because *NoC-sprinting* can adapt the network topology according to workload characteristics and only operates on a subset of nodes. In comparison, *full-sprinting* activates a fully functional network and loses opportunities for power gating.

More Analysis with Synthetic Traffic Furthermore, we construct some synthetic traffic on a network simulator booksim 2.0 [11] to test *NoC-sprinting* under different traffic scenarios. For *full-sprinting*, we consider traffic to be *randomly mapped* in the fully functional network and results are averaged over ten samples. We implemented these sprinting algorithms to generate traffic from a selected number of nodes while others nodes only used for packet forwarding. This can mimic the scenario of an over-designed full-sprinting scheme that actually only uses a few cores for

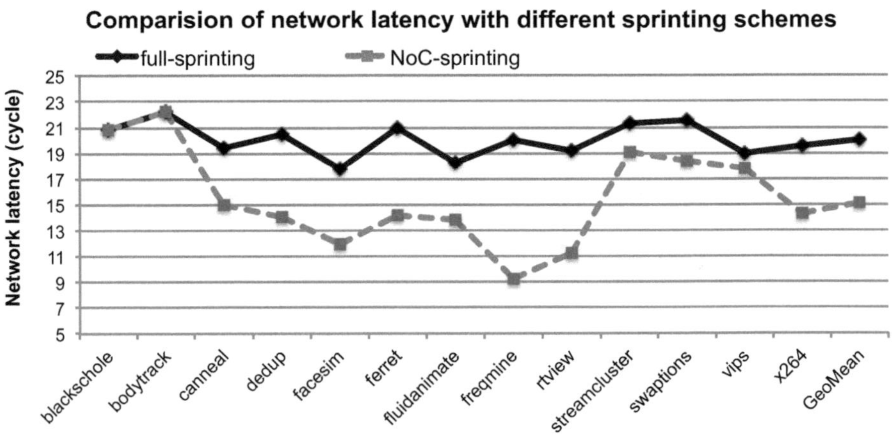

Fig. 12.9 Comparisons of average network latency after running PARSEC with full-sprinting and NoC-sprinting

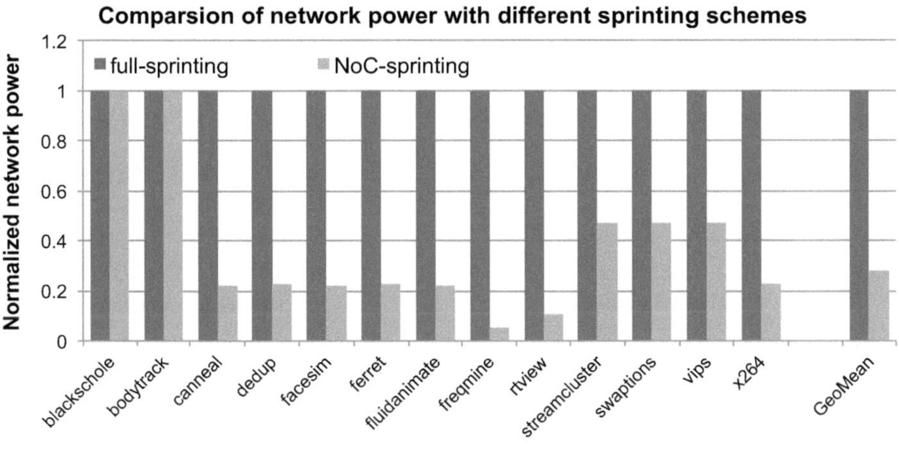

Fig. 12.10 Comparisons of total network power after running PARSEC with full-sprinting and NoC-sprinting

actual computation. Furthermore, we extract dynamic and leakage power numbers from DSENT [22] and thus enable booksim to generate network power results. We compare *full-sprinting* with *NoC-sprinting* and observe the differences in performance and power while varying the network load. As an example, Fig. 12.11 shows the results of 4-core and 8-core sprinting for a 16-core system under uniform-random traffic. There are a few key observations:

- As shown in Fig. 12.11a, c, *NoC-sprinting* cuts down the average flit latency by 45.1 and 16.1 % before saturation for 4-core and 8-core sprinting, respectively, because it uses a dedicated region of network for more efficient communication without traversing the dark region. The latency benefit drops when switching to a higher level of sprinting because less routers/links are wasted as intermediate forwarding stations like *full-sprinting*.
- Correspondingly, *NoC-sprinting* decreases the total network power consumption by 62.1 and 25.9 % for 4-core and 8-core sprinting, respectively, as indicated by the gap between the two power curves in both Fig. 12.11b, d. The extra routers/links used in *full-sprinting* not only consume leakage power but also generate dynamic power from packet traversals. As expected, the lower the sprint level, the more power saving the NoC-sprinting can achieve.
- The downside of *NoC-sprinting* is that the network saturates earlier than that of *full-sprinting*. This is because *NoC-sprinting* uses a subset of network where each node is generating and accepting packets. Differently, *full-sprinting* spreads the same amount of traffic among a fixed fully functional network where some nodes are simply used for intermediate packet forwarding. However, this usually would not affect the network performance in real cases. For example, in the PARSEC benchmarks we have evaluated, the average network injection rate never exceeds 0.3 flits/cycle, which is far from saturating the network.

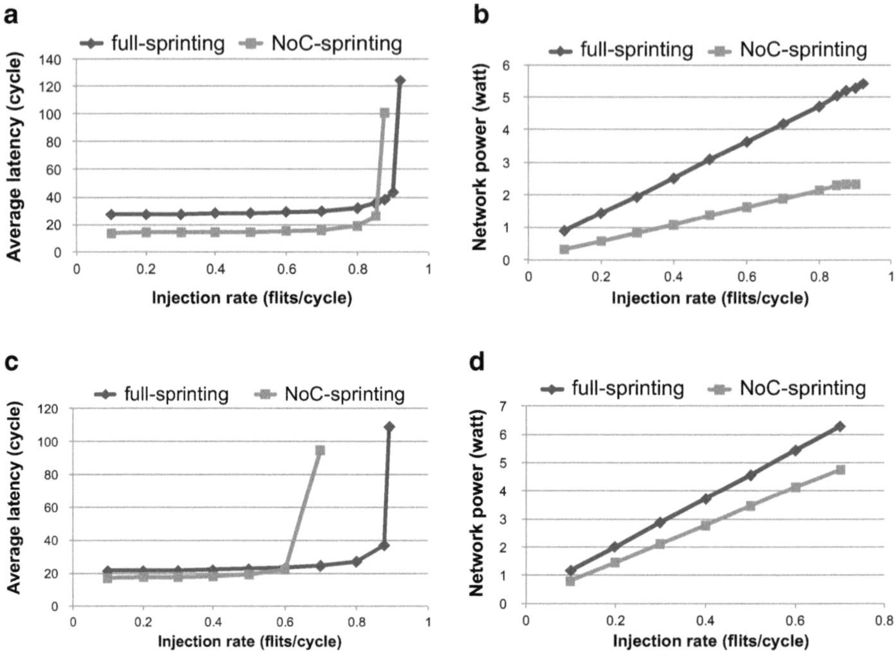

Fig. 12.11 Performance and power analysis on *full-sprinting* and *NoC-sprinting* with synthetic uniform-random traffic. For a specific sprint level (4-core or 8-core), *full-sprinting* activates all the routers/links although only a few nodes are generating traffic, whereas *NoC-sprinting* triggers a subnet of nodes and power-off all the other idle network components. (**a**) Latency on 4-core sprint, (**b**) power on 4-core sprint, (**c**) latency on 8-core sprint, (**d**) power on 8-core sprint

Thermal Analysis

NoC-sprinting heavily relies on the sprint duration to sustain the desired parallelism. Figure 12.1 in section "Challenges and Opportunities" demonstrated the sprinting process which includes three phases. The duration of each phase is dependent on the property of the phase change material placed close to the die. However, we can still conduct some qualitative analyses to evaluate how NoC-sprinting affects the sprint duration.

Phase 1 indicates that the temperature rises abruptly when sprinting starts, and so does the phase 3 after the melting phase ends. Intuitively, the more the power-on components, the faster the temperature will increase. Therefore, *NoC-sprinting* can slow down the heating process by allocating just enough power for the maximum performance speedup. As an example, we analyze *dedup* (one of the PARSEC benchmarks) whose optimal level of sprinting is 4. We collect the power densities using McPAT and feed them into a thermal modeling tool HotSpot [10] as the power trace. As for the floorplan, we abstract the 16-core CMP system as 16 blocks placed in a 2D grid, where each block comprises the Alpha CPU, local caches, and

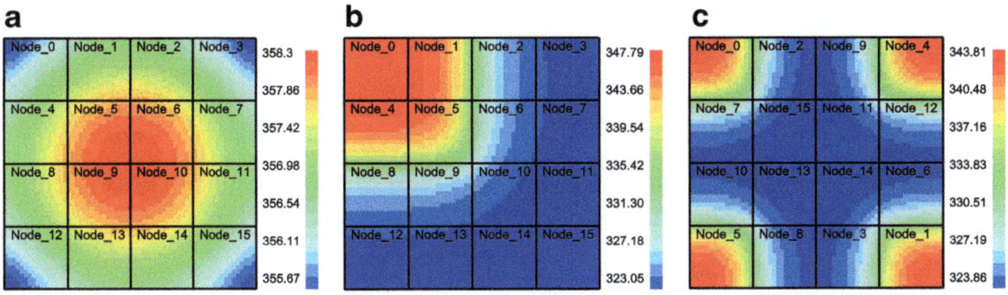

Fig. 12.12 Heat maps for running *dedup* with *full-sprinting* and *NoC-sprinting*. (**a**) Full-sprinting, (**b**) 4-core sprinting with aggregated nodes, (**c**) thermal-aware NoC-sprinting

other network resources. We use a fine-grained grid model to observe the stable temperatures of the whole chip. Figure 12.12 shows the heat maps for *full-sprinting* and *NoC-sprinting*.

As shown in Fig. 12.12a, though power is almost uniformly distributed across the chip, *full-sprinting* results in an overheated spot in the center (358.3 K). In contrast, *fine-grained sprinting* only activates four nodes as shown in Fig. 12.12b and the corresponding peak temperature drops (347.79 K). Furthermore, our thermal-aware floorplanning generates better temperature profile (343.81 K) as shown in Fig. 12.12c. This example implies that *NoC-sprinting* can slow down the heating process by sprinting with a less number of cores/interconnect compared to *full-sprinting*.

Phase 2 is the most critical phase that determines the capability of sprinting. Placing phase change materials close to the die elongates this period by increasing the thermal capacitance. Temperature remains stable during melting and the duration of melting is mostly determined by its *latent heat of fusion*—the amount of energy to melt a gram of such material. As such, based on the power results we collected from PARSEC, *NoC-sprinting* increases the duration by 55.4 % on average.

As a summary, *NoC-sprinting* reduces the slopes of temperature rise in phase 1 & 3, and enhances the melting duration in phase 2 by slowing down thermal capacitance depletion. Thus, it guarantees a longer sprint for intense parallel computation and further increases the performance.

Conclusion

In this work, we reveal the challenges and opportunities in designing NoC in the dark silicon age. We present *NoC-sprinting*: it provides topology/routing support, thermal-aware floorplanning, and network power gating for fine-grained sprinting. Our experiments show that *NoC-sprinting* outperforms conventional *full-sprinting* which always activates all the dark cores. It is able to provide tremendous performance speedup, longer sprint duration, and reduces the chip power significantly.

References

1. N. Agarwal, T. Krishna, L.-S. Peh, N.K. Jha, Garnet: a detailed on-chip network model inside a full-system simulator, in *International Symposium on Performance Analysis of Systems and Software (ISPASS)* (2009), pp. 33–42
2. C. Bienia, Benchmarking modern multiprocessors, PhD thesis, Princeton University, 2011
3. N. Binkert et al., The gem5 simulator. ACM SIGARCH Comput. Archit. News **39**(2), 1–7 (2011)
4. L. Chen, T.M. Pinkston, Nord: Node-router decoupling for effective power-gating of on-chip routers, in *MICRO-45* (2012), pp. 270–281
5. H.-Y. Cheng, J. Zhan, J. Zhao, Y. Xie, J. Sampson, M. Irwin, Core vs. uncore: the heart of darkness, in *52nd Design Automation Conference (DAC)* (2015), pp. 1–6
6. R. Das, S. Narayanasamy, S. Satpathy, R. Dreslinski, Catnap: energy proportional multiple network-on-chip, in *Proceedings of the 40th Annual International Symposium on Computer Architecture (ISCA)* (2013), pp. 320–331
7. Y. Ding, M. Kandemir, P. Raghavan, M.J. Irwin, A helper thread based EDP reduction scheme for adapting application execution in CMPs, in *International Symposium on Parallel and Distributed Processing (IPDPS)* (2008), pp. 1–14
8. J. Flich, S. Rodrigo, J. Duato, An efficient implementation of distributed routing algorithms for NoCs, in *Second ACM/IEEE International Symposium on Networks-on-Chip (NoCs)* (2008), pp. 87–96
9. Y. Hoskote, S. Vangal, A. Singh, N. Borkar, S. Borkar, A 5-GHz mesh interconnect for a teraflops processor. IEEE Micro **27**(5), 51–61 (2007)
10. W. Huang et al., HotSpot: a compact thermal modeling methodology for early-stage VLSI design. IEEE Trans. Very Large Scale Integr. Syst. **14**(5), 501–513 (2006)
11. N. Jiang et al., A detailed and flexible cycle-accurate network-on-chip simulator, in *IEEE International Symposium on Performance Analysis of Systems and Software (ISPASS)* (2013), pp. 86–96
12. T. Krishna et al., Single-cycle multihop asynchronous repeated traversal: a SMART future for reconfigurable on-chip networks. Computer **46**(40), 48–55 (2013)
13. J. Li, J.F. Martinez, Dynamic power-performance adaptation of parallel computation on chip multiprocessors, in *The 12th International Symposium on High-Performance Computer Architecture* (2006), pp. 77–87
14. S. Li et al., McPAT: an integrated power, area, and timing modeling framework for multicore and manycore architectures, in *42nd Annual IEEE/ACM International Symposium on Microarchitecture (MICRO-42)* (2009), pp. 469–480
15. M.M. Martin et al., Multifacet's general execution-driven multiprocessor simulator (GEMS) toolset. ACM SIGARCH Comput. Archit. News **33**(4), 92–99 (2005)
16. H. Matsutani, M. Koibuchi, D. Wang, H. Amano, Run-time power gating of on-chip routers using look-ahead routing, in *Proceedings of the 2008 Asia and South Pacific Design Automation Conference (ASP-DAC)* (2008), pp. 55–60
17. T.G. Mattson et al., The 48-core SCC Processor: the programmer's view, in *Proceedings of the 2010 ACM/IEEE International Conference for High Performance Computing, Networking, Storage and Analysis* (2010), pp. 1–11
18. R. Merritt, ARM CTO: power surge could create 'dark silicon'. EE Times, 2009
19. U.G. Nawathe et al., Implementation of an 8-core, 64-thread, power-efficient sparc server on a chip. IEEE J. Solid State Circuits **43**(1), 6–20 (2008)
20. A. Raghavan et al., Computational sprinting, in *IEEE 18th International Symposium High Performance Computer Architecture (HPCA)* (2012), pp. 1–12
21. A. Samih et al., Energy-efficient interconnect via router parking, in *IEEE 18th International Symposium High Performance Computer Architecture (HPCA)* (2013), pp. 508–519
22. C. Sun et al., DSENT-a tool connecting emerging photonics with electronics for opto-electronic networks-on-chip modeling, in *2012 Sixth IEEE/ACM International Symposium on Networks on Chip (NoCS)* (2012), pp. 201–210

23. M.B. Taylor et al., The Raw microprocessor: a computational fabric for software circuits and general-purpose programs. IEEE Micro **22**(2), 25–35 (2002)
24. G. Venkatesh et al., Conservation cores: reducing the energy of mature computations, in *Proceedings of the Fifteenth Edition of ASPLOS on Architectural Support for Programming Languages and Operating Systems*, vol. 38, issue 1 (2010), pp. 205–218
25. J. Zhan, Y. Xie, G. Sun, NoC-sprinting: interconnect for fine-grained sprinting in the dark silicon era, in *51nd Design Automation Conference (DAC)* (2014), pp. 1–6
26. J. Zhan, J. Ouyang, F. Ge, J. Zhao, Y. Xie, DimNoC: a dim silicon approach towards power-efficient on-chip network, in *52nd Design Automation Conference (DAC)* (2015), pp. 1–6

MIX
Papier aus verantwortungsvollen Quellen
Paper from responsible sources
FSC® C105338

If you have any concerns about our products,
you can contact us on
ProductSafety@springernature.com

In case Publisher is established outside the EU,
the EU authorized representative is:
Springer Nature Customer Service Center GmbH
Europaplatz 3, 69115 Heidelberg, Germany

Printed by Libri Plureos GmbH
in Hamburg, Germany